Challenges and New Strategies on Rabbit Breeding

Challenges and New Strategies on Rabbit Breeding

Editors

Rosa María García-García
María Arias Álvarez

MDPI • Basel • Beijing • Wuhan • Barcelona • Belgrade • Manchester • Tokyo • Cluj • Tianjin

Editors
Rosa María García-García
Complutense University of Madrid
Spain

María Arias Álvarez
Complutense University of Madrid
Spain

Editorial Office
MDPI
St. Alban-Anlage 66
4052 Basel, Switzerland

This is a reprint of articles from the Special Issue published online in the open access journal *Animals* (ISSN 2076-2615) (available at: https://www.mdpi.com/journal/animals/special_issues/Challenges_and_New_Strategies_on_Rabbit_Breeding).

For citation purposes, cite each article independently as indicated on the article page online and as indicated below:

LastName, A.A.; LastName, B.B.; LastName, C.C. Article Title. *Journal Name* **Year**, *Volume Number*, Page Range.

ISBN 978-3-0365-4497-7 (Hbk)
ISBN 978-3-0365-4498-4 (PDF)

Cover image courtesy of Rosa María García-García and María Arias Álvarez

Contents

About the Editors

Rosa María García-García

Dr. Rosa María García-García, Ph.D., DVM. She graduated in Veterinary Medicine in 1996. In 2002, she obtained her Ph.D. in Veterinary Sciences. She has been Associate Professor at the Department of Physiology of the Faculty of Veterinary Medicine, Complutense University of Madrid, Spain since 2011. She is a member of the UCM research group "Physiology of the Reproduction in Lagomorphs" (CONEREPRO). According to Scopus, she is author/co-author of 56 indexed research papers (h-index: 18). Her main research topics have been developed in the area of reproductive physiology in the female, always focusing on the improvement of reproductive performance both in small ruminants (sheep and goat) and in rabbits, with a physiological basis.

María Arias Álvarez

Dr. María Arias Álvarez, Ph.D., DVM with European mention, "cum laude" qualification and doctorate extraordinary prize (2009). Currently Associate Professor in Animal Production Department and Vice-dean for Research in the Faculty of Veterinary Medicine, Complutense University of Madrid. Her research activity has been funded continuously with a total of 14 competitive research projects. She is a member of the UCM research group "Physiology of the Reproduction in Lagomorphs" (CONEREPRO). She is co-author of 39 indexed research papers (91% in Q1) (h-index: 14) according to Thomson Reuters Web of Science. Her studies have been focused on the effect of different management strategies on follicular development, oocyte maturation and embryo preimplantation development. In particular, she studies the maternal—oocyte—embryo interaction using in vitro and in vivo systems (IVM, IVF and IVC) in different species such as rabbit, cattle and mouse.

Preface to "Challenges and New Strategies on Rabbit Breeding"

Rabbit meat is a healthy source of protein for people: it is low in fat, cholesterol, and sodium and can meet the growing population's requirements in the coming decades. In addition, rabbits are less costly and have less environmental impact than other livestock species. Therefore, this valuable livestock species can potentially help to satisfy the nutritional needs of humans thanks to its sustainable husbandry system. However, intensive rabbit production shows some difficulties that negatively impair the reproduction and lifespan of rabbit females, which ultimately affect productive outcomes; then, there is an increase in the use of antibiotics and decrease in animal welfare that subsequently reduce the farmer's profitability, and negatively prejudices consumer perception of rabbit meat production. In addition, new breeding selection tools and improvement in nutritional strategies should arise to balance animal physiological demands and sustainability. Consequently, rabbit production should adapt to the new green challenges promoting circular bioeconomy, benefiting the rabbits' welfare, and promoting both human and rabbit health in a One Health approach to be a competitive and sustainable production system.

Rosa María García-García and María Arias Álvarez
Editors

Review

Current Knowledge on the Multifactorial Regulation of Corpora Lutea Lifespan: The Rabbit Model

Massimo Zerani, Angela Polisca *, Cristiano Boiti and Margherita Maranesi

Dipartimento di Medicina veterinaria, Università di Perugia, via San Costanzo 4, 06126 Perugia, Italy; massimo.zerani@unipg.it (M.Z.); boiti.cristiano@gmail.com (C.B.); margherita.maranesi@unipg.it (M.M.)
* Correspondence: angela.polisca@unipg.it

Simple Summary: Corpora lutea (CL) are temporary endocrine structures that secrete progesterone, which is essential for maintaining a healthy pregnancy. A variety of regulatory factors come into play in modulating the functional lifespan of CL, with luteotropic and luteolytic effects. Many aspects of luteal phase physiology have been clarified, yet many others have not yet been determined, including the molecular and/or cellular mechanisms that maintain the CL from the beginning of luteolysis during early CL development. This paper summarizes our current knowledge of the endocrine and cellular mechanisms involved in multifactorial CL lifespan regulation, using the pseudopregnant rabbit model.

Abstract: Our research group studied the biological regulatory mechanisms of the corpora lutea (CL), paying particular attention to the pseudopregnant rabbit model, which has the advantage that the relative luteal age following ovulation is induced by the gonadotrophin-releasing hormone (GnRH). CL are temporary endocrine structures that secrete progesterone, which is essential for maintaining a healthy pregnancy. It is now clear that, besides the classical regulatory mechanism exerted by prostaglandin E2 (luteotropic) and prostaglandin F2α (luteolytic), a considerable number of other effectors assist in the regulation of CL. The aim of this paper is to summarize our current knowledge of the multifactorial mechanisms regulating CL lifespan in rabbits. Given the essential role of CL in reproductive success, a deeper understanding of the regulatory mechanisms will provide us with valuable insights on various reproductive issues that hinder fertility in this and other mammalian species, allowing to overcome the challenges for new and more efficient breeding strategies.

Keywords: rabbit; corpus luteum; reproduction

Citation: Zerani, M.; Polisca, A.; Boiti, C.; Maranesi, M. Current Knowledge on the Multifactorial Regulation of Corpora Lutea Lifespan: The Rabbit Model. *Animals* **2021**, *11*, 296. https://doi.org/10.3390/ani11020296

Academic Editor: Rosa María García-García and Maria Arias Alvarez
Received: 30 November 2020
Accepted: 21 January 2021
Published: 25 January 2021

Publisher's Note: MDPI stays neutral with regard to jurisdictional claims in published maps and institutional affiliations.

1. Introduction

Corpora lutea (CL) are temporary endocrine structures that secrete progesterone, which is essential for a healthy pregnancy in most species. In rabbits, the CL develop rapidly following ovulation and reach their maximum size and functional capacity within nine to ten days. This process shows the intense angiogenesis and active granulosa or theca cell luteinization of preovulatory follicles, due to the effects of several local angiogenic growth factors, gonadotropins and other hormones [1,2]. In pregnant rabbits, the CL lifespan lasts for about 30 days [3]; however, if pregnancy does not occur, the lifespan of the CL is much shorter, and luteal regression starts around day 12 and ends 16 days after ovulation when the peripheral plasma progesterone concentrations drop to the baseline values [4,5]. Therefore, the absence of embryonic signals or the end of gestation activates luteolysis, a comprehensive regressive process that leads to total functional and structural CL demise, in which prostaglandin (PG) F2 α (PGF2α) plays a central role [6].

Many regulatory factors, including cytokines, growth factors, prostaglandin E2 (PGE2) and PGF2α released by different CL cell types, including endothelial and local immune cells and fibroblasts, as well as progesterone and 17β-estradiol released by luteal and follicular cells and hormones, control the functional lifespan of the CL, with luteotrophic and luteolytic effects [7]. However, the overall balance between these contrasting actions varies considerably with the age of the CL and/or in the presence/absence of an embryo [8]. Many facets of luteal physiology have been clarified, but others are still poorly understood, including the molecular and/or cellular mechanisms that protect the CL from luteolysis from the early luteal phase. Moreover, the mechanisms that are induced by the administration of exogenous PGF2α have been extensively investigated in rabbits [9–11] in order to evaluate PG paracrine and/or autocrine functions and other possible regulators that switch on (luteotropic)/off (luteolyic) progesterone production by the CL at a specific stage of its life cycle. However, there are few data on the mechanisms that protect the developing CL from functional luteolysis in the early luteal phase, which starts on day six of a pseudopregnancy, when the luteal cells acquire the ability to respond to the luteolytic effects of exogenous PGF2α (luteolytic capacity) [9]. Luteolysis is a key event in reproduction for spontaneously ovulating species, as well as for rabbits, whose mating activity triggers a neuroendocrine reflex, which, combined with GnRH or exogenous human chorionic gonadotropin (hCG) exogenous administration, induces ovulation [12,13].

This paper provides a summary of our current knowledge on the endocrine and cellular mechanisms of multifactorial CL lifespan regulation, acquired using the pseudopregnant rabbit model, which was able to determine the relative luteal phase following GnRH-induced ovulation. Most of the mechanisms described in this review were observed during our studies on the progressive age-dependent response of the CL to PGF2α conducted over a 20-year period [9]. A better understanding of these mechanisms may provide us with valuable insights in the challenge to find more efficient breeding strategies for rabbits, as well as for other species.

2. Prostaglandins

Prostaglandins (PGs) play a key regulatory role in CL function and the lifespan: PGF2α is the main luteolytic agent produced by the uterine endometrium of numerous mammals, including rabbits, but not by primates [14–16], while PGE2 plays a crucial luteoprotective role, with luteotrophic and/or antiluteolytic effects [6]. In some species, PGF2α and PGE2 are produced by the CL [17–21].

An essential step in PG biosynthesis is the cyclooxygenase (COX) 1 (COX1) and/or COX2 enzymatic conversion of arachidonic acid (AA)—produced by phospholipase A2 (PLA2) activity—into PGH2 [22–24]. This latter PG is then transformed into four structurally active PGs (PGE2, PGF2α, PGD2 and PGI2) by specific PG synthases [25]. PGF2α biosynthesis is particular, since three specific ketoreductases catalyze this PG from PGH2, PGD2 or PGE2, respectively [26]. PGE2-9-ketoreductase (PGE2-9-K) is present in the rabbit ovary [27] and CL [28]. This ketoreductase also converts progesterone into its inactive metabolite through its 20α-hydroxysteroid dehydrogenase (HSD) catalytic activity.

We previously reported [21] that, in rabbits, intra-luteal PGF2α activates luteolysis with an auto-amplification loop: during the mid- and late-luteal phases, it activates COX2 and PGE2-9-K; the former converts AA into PGH2, which is then transformed into PGF2α and PGE2, while the latter is converted into PGF2α through PGE2-9-K activation. Moreover, this enzyme significantly reduces PGF2α-induced progesterone through its 20α-hydroxysteroid dehydrogenase (20α-HSD) activity that converts progesterone into 20α-OH-progesterone. Late-luteal phase PGE2 production plays another essential role: PGE2-9-K enzymatic activity make this PG the main source of PGF2α synthesis.

Arosh et al. [29] suggested that CL PG biosynthesis is mainly directed toward PGE2 production rather than PGF2α. In fact, PGH2 conversion into PGE2 (PGE synthase) is 150-fold higher than that of PGH2 into PGF2α (PGF synthase) [30]. These results [29,30], combined with our data [21,31], allow us to hypothesize [31] that rabbit CL in the early and mid-luteal phases use the same cellular enzymatic pathways (PLA2/AA/COX2/PGH2/PGE synthase/PGE2) to produce an initial PGE2 amount, while the final luteal production of PGE2 (early CL) or PGF2α (mid-CL) is regulated by PGE2-9-K inactivation or activation, respectively (Figure 1, upper, functional luteolysis).

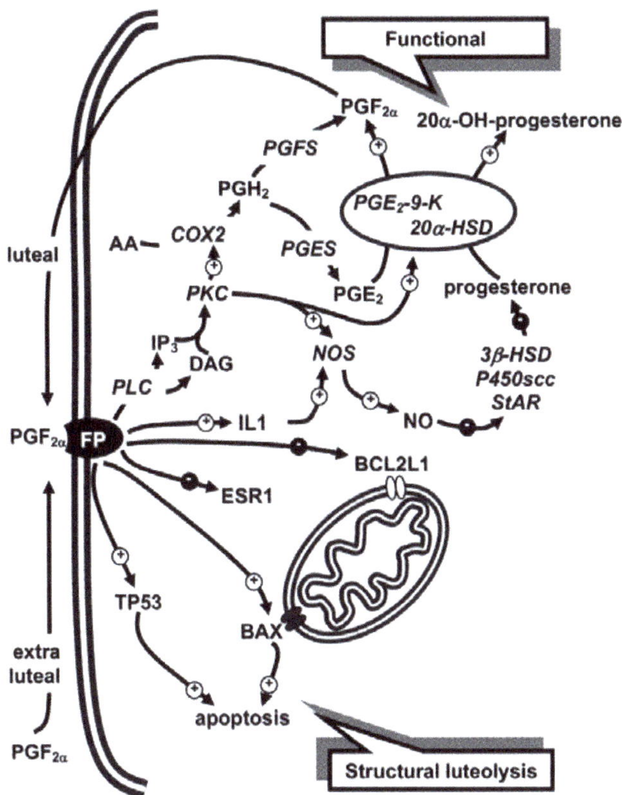

Figure 1. Schematic model reporting the functional (upper) and structural (lower) luteolytic pathways induced by prostaglandin F2 α (PGF2α) in rabbit mid-corpora lutea (CL) (day 9 of pseudopregnancy). Since prostaglandin E2 (PGE2)-9-K and 20α-hydroxysteroid dehydrogenase (HSD) represent two different activities of a single enzyme, they are joined. Figure from the study by Maranesi et al. 2010 [31]. For acronyms, see the list of abbreviations in the text.

Several studies have investigated the possible factors involved in PGF2α-induced luteolytic capacity during the mid-luteal phase [7,9,32–36]. Interleukin 1 (IL1), with other cytokines that are normally present in rabbit luteal cells [32,33], are locally involved in the CL function control leading to apoptosis as proinflammatory mediators [34]. Moreover, locally acting hormones and pro- and antiapoptotic intra-luteal factors may interact dynamically. 17β-Estradiol is one of the main luteotropic effectors, since its absence leads to luteolysis through apoptosis activation [7]. Nitric oxide synthase (NOS) and its product nitric oxide (NO) are also known to have pro- and antiapoptotic properties that modulate various intracellular pathways—in particular B-cell CLL/lymphoma 2 (BCL2)-like 1 (BCL2L1) and tumor protein p53 (TP53) proteins [35]. In rabbits, NOS luteal inhibition favors apoptosis [36].

Our study [31] on the key protein-encoding genes involved in apoptotic mechanism control revealed that PGF2α induces luteolysis in luteal cells with an acquired luteolytic capacity through the upregulation of luteal IL1B and TP53 gene transcripts and the downregulation of the estrogen receptor 1 (ESR1) and BCL2L1 receptors. This PGF2α-induced CL regression seems to be the result of two distinct mechanisms: the steroidogenic pathway, by ESR1 downregulation, and the apoptotic pathway, by the dynamic changes of the TP53 and BCL2L1 proteins and gene transcripts (Figure 1, lower, structural luteolysis). Finally, aglepristone (RU534), an antiprogestinic, increases progesterone release in rabbit mid- and late-CL, whereas this antiprogestinic reduces PGF2α and enhances PGE2 only during the late-luteal stage [37].

3. Nitric Oxide

Nitric oxide is a potent vasodilator factor involved in several biological processes, such as neurotransmissions and cytotoxicity, under both physiological and pathological conditions [38,39]. NO is produced by the enzymatic action of NOS, which converts L-arginine into NO and L-citrulline. There are three forms of NOS: two constitutive Ca^{2+}-dependent forms neuronal NOS (nNOS) and endothelial NOS (eNOS) and an inducible Ca^{2+}-independent form (iNOS) [38,40]. With the exception of neuronal and endothelial cells, constitutive eNOS and nNOS are normally expressed in various cell types and produce low levels of NO. Contrastingly, the inducible form only produces large quantities of NO when the expression is activated [38,40]. NOS is present in both ovarian stroma and follicular granulosa cells of several mammalian species, including rabbit ovaries, where it regulates steroidogenesis [17,41–44]. The NO/NOS system present in rabbit, rat and mare ovaries is also involved in ovulation [43–49]. All of these studies suggest that NO regulates the key mechanisms of ovarian physiology.

In rabbits, NO has a direct antisteroidogenic effect at the luteal level. Numerous in vivo and in vitro experiments have found that NO and NOS are the main targets of PGF2α and effectors of PGF2α-induced luteolysis in competent CL [10,11,17,18,33,50]. Ovarian NO is known to be a mediator of the luteolytic action induced by PGF2α in rabbits and other mammalian species [17,51–55]. Ovarian NO might also control the CL lifespan by regulating 17β-estradiol and progesterone concentrations. However, in contrast to earlier findings in rat and human in vitro cultured CL [41,56], NO did not affect the total androgens and 17β-estradiol production in rabbit CL [17]. Contrastingly, in rabbit CL cultured in vitro, the NO donor, sodium nitroprusside, greatly reduced progesterone secretion in all luteal developmental stages [17]. Luteal NOS activity decreases between the early-to mid-luteal phases with elevated steroidogenesis levels [17,57], which increase again in late-CL when the progesterone levels drop and natural luteolysis initiates [5,57].

4. Leptin

Leptin is a cytokine secreted mainly by adipocytes and encoded by the obese gene [58]. Leptin regulates the hypothalamic centers of satiety and energy metabolism through the modulation of various neurotransmitters [59,60].

The leptin receptor (ObR) has six isoforms (a–f) resulting from mRNA splice variants [61,62]. ObRa–d and ObRf have identical extracellular and transmembrane domains [62,63]. A long intracellular domain of ObRb activates the Janus kinase (JAK)/signal transducer with the subsequent signal transducer and activator of transcription (STAT) phosphorylation [64]. Contrastingly, the short intracellular domain of ObRa, ObRc,d and ObRf activates the mitogen-activated protein kinase (MAPK) pathway [61,65].

Several studies have found that various key mammalian reproductive processes are modulated by leptin [66], including steroidogenesis [67,68], ovulation [69,70], pregnancy [71,72] and menstrual cycles [73,74]. Moreover, leptin is the crucial link between adipose tissue and the reproductive system, since it provides information on whether energy reserves are adequate for normal reproductive function [75].

Leptin receptors are present in several tissues of the hypothalamic–pituitary–gonadal (HPG) axis and in pituitary [76], granulosa, theca and interstitial ovary cells [77]. Various studies have reported that leptin directly inhibits steroidogenesis in intracellular signaling pathways in theca, granulosa and luteinized granulosa cells of rodents, bovines and primates [67,68,77–79].

Our studies on rabbit CL [80] show that leptin affects progesterone and PGF2α release with different intracellular signaling pathways through different receptors (long ObR and short ObR). More specifically, leptin inhibits progesterone release through the MAPK cascade (short ObR) and stimulates PGF2α release through the JAK pathway (long ObR) (Figure 2).

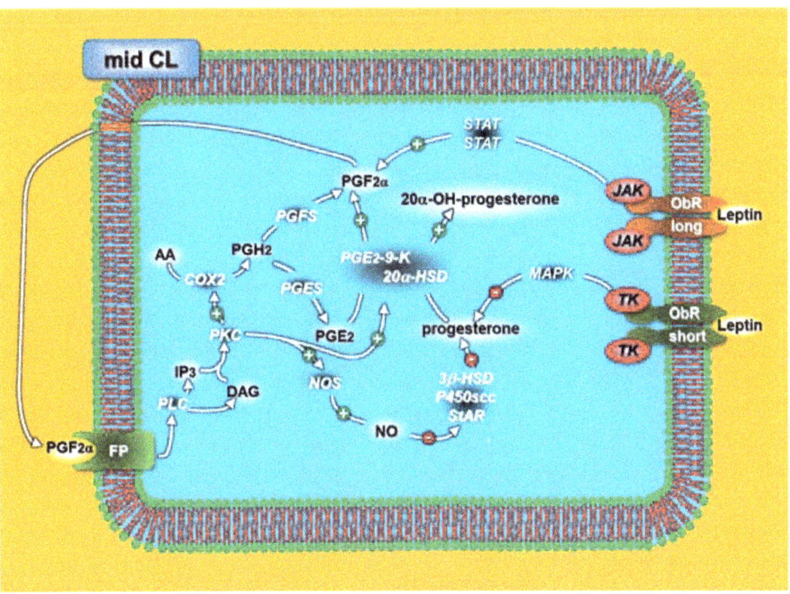

Figure 2. Schematic representation of the leptin mechanisms regulating progesterone release in rabbit mid-CL. For acronyms, see the list of abbreviations in the text.

5. Gonadotropin-Releasing Hormone (GnRH)

Gonadotropin-Releasing Hormone (GnRH) is a hypothalamic-releasing decapeptide and a key regulator of the mammalian reproductive system. GnRH regulatory action on the reproductive functions is exerted largely via luteinizing hormone (LH) and follicle-stimulating hormone (FSH) secretion, which also affect steroidogenesis and germ cell development [81]. Although the hypothalamus and pituitary gland are the main GnRH synthesis and action sites, several studies have reported an extra-hypothalamic presence of GnRH and its cognate receptor (GnRHR) in numerous peripheral tissues, including reproductive organs such as the gonads, prostate, uterine tube, placenta and mammary glands [82]. Previous studies have highlighted that GnRH regulates the ovarian steroid hormones [82]. In rabbit CL, GnRH administration was found to be associated with CL regression with decreased levels of serum progesterone [83]. Contrastingly, no GnRH effects were observed on ovarian tissue steroid production by other authors [84].

The studies conducted in our laboratory [85] highlighted that the autocrine, paracrine and/or endocrine roles of GnRH type I (GnRH-I) directly diminished the progesterone secretion in rabbit CL that had acquired luteolytic competence (Figure 3): GnRH-I acts via GnRHR-I by activating phospholipase C (PLC) and stimulating the inositol trisphosphate (IP3) and diacylglycerol (DAG) pathways. Through the activation of protein kinase C (PKC), these two intracellular messengers stimulate COX2 activity and PGF2α release. This PG induces (via paracrine, autocrine and/or intracrine mechanisms) an increase in NOS activity and NO levels [11], which downregulates the progesterone levels [18,31] (Figure 1, upper, functional luteolysis).

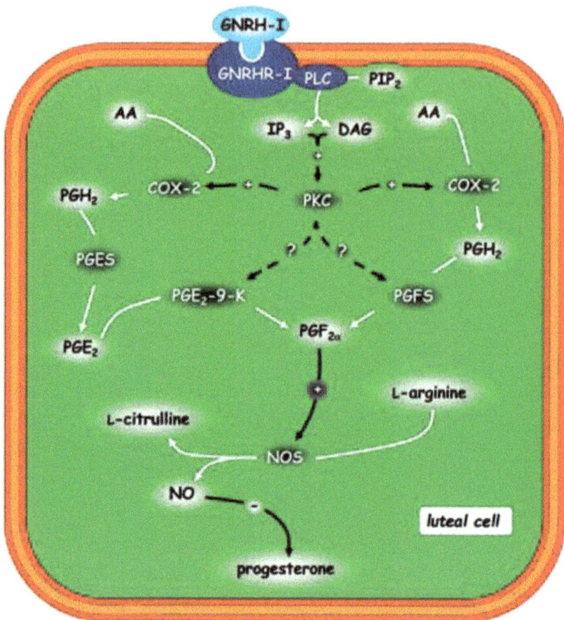

Figure 3. Schematic representation of the post-receptorial mechanism of GnRH-I regulating the progesterone release in rabbit CL. The other possible protein kinase C (PKC) targets are represented by hatched lines. Figure from the study by Zerani et al. 2010 [85]. For acronyms, see the list of abbreviations in the text.

6. Endothelin 1

Endothelin 1 (ET1), a 21-amino acid peptide, is a potent vasoconstrictor secreted by vascular endothelial cells [86,87]. Many tissues other than the vascular endothelium are known to express ET1, including follicular granulosa cells [88–92].

In rabbit CL, ET1 receptors are expressed in the vascular compartments and luteal cells, thus evidencing that the ET1 system is related to ovarian blood flow and steroid hormone production [91,92]. Moreover, ET1-induced luteolysis in rabbits on day nine of the pseudopregnancies was prevented by administering captopril, the angiotensin-converting enzyme inhibitor (ACE). It is important to note that PGF2α-induced luteolysis was not influenced by captopril. These findings indicate that the cascade mechanism triggered by PGF2α does not require the renin–angiotensin system for inducing luteolysis in rabbits [92], which is in good agreement with the data obtained for cows [93]. Strict cooperation between endothelin and NO is required for endothelial cell migration and angiogenesis [94]. ET1 was found to stimulate endothelial NOS under different physio-pathological conditions [95], while NO/NOS is a recognized system involved in both PGF2α [11] and ET1 [96]-induced luteal regression.

7. Adrenocorticotropic Hormone

Adrenocorticotropic hormone (ACTH) is a major component of the hypothalamic–pituitary–adrenal (HPA) axis, which is synthesized and secreted by the anterior pituitary gland in response to stress. This response is activated by the hypothalamic corticotropin-releasing hormone (CRH), which stimulates pituitary ACTH release, with subsequent glucocorticoid secretion from the adrenal glands.

There is strong evidence that female reproduction can be impaired by stress [97]. In fact, CRH, ACTH and glucocorticoid negatively affect fertility by targeting the hypothalamic GnRH neurons [98], as well as pituitary LH and/or FSH production and sex steroid synthesis by ovarian follicles and CL. However, the mechanisms by which hormones released during stress may inhibit reproductive mechanisms have yet to be clarified; however, any direct action of ACTH on ovarian functions requires the activation of melanocortin receptor 2 (MC2R) [99], while any indirect action requires glucocorticoid receptor (GR) activation.

The presence of ACTH and glucocorticoid receptors in the luteal cells of rabbit CL [100] supports the hypothesis that ACTH affects ovarian functions both directly and indirectly. During the early and mid-luteal phases (days four and nine of the pseudopregnancies), ACTH increased the in vitro progesterone and PGE2 releases but reduced the PGF2α release. Contrastingly, ACTH increased the in vivo plasmatic cortisol levels within four hours, while the progesterone levels dropped 24 h later and for the following 48 h. Daily injections of ACTH did not affect the progesterone profile following ovulation. Taken together, these findings indicate that ACTH directly induces the upregulation of luteal progesterone synthesis through MC2R (Figure 4), while it indirectly blocks CL functions through the cortisol/GR system.

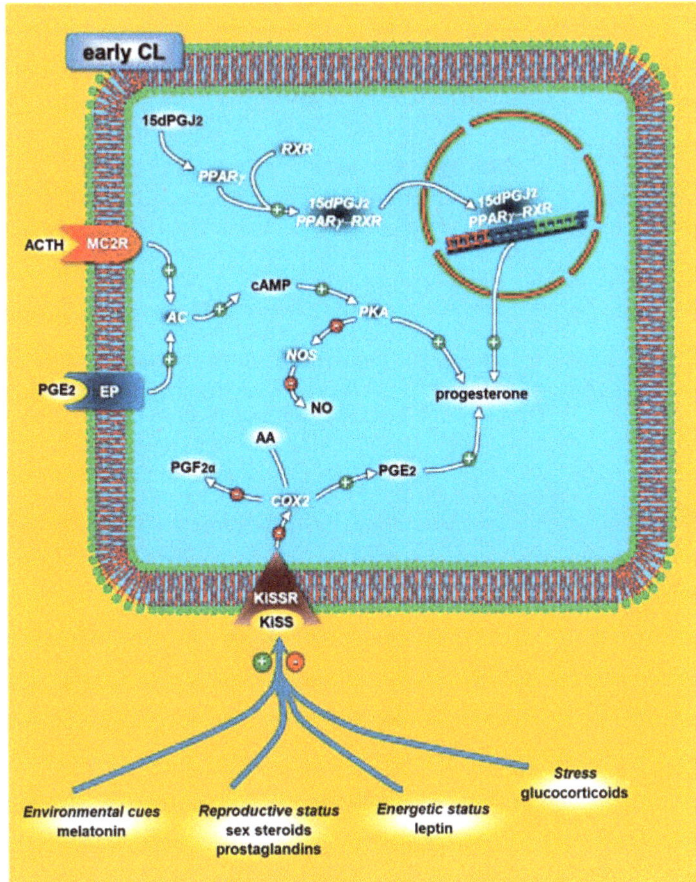

Figure 4. Schematic diagram of the adrenocorticotropic hormone (ACTH), kisspeptins (KiSS) and peroxisome proliferator-activated receptor (PPAR) mechanisms modulating progesterone release in early rabbit CL. The effectors that could directly modulate the KiSS/KiSSR (receptor) system at the CL level are represented by blue lines. For acronyms, see the list of abbreviations in the text.

8. Immunity Mediators

It is now widely accepted that luteolysis is an event mediated by immune effectors in rabbits and other species, as demonstrated by the presence of immune cells during spontaneous luteal regression [32]. Luteal immune cells are key modulators of CL activity, affecting the luteal, endothelial and stromal cells through several cytokines, including IL1, tumor necrosis factor (TNF)α, monocyte chemoattractant protein-1 (MCP1) and interleukin 2 (IL2) [33,101,102]. In rabbits, during spontaneous luteolysis, the expression levels of MCP1 and IL1β increased on day 15 of the pseudopregnancies [33]. These findings show the greater influx of macrophages and immune cells observed during luteal regression [103]. The IL2 transcript increases earlier (day 13 of the pseudopregnancies) than the other cytokines [33]; in fact, T lymphocytes were detected in rabbit CL before the macrophages [103].

The IL-1 cytokine is present in the ovaries of various species, including rabbits [104,105]. IL1β has various effects on the ovaries [106]: it inhibits progesterone production, increases PG synthesis and PGF2 receptor expression, it inhibits COX2 mRNA degradation [107], enhances NO production and induces the activation of constitutive and inducible NOS [108].

Our studies report [21] that injecting pseudopregnant rabbits with PGF2α markedly upregulated COX2 and IL1β mRNA expression and increased PGF2α release and COX2 activity only in CL with acquired luteolytic capacity [31]. These data suggest that IL1β enhances intra-luteal PGF2α synthesis by upregulating the luteal function of COX2 and NOS, thus promoting functional regression in luteal cells that have achieved luteolytic capacity.

9. Peroxisome Proliferator-Activated Receptor

The peroxisome proliferator-activated receptors (PPARs) include a family of three (a, d and c) nuclear receptor/transcription factors, which regulate steroidogenesis, angiogenesis, tissue remodeling, cell cycle and apoptosis [109], which are all essential processes for normal ovarian function [110]. All three PPARs have been detected in the ovaries of numerous species [111], including rats [110,112], mice [113], pigs [114], sheeps [115], cows [116–118], rabbits [119] and humans [120,121].

Komar [110] reported that PPARc activation affected the progesterone synthesis in ovarian cells. In particular, an endogenous activator of PPARc 15d-PGJ2 inhibited both the basal and gonadotropin-induced production of progesterone in human granulosa cells [122], while 15d-PGJ2 and ciglitazone, a synthetic PPARc activator, increased progesterone production by granulosa cells in equine chorionic gonadotropin (eCG)-primed immature rats [123]. PPARc activation by 15d-PGJ2, ciglitazone or another synthetic activator, troglitazone, also increased progesterone release by porcine theca and bovine luteal cells [114,124]. Taken together, these findings indicate that the cell type, stage of cell differentiation, stage of the ovarian cycle and/or animal species influence the effects of PPARc on progesterone production [110].

Our study [125,126] suggests that PPARc may play a luteotropic role in rabbit CL through a mechanism that upregulates 3β-hydroxysteroid dehydrogenase (3β-HSD) and increases progesterone while it downregulates PGF2α and its correlated enzyme COX2 [21] (Figure 4). Moreover, the significant decrease in PPARc in the luteal cell nucleus during the late-luteal stage supports the aforementioned mechanism, thus suggesting that this reduction may be required for luteolysis to take place.

10. Dopamine

The catecholamine dopamine (DA) is a neurotransmitter widely distributed in the brain and in various peripheral organs of numerous species [127]. DA exerts its physiological actions by binding to specific receptors (DR). In mammals, there are five dopamine receptor subtypes, which are grouped into the D1R-like and D2R-like receptor superfamilies [127,128].

D1R-like receptors stimulate the production of the second messenger cyclic adenosine monophosphate (cAMP); contrastingly, D2R-like receptors inhibit cAMP synthesis, which decreases the protein kinase A (PKA) activity [128]. In mammals, dopamine receptors are widely expressed in many organs and tissues, including the reproductive system [128]. D1R has been detected in the luteal cells of humans [129,130], horses [131], rats [132], cows [118] and rabbits [133], suggesting that DA might be directly involved in the physiological pathways regulating the CL function.

Our studies [133] provide evidence that CL produce DA and that the DA/D1R-D3R system regulates the CL lifespan by exerting either luteotrophic or luteolytic actions depending on the luteal stage. In fact, the DA/D1R-D3R system stimulated PGE2 and progesterone synthesis by early CL, while it increased PGF2α production and decreased progesterone production by mid- and late-CL (Figure 5).

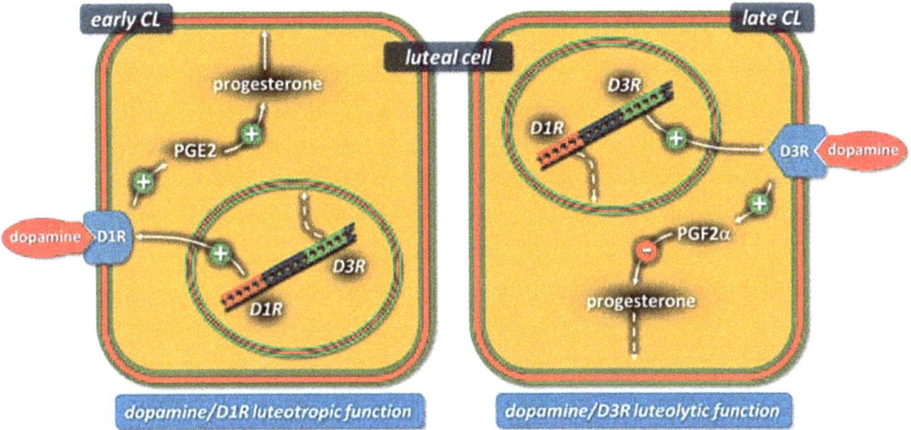

Figure 5. Schematic representation of the dopamine receptor-dependent mechanism modulating progesterone production in early (Day 4) and late (Day 9) rabbit CL. D1R: dopamine receptor subtype 1 (cAMP increase) and D3R: dopamine receptor subtype 3 (cAMP decrease). Italic: D1R and D3R genes. Figure from the study by Parillo et al. 2014 [132].

A multi-synaptic neural pathway connects the ovaries to the central nervous system in mammals [134]. Moreover, the ovarian interstitial stroma is composed of many different cell types, including neuron-like or neuroendocrine cells [135]. These data suggest that extrinsic and intrinsic neurons are another paracrine source of DA that can bind its cognate receptors D1R and D3R in the CL, thus supporting the hypothesis that the DA/DR system plays a physiological role in regulating the CL lifespan and functions.

11. Kisspeptin

The hypothalamic neuropeptide kisspeptins (KiSS) are greatly involved in mammalian reproduction. In fact, they regulate the synthesis and production of GnRH that are required to initiate puberty and sustain normal reproductive function [136].

KiSS and its receptor KiSS1R are expressed in various ovarian structures, including the CL of several mammalian species [137–139], supporting the hypothesis that these neuropeptides can regulate the CL lifespan by modulating the steroidogenic enzymes controlling progesterone synthesis. Moreover, Laoharatchatathanin et al. [140] suggested that KiSS is involved in the luteinization of rat granulosa cells.

Based on data obtained in our laboratory [141], we hypothesize that, besides the well-known hypothalamic mechanism, the KiSS/KiSS1R system may also directly control the rabbit CL lifespan via local mechanisms. In fact, KiSS was found to exert a luteotrophic action by increasing luteal progesterone synthesis, likely through autocrine and/or paracrine mechanisms that simultaneously reduce PGF2α production and increase PGE2 production by blocking COX2 activity (Figure 4). The lack of KiSS1R expression in late-CL suggests that the functional activity of the KiSS/KiSS1R system is mainly regulated by the gene and/or protein expression of the receptor.

Interestingly, there is sufficient evidence to suggest that the hypothalamic KiSS-1 gene expression is regulated by several factors, including melatonin, gonadal steroids and leptin, which convey environmental cues and reproductive and metabolic conditions, respectively [142,143]. The theory that these factors could modulate the luteal KiSS/KiSS1R system cannot be ruled out (Figure 4).

12. Nerve Growth Factor

The nerve growth factor (NGF), together with brain-derived growth factor and other neurotrophins, belong to the neurotrophin family [144]. These neurotrophins maintain normal physiological functions in the central and peripheral nervous systems, including neural development, differentiation and synaptic plasticity [145,146]. NGF and its receptors neurotrophic receptor tyrosine kinase 1 (NTRK1) and nerve growth factor receptor (NGFR) have been found in rabbit ovaries [147,148]. In particular, our studies [149] have evidenced that NGF from seminal plasma supports the neuroendocrine ovulatory reflex induced by mating and/or vaginal stimulation through a novel mechanism exerted on the uterus and/or cervix.

Although there is sufficient experimental evidence suggesting that seminal plasma NGF is able to induce ovulation in rabbits [147], its potential role in regulating the CL lifespan has not yet been thoroughly explored. To date, we only know that NGF and its cognate receptor NTRK1 are expressed in rabbit CL at various stages of a pseudopregnancy [149]. Contrastingly, using purified NGF obtained from seminal plasma, Silva et al. [150,151] observed that, in llamas, CL increased vascularization, upregulated cytochrome P450, family 11, subfamily A, member 1/P450 side chain cleavage and steroidogenic acute regulatory protein transcripts and increased progesterone secretion. All of these findings support the hypothesis that NGF positively affects CL development. Tribulo et al. [152] and Stewart et al. [153] obtained similar results in heifers; however, no luteotrophic effect was observed in alpaca CL using recombinant human NGF [154,155].

13. Conclusions

In conclusion, it is now well-documented that the progressive acquisition of luteolytic competence by rabbit CL is not only due to their increased sensitivity to PGF2 induced by the upregulation of PGF2α and its receptors and to the decrease of the luteotropic factors (E2, PGE2 and ACTH), but it is also caused by several antisteroidogenic factors. These include, among others, GnRH, ET1 and leptin, which influence the inflammatory, vascular and apoptotic processes involved in the luteolytic process through interaction with PGF2α and the NO/NOS system. During PGF2α-induced CL regression with luteolytic competence, all these factors concomitantly induce the upregulation of NOS, COX2 and PGE2-9-K activities and gene transcripts for ETI, COX2, IL1B and TP53, as well as the downregulation of several other transcripts, including ESR1 and BCLXL. Therefore, the luteolytic effect of PGF2α is the result of its influence on distinct processes involving the regulation of vasoactive peptides, steroidogenic pathways and apoptotic pathways. However, despite the increased knowledge on the physiology of rabbit CL, it is recommended that further research should be undertaken in the near future by a younger generation of researchers who will be able to apply these new discoveries in the challenge for new rabbit breeding strategies.

Author Contributions: Conceptualization, C.B., M.M. and M.Z.; infographic, M.Z.; writing—original draft preparation, M.M. and A.P. and writing—review and editing, C.B. and M.Z. All authors have read and agreed to the published version of the manuscript.

Funding: This research was partially funded by the BC Red-water (Perugia, Italy), PFZM Kitchenbrown (Matelica, Italy) and MMZM Bighead (Settevalli, PG, Italy) trusts.

Acknowledgments: In memory of Francesco Parillo (1964–2018), friend and colleague, who substantially contributed to the results of our research group.

Conflicts of Interest: The authors declare no conflict of interest.

Abbreviations

3β-HSD	3β-hydroxysteroid dehydrogenase
15d-PGJ2	15-deoxy-Δ12,14-prostaglandin J2
20α-HSD	20α-hydroxysteroid dehydrogenase
AA	arachidonic acid
ACE	angiotensin converting enzyme
ACTH	adrenocorticotropic hormone
BAX	BCL2-associated X protein
BCL2L1	B-cell CLL/lymphoma 2 (BCL2)-like 1
cAMP	cyclic adenosine monophosphate
CL	corpora lutea; COX1
COX1	cyclooxygenase 1
COX2	cyclooxygenase 2
CRH	corticotropin-releasing hormone
DA	dopamine
DAG	diacylglycerol
DR	dopamine receptor
eCG	equine chorionic gonadotropin
eNOS	endothelial NOS
ESR1	estrogen receptor subtype-1
ET1	endothelin 1
GnRH	gonadotropin-releasing hormone
GnRH-I	gonadotropin-releasing hormone type I
GnRHR	gonadotropin-releasing hormone receptor
GR	glucocorticoid receptor
hCG	human chorionic gonadotropin
HPA	hypothalamic-pituitary-adrenal
HPG	hypothalamic–pituitary–gonadal
IL1B	interleukin 1 Beta
iNOS	inducible NOS
IP3	inositol trisphosphate
JAK	Janus kinase
KiSS	kisspeptin
KiSSR	kisspeptin receptor
MAPK	mitogen-activated protein kinase
MC2R	melanocortin receptor type 2
MCP1	monocyte chemoattractant protein-1
nNOS	neuronal NOS
NGF	nerve growth factor
NGFR	nerve growth factor receptor
NO	nitric oxide
NOS	nitric oxide synthase
NTRK1	neurotrophic receptor tyrosine kinase 1
ObR	leptin (obesity) receptor
P450scc	P450 side-chain cleavage
PG	prostaglandin
PGD2	prostaglandin D2
PGE2-9-K	PGE2-9-ketoreductase

PGE2	prostaglandin E2
PGF2α	prostaglandin F2α
PGH2	prostaglandin H2
PGI2	prostaglandin I2
PKA	protein kinase A
PKC	protein kinase C
PLA2	phospholipase A2
PLC	phospholipase C
PPAR	peroxisome proliferator-activated receptor
RXR	retinoid X receptor
StAR	steroidogenic acute regulatory protein
STAT	signal transducer and activator of transcription
TK	tyrosine kinase
TNFα	tumor necrosis factor α
TP53	tumor protein p53

References

1. Webb, R.; Woad, K.J.; Armstrong, D.G. Corpus luteum (CL) function: Local control mechanisms. *Domest. Anim. Endocrinol.* **2002**, *23*, 277–285. [CrossRef]
2. Acosta, T.J.; Miyamoto, A. Vascular control of ovarian function: Ovulation, corpus luteum formation and regression. *Anim. Reprod. Sci.* **2004**, *82–83*, 127–140. [CrossRef] [PubMed]
3. Spies, H.G.; Hilliard, J.; Sawyer, C.H. Maintenance of corpora lutea and pregnancy in hypophysectomized rabbits. *Endocrinology* **1968**, *83*, 354–356. [CrossRef] [PubMed]
4. Scott, R.S.; Rennie, P.I. Factors controlling the life-span of the corpora lutea in the pseudopregnant rabbit. *Reprod. Fertil.* **1970**, *23*, 415–422. [CrossRef] [PubMed]
5. Browning, J.Y.; Keyes, P.F.; Wolf, R.C. Comparison of serum progesterone, 20α-dihydroprogesterone, and estradiol-17β in pregnant and pseudopregnant rabbits: Evidence for postimplantation recognition of pregnancy. *Biol. Reprod.* **1980**, *23*, 1014–1019. [CrossRef]
6. Niswender, G.D.; Juengel, J.L.; Silva, P.J.; Rollyson, M.K.; McIntush, E.W. Mechanisms controlling the function and life span of the corpus luteum. *Physiol. Rev.* **2000**, *80*, 1–29. [CrossRef] [PubMed]
7. Goodman, S.B.; Kugu, K.; Chen, S.H.; Preutthipan, S.; Tilly, K.I.; Tilly, J.L.; Dharmarajan, A.M. Estradiol-mediated suppression of apoptosis in the rabbit corpus luteum with a shift in expression of Bcl-2 family members favoring cellular survival. *Biol. Reprod.* **1998**, *59*, 820–827. [CrossRef]
8. Stocco, C.; Telleria, C.; Gibori, G. The molecular control of corpus luteum formation, function, and regression. *Endocr. Rev.* **2007**, *28*, 117–149. [CrossRef]
9. Boiti, C.; Canali, C.; Zerani, M.; Gobbetti, A. Changes in refractoriness of rabbit Corpora lueta to a Prostaglandin F2α analogue, alfaprostol, during pseudopregnancy. *Prostag. Other Lipid. Mediat.* **1998**, *56*, 255–264. [CrossRef]
10. Boiti, C.; Zampini, D.; Zerani, M.; Guelfi, G.; Gobbetti, A. Prostaglandin receptors and role of G protein-activated pathways on corpora lutea of pseudopregnant rabbit in vitro. *J. Endocrinol.* **2001**, *168*, 141–151. [CrossRef]
11. Boiti, C.; Guelfi, G.; Zampini, D.; Brecchia, G.; Gobbetti, A.; Zerani, M. Regulation of nitric-oxide synthase isoforms and role of nitric oxide during prostaglandin $F_{2\alpha}$-induced luteolysis in rabbits. *Reproduction* **2003**, *125*, 807–816. [CrossRef] [PubMed]
12. Mehaisen, G.M.; Vicente, J.S.; Lavara, R.; Viudes-de-Castro, M.P. Effect of eCG dose and ovulation induction treatments on embryo recovery and in vitro development post-vitrification in two selected lines of rabbit does. *Anim. Reprod. Sci.* **2005**, *90*, 175–184. [CrossRef] [PubMed]
13. Dal Bosco, A.; Rebollar, P.G.; Boiti, C.; Zerani, M.; Castellini, C. Ovulation induction in rabbit does: Current knowledge and perspectives. *Anim. Reprod. Sci.* **2011**, *129*, 106–117. [CrossRef] [PubMed]
14. O'Grady, J.P.; Caldwell, B.V.; Auletta, F.J.; Speroff, L. The effects of an inhibitor of prostaglandin synthesis (indomethacin) on ovulation, pregnancy, and pseudopregnancy in the rabbit. *Prostaglandins* **1972**, *1*, 97–106. [CrossRef]
15. Keyes, P.L.; Bullock, D.W. Effects of prostaglandin $F_{2\alpha}$ on ectopic and ovarian corpora lutea of the rabbit. *Biol. Reprod.* **1974**, *10*, 519–525. [CrossRef]
16. Lytton, F.D.; Poyser, N.L. Prostaglandin production by the rabbit uterus and placenta in vitro. *J. Reprod. Fertil.* **1982**, *66*, 591–599. [CrossRef]
17. Gobbetti, A.; Boiti, C.; Canali, C.; Zerani, M. Nitric oxide synthase acutely regulates progesterone production by in vitro cultured rabbit corpora lutea. *J. Endocrinol.* **1999**, *160*, 275–283. [CrossRef]
18. Boiti, C.; Zerani, M.; Zampini, D.; Gobbetti, A. Nitric oxide synthase activity and progesterone release by isolated corpora lutea of rabbits in early- and mid-luteal phase of pseudopregnancy are differently modulated by prostaglandin E-2 and prostaglandin F-2α via adenylate cyclase and phospholipase C. *J. Endocrinol.* **2000**, *164*, 179–186. [CrossRef]

19. Parillo, F.; Catone, G.; Maranesi, M.; Gobbetti, A.; Gasparrini, B.; Russo, M.; Boiti, C.; Zerani, M. Immunolocalization, gene expression, and enzymatic activity of cyclooxygenases, prostaglandin E_2-9-ketoreductase, and nitric oxide synthases in Mediterranean buffalo (*Bubalus bubalis*) corpora lutea during diestrus. *Microsc. Res. Tech.* **2012**, *75*, 1682–1690. [CrossRef]
20. Diaz, F.J.; Anderson, L.E.; Wu, Y.L.; Rabot, A.; Tsai, S.J.; Wiltbank, M.C. Regulation of progesterone and prostaglandin $F_{2\alpha}$ production in the CL. *Mol. Cell Endocrinol.* **2002**, *191*, 65–80. [CrossRef]
21. Zerani, M.; Dall'Aglio, C.; Maranesi, M.; Gobbetti, A.; Brecchia, G.; Mercati, F.; Boiti, C. Intraluteal regulation of prostaglandin $F_{2\alpha}$-induced prostaglandin biosynthesis in pseudopregnant rabbits. *Reproduction* **2007**, *133*, 1005–1116. [CrossRef] [PubMed]
22. Smith, W.L.; Garavito, R.M.; De Witt, D.L. Prostaglandin endoperoxide H synthase (cyclooxygenase)-1 and -2. *J. Biol. Chem.* **1996**, *271*, 33157–33160. [CrossRef] [PubMed]
23. Sakurai, T.; Tamura, K.; Okamoto, S.; Hara, T.; Kogo, H. Possible role of cyclooxygenase 2 in the acquisition of ovarian luteal function in rodents. *Biol. Reprod.* **2003**, *69*, 835–842. [CrossRef] [PubMed]
24. Simmons, D.L.; Botting, R.M.; Hla, T. Cyclooxygenase isozymes: The biology of prostaglandin synthesis and inhibition. *Pharmacol. Rev.* **2004**, *56*, 387–437. [CrossRef]
25. Helliwell, R.J.A.; Adams, L.F.; Mitchell, M.D. Prostaglandin synthases: Recent developments and a novel hypothesis. *Prostaglandins Leukot. Essent. Fat. Acids* **2004**, *70*, 101–113. [CrossRef]
26. Watanabe, K. Prostaglandin F synthase. *Prostaglandins Other Lipid Mediat.* **2002**, *68–69*, 401–407. [CrossRef]
27. Schlegel, W.; Daniels, D.; Kruger, S. Partial purification of prostaglandin E_2-9 ketoreductase and prostaglandin-15-hydroxydehydrogenase from ovarian tissues of rabbits. *Clin. Physiol. Biochem.* **1987**, *5*, 336–342.
28. Wintergalen, N.; Thole, H.H.; Galla, H.J.; Schlegel, W. Prostaglandin-E_2-9-reductase from corpus-luteum of pseudopregnant rabbit is a member of the aldo-keto reductase superfamily featuring 20-alphahydroxysteroid dehydrogenase-activity. *Eur. J. Biochem.* **1995**, *234*, 264–270. [CrossRef]
29. Arosh, J.A.; Banu, S.K.; Chapdelaine, P.; Madore, E.; Sirois, J.; Fortier, M.A. Prostaglandin biosynthesis, transport and signaling in corpus luteum: A basis for autoregulation of luteal function. *Endocrinology* **2004**, *145*, 2551–2560. [CrossRef]
30. Madore, E.; Harvey, N.; Parent, J.; Chapdelaine, P.; Arosh, J.A.; Fortier, M.A. An aldose reductase with 20a-hydroxysteroid dehydrogenase activity is most likely the enzyme responsible for the production of prostaglandin F2a in the bovine endometrium. *J. Biol. Chem.* **2003**, *278*, 11205–11212. [CrossRef]
31. Maranesi, M.; Zerani, M.; Lilli, L.; Dall'Aglio, C.; Brecchia, G.; Gobbetti, A.; Boiti, C. Expression of luteal estrogen receptor, interleukin-1, and apoptosis-associated genes after $PGF_{2\alpha}$ administration in rabbits at different stages of pseudopregnancy. *Domest. Anim. Endocrinol.* **2010**, *39*, 116–130. [CrossRef] [PubMed]
32. Krusche, C.A.; Vloet, T.D.; Herrier, A.; Black, S.; Beier, H.M. Functional and structural regression of the rabbit corpus luteum is associated with altered luteal immune cell phenotypes and cytokine expression patterns. *Histochem. Cell Biol.* **2002**, *118*, 479–489. [CrossRef] [PubMed]
33. Boiti, C.; Guelfi, G.; Zerani, M.; Zampini, D.; Brecchia, G.; Gobbetti, A. Expression patterns of cytokines, p53, and nitric oxide synthase isoenzymes in corpora lutea of pseudopregnant rabbits during spontaneous luteolysis. *Reproduction* **2004**, *127*, 229–238. [CrossRef] [PubMed]
34. Del Vecchio, R.P.; Sutherland, W.D. Prostaglandin and progesterone production by bovine luteal cells incubated in the presence or absence of the accessory cells of the corpus luteum and treated with interleukin-1beta, indomethacin and luteinizing hormone. *Reprod. Fertil. Dev.* **1997**, *9*, 651–658. [CrossRef]
35. Leon, L.; Jeannin, J.F.; Bettaieb, A. Post-translational modifications induced by nitric oxide (NO): Implication in cancer cells apoptosis. *Nitric Oxide* **2008**, *19*, 77–83. [CrossRef]
36. Preutthipan, S.; Chen, S.H.; Tilly, J.L.; Kugu, K.; Lareu, R.R.; Dharmarajan, A.M. Inhibition of nitric oxide synthesis potentiates apoptosis in the rabbit corpus luteum. *Reprod. Biomed. Online* **2004**, *9*, 264–270. [CrossRef]
37. Parillo, F.; Dall'Aglio, C.; Brecchia, G.; Maranesi, M.; Polisca, A.; Boiti, C.; Zerani, M. Aglepristone (RU534) effects on luteal function of pseudopregnant rabbits: Steroid receptors, enzymatic activities, and hormone productions in corpus luteum and uterus. *Anim. Reprod. Sci.* **2013**, *138*, 118–132. [CrossRef]
38. Moncada, S.; Palmer, R.M.J.; Higgs, E.A. Nitric oxide: Physiology, pathophysiology, and pharmacology. *Pharmacol. Rev.* **1991**, *42*, 109–142.
39. Schmidt, H.H.; Walter, U. NO at work. *Cell* **1994**, *78*, 919–925. [CrossRef]
40. Xie, Q.W.; Nathan, C. The high-output nitric oxide pathway: Role and regulation. *J. Leukoc. Biol.* **1994**, *56*, 576–582. [CrossRef]
41. Van Voorhis, B.J.; Dunn, M.S.; Snyder, G.D.; Weiner, C.P. Nitric oxide: An autocrine regulator of human granulosa-luteal cell steroidogenesis. *Endocrinology* **1994**, *135*, 1799–1806. [CrossRef] [PubMed]
42. Chatterjee, S.; Gangula, P.R.; Dong, Y.L.; Yallampalli, C. Immunocytochemical localization of nitric oxide synthase-III in reproductive organs of female rats during the oestrous cycle. *Histochem. J.* **1996**, *28*, 715–723. [CrossRef] [PubMed]
43. Hesla, J.S.; Preutthipan, S.; Maguire, M.P.; Chang, T.S.; Wallach, E.E.; Dharmarajan, A.M. Nitric oxide modulates human chorionic gonadotropin-induced ovulation in the rabbit. *Fertil. Steril.* **1997**, *67*, 548–552. [CrossRef]
44. Jablonka-Shariff, A.; Olson, L.M. Hormonal regulation of nitric oxide synthases and their cell-specific expression during follicular development in the rat ovary. *Endocrinology* **1997**, *138*, 460–468. [CrossRef] [PubMed]
45. Yamauchi, J.; Miyazaki, T.; Iwasaki, S.; Kishi, I.; Kuroshima, M.; Tei, C.; Yoshimura, Y. Effects of nitric oxide on ovulation and ovarian steroidogenesis and prostaglandin production in the rabbit. *Endocrinology* **1997**, *138*, 3630–3637. [CrossRef]

46. Pinto, C.R.F.; Paccamonti, D.L.; Eilts, B.E.; Short, C.R.; Godke, R.A. Effect of nitric oxide synthase inhibitors on ovulation in hCG stimulated mares. *Theriogenology* **2002**, *58*, 1017–1026. [CrossRef]
47. Shukovski, L.; Tsafriri, T. The involvement of nitric oxide in the ovulatory process in the rat. *Endocrinology* **1995**, *135*, 2287–2290. [CrossRef]
48. Bonello, N.; McKie, K.; Jasper, M.; Andrew, L.; Ross, N.; Braybon, E.; Brannstrom, M.; Norman, R.J. Inhibition of nitric oxide: Effects on interleukin-1á-enhanced ovulation rate, steroid hormone, and ovarian leukocyte distribution at ovulation in the rat. *Biol. Reprod.* **1996**, *54*, 436–445. [CrossRef]
49. Zackrisson, U.; Mikuni, M.; Wallin, A.; Delbro, D.; Hedin, L.; Brannstrom, M. Cell-specific localization of nitric oxide synthases (NOS) in the rat ovary during follicular development, ovulation and luteal formation. *Hum. Reprod.* **1996**, *11*, 2667–2673. [CrossRef]
50. Boiti, C.; Zampini, D.; Guelfi, G.; Paolocci, F.; Zerani, M.; Gobbetti, A. Expression patterns of endothelial and inducible isoforms in corpora lutea of pseudopregnant rabbits at different luteal stages. *J. Endocrinol.* **2002**, *173*, 285–296. [CrossRef]
51. Korzekwa, A.; Woclawek-Potocka, I.; Okuda, K.; Acosta, T.J.; Skarzynski, D.J. Nitric oxide in bovine corpus luteum: Possible mechanisms of action in luteolysis. *J. Anim. Sci.* **2007**, *78*, 233–242. [CrossRef]
52. Korzekwa, A.; Jaroszewski, J.J.; Bogacki, M.; Deptula, K.M.; Maslanka, T.S.; Acosta, T.J.; Okuda, K.; Skarzynski, D.J. Effects of prostaglandin $F_{2\alpha}$ and nitric oxide on the secretory function of bovine luteal cells. *J. Reprod. Dev.* **2004**, *50*, 411–417. [CrossRef] [PubMed]
53. Jaroszewski, J.J.; Skarzynski, D.J.; Hansel, W. Nitric oxide as a local mediator of prostaglandin F2-induced regression in bovine corpus luteum: An in vivo study. *Exp. Biol. Med.* **2003**, *228*, 1057–1062. [CrossRef] [PubMed]
54. Roberto da Costa, R.P.; Costa, A.S.; Korzekwa, A.; Platek, R.; Siemieniuch, M.; Galvão, A.; Redmer, D.A.; Robalo Silva, J.; Skarzynski, D.J.; Ferreira-Dias, G. Actions of a nitric oxide donor on prostaglandins production and angiogenic activity in the equine endometrium. *Reprod. Fertil. Dev.* **2008**, *20*, 674–683. [CrossRef]
55. Ferreira-Dias, G.; Costa, A.S.; Mateus, L.; Korzekwa, A.J.; Galvão, A.; Redmer, D.A.; Lukasik, K.; Szóstek, A.Z.; Woclawek-Potocka, I.; Skarzynski, D.J. Nitric oxide stimulates progesterone and prostaglandin E_2 secretion as well as angiogenic activity in the equine corpus luteum. *Domest. Anim. Endocrinol.* **2010**, *40*, 1–9. [CrossRef]
56. Olson, L.M.; Jones-Burton, C.M.; Jablonka-Shariff, A. Nitric oxide decreases estradiol synthesis of rat luteinized ovarian cells: Possible role for nitric oxide in functional luteal regression. *Endocrinology* **1996**, *137*, 3531–3539. [CrossRef]
57. Holt, J.A. Regulation of progesterone production in the rabbit corpus luteum. *Biol. Reprod.* **1989**, *40*, 201–208. [CrossRef]
58. Zhang, Y.; Proenca, R.; Maffei, M.; Barone, M.; Leopold, L.; Friedman, J.M. Positional cloning of the mouse obese gene and its human homologue. *Nature* **1994**, *372*, 425–432. [CrossRef]
59. Morash, B.; Li, A.; Murphy, P.R.; Wilkinson, M.; Ur, E. Leptin gene expression in the brain and pituitary gland. *Endocrinology* **1999**, *140*, 5995–5998. [CrossRef]
60. Harris, R.B. Leptin–much more than a satiety signal. *Annu. Rev. Nutr.* **2000**, *20*, 45–75. [CrossRef]
61. Houseknecht, K.L.; Portocarrero, C.P. Leptin and its receptors: Regulators of whole-body energy homeostasis. *Domest. Anim. Endocrinol.* **1998**, *15*, 457–475. [CrossRef]
62. Sweeney, G. Leptin signalling. *Cell. Signal.* **2002**, *14*, 655–663. [CrossRef]
63. Zabeau, L.; Lavens, D.; Peelman, F.; Eycherman, S.; Vandekerckhove, J.; Tavernier, J. The ins and out of leptin receptor activation. *FEBS Lett.* **2003**, *546*, 45–50. [CrossRef]
64. Baumann, H.; Morella, K.K.; White, D.W.; Dembski, M.; Bailon, P.S.; Kim, H.; Lai, C.F.; Tartaglia, L.A. The full-length leptin receptor has signaling capabilities of interleukin 6-type cytokine receptors. *Proc. Nat. Acad. Sci. USA* **1996**, *9*, 8374–8378. [CrossRef] [PubMed]
65. Bjørbæk, A.S.; Uotani, S.; da Silva, B.; Flier, J.S. Divergent signaling capacities of the long and short isoforms of the leptin receptor. *J. Biol. Chem.* **1997**, *272*, 32686–32695. [CrossRef] [PubMed]
66. Mantzoros, C.S. Role of leptin in reproduction. *Ann. N. Y. Acad. Sci.* **2000**, *900*, 174–183. [CrossRef]
67. Spicer, L.J.; Francisco, C.C. The adipose obese gene product, leptin: Evidence of a direct inhibitory role in ovarian function. *Endocrinology* **1997**, *138*, 3374–3379. [CrossRef]
68. Agarwal, S.K.; Vogel, K.; Weitsman, S.R.; Magoffin, D.A. Leptin antagonizes the insulin-like growth factor-I augmentation of steroidogenesis in granulosa and theca cells of the human ovary. *J. Clin. Endocrinol. Metab.* **1999**, *84*, 1072–1076. [CrossRef]
69. Cunningham, M.J.; Clifton, D.K.; Steiner, R.A. Leptin's actions on the reproductive axis: Perspectives and mechanisms. *Biol. Reprod.* **1999**, *60*, 216–222. [CrossRef]
70. Ryan, N.K.; Woodhouse, C.M.; Van Der Hoeck, K.H.; Gilchrist, R.B.; Armstrong, D.T.; Norman, R.J. Expression of leptin and its receptor in the murine ovary: Possible role in the regulation of oocyte maturation. *Biol. Reprod.* **2002**, *66*, 1548–1554. [CrossRef]
71. Mounzih, K.; Qiu, J.; Ewart-Toland, A.; Chehab, F.F. Leptin is not necessary for gestation and parturition but regulates maternal nutrition via a leptin resistance state. *Endocrinology* **1998**, *139*, 5259–5262. [CrossRef] [PubMed]
72. Mukherjea, R.; Castonguay, T.W.; Douglass, L.W.; Moser-Veillon, P. Elevated leptin concentrations in pregnancy and lactation: Possible role as a modulator of substrate utilization. *Life Sci.* **1999**, *65*, 1183–1193. [CrossRef]
73. Quinton, N.D.; Laird, S.M.; Kocon, M.A.; Li, T.C.; Smith, R.F.; Ross, R.J.; Blakemore, A.I. Serum leptin levels during the menstrual cycle of healthy fertile women. *Br. J. Biomed. Sci.* **1999**, *56*, 16–19. [PubMed]

74. Ludwig, M.; Klein, H.H.; Diedrich, K.; Ortmann, O. Serum leptin concentrations throughout the menstrual cycle. *Arch. Gynecol. Obstet.* **2000**, *263*, 99–101. [CrossRef] [PubMed]
75. Moschos, S.; Chan, J.L.; Mantzoros, C.S. Leptin and reproduction: A review. *Fertil. Steril.* **2002**, *77*, 433–444. [CrossRef]
76. Jin, L.; Zhang, S.; Burguera, B.G.; Couce, M.E.; Osamura, R.Y.; Kulig, E.; Lloyd, R.V. Leptin and leptin receptor expression in rat and mouse pituitary cells. *Endocrinology* **2000**, *141*, 333–339. [CrossRef] [PubMed]
77. Karlsson, C.; Lindell, K.; Svensson, E.; Bergh, C.; Lind, P.; Billig, H.; Carlsson, L.M.; Carlsson, B. Expression of functional leptin receptors in the human ovary. *J. Clin. Endocrinol. Metab.* **1997**, *82*, 4144–4148. [CrossRef]
78. Zachow, R.J.; Magoffin, D.A. Direct intraovarian effects of leptin: Impairment of the synergistic action of insulin-like growth factor-I on follicle-stimulating hormone-dependent estradiol-17 beta production by rat ovarian granulosa cells. *Endocrinology* **1997**, *138*, 847–850. [CrossRef]
79. Brannian, J.D.; Zhao, Y.; McElroy, M. Leptin inhibits gonadotrophin-stimulated granulosa cell progesterone production by antagonizing insulin action. *Hum. Reprod.* **1999**, *14*, 1445–1448. [CrossRef]
80. Zerani, M.; Boiti, C.; Zampini, D.; Brecchia, G.; Dall'Aglio, C.; Ceccarelli, P.; Gobbetti, A. Ob receptor in rabbit ovary and leptin in vitro regulation of corpora lutea. *J. Endocrinol.* **2004**, *183*, 279–288. [CrossRef]
81. Conn, P.M.; Crowley, W.F., Jr. Gonadotropin-releasing hormone and its analogues. *Ann. Rev. Med.* **1994**, *45*, 391–405. [CrossRef] [PubMed]
82. Ramakrishnappa, N.; Rajamahendran, R.; Lin, Y.M.; Leung, P.C.K. GnRH in non- hypothalamic reproductive tissues. *Anim. Reprod. Sci.* **2005**, *88*, 95–113. [CrossRef] [PubMed]
83. Rippel, R.H.; Johnson, E.S.; Kimura, E.T. Regression of corpora lutea in the rabbit after injection of a gonadotropin-releasing peptide. *Proc. Soc. Experim. Biol. Med.* **1976**, *152*, 29–32. [CrossRef] [PubMed]
84. Eisenberg, E.; Kitai, H.; Kobayashi, Y.; Santulli, R.; Wallach, E.E. Gonadotropin releasing hormone: Effects on the in vitro perfused rabbit ovary. *Biol. Reprod.* **1984**, *30*, 1216–1221. [CrossRef] [PubMed]
85. Zerani, M.; Parillo, F.; Brecchia, G.; Guelfi, G.; Dall'Aglio, C.; Lilli, L.; Maranesi, M.; Gobbetti, A.; Boiti, C. Expression of type I GnRH receptor and in vivo and in vitro GnRH-I effects in corpora lutea of pseudopregnant rabbits. *J. Endocrinol.* **2010**, *207*, 289–300. [CrossRef]
86. Meidan, R.; Milvae, R.A.; Weiss, S.; Levy, N.; Friedman, A. Intra-ovarian regulation of luteolysis. *J. Reprod. Fertil.* **1999**, *54*, 217–228.
87. Yanagisawa, M.; Kurihara, H.; Kimura, S.; Tomobe, Y.; Kobayashi, M.; Mitsui, Y.; Yazaki, Y.; Goto, K.; Masaki, T. A novel potent vasoconstrictor peptide produced by vascular endothelial cells. *Nature* **1988**, *332*, 411–415. [CrossRef]
88. Iwai, M.; Hasegawa, M.; Taii, S.; Sagawa, N.; Nakao, K.; Imura, H.; Nakanishi, S.; Mori, T. Endothelins inhibit luteinization of cultured porcine granulosa cells. *Endocrinology* **1991**, *129*, 1909–1914. [CrossRef]
89. Bagavandoss, P.; Wilks, J.W. Isolation and characterization of microvascular endothelial cells from developing corpus luteum. *Biol. Reprod.* **1991**, *44*, 1132–1139. [CrossRef]
90. Usuki, S.; Suzuki, N.; Matsumoto, H.; Yanagisawa, M.; Masaki, T. Endothelin-1 in luteal tissue. *Mol. Cell Endocrinol.* **1991**, *80*, 147–151. [CrossRef]
91. Boiti, C.; Guelfi, G.; Brecchia, G.; Dall'Aglio, C.; Ceccarelli, P.; Maranesi, M.; Mariottini, C.; Zampini, D.; Gobbetti, A.; Zerani, M. Role of the endothelin-1 system in the luteolytic process of pseudopregnant rabbits. *Endocrinology* **2005**, *146*, 1293–1300. [CrossRef] [PubMed]
92. Boiti, C.; Maranesi, M.; Dall'aglio, C.; Pascucci, L.; Brecchia, G.; Gobbetti, A.; Zerani, M. Vasoactive peptides in the luteolytic process activated by $PGF_{2\alpha}$ in pseudopregnant rabbits at different luteal stages. *Biol. Reprod.* **2007**, *77*, 156–164. [CrossRef] [PubMed]
93. Schams, D.; Berisha, B.; Neuvians, T.; Amselgruber, W.; Kraetzl, W.D. Real-time changes of the local vasoactive peptide systems (angiotensin, endothelin) in the bovine corpus luteum after induced luteal regression. *Mol. Reprod. Dev.* **2003**, *65*, 57–66. [CrossRef] [PubMed]
94. Goligorsky, M.S.; Budzikowski, A.S.; Tsukahara, H.; Noiri, E. Co-operation between endothelin and nitric oxide in promoting endothelial cell migration and angiogenesis. *Clin. Exp. Pharm. Physiol.* **1999**, *26*, 269–271. [CrossRef]
95. Zhang, M.; Luo, B.; Chen, S.J.; Abrams, G.A.; Fallon, M.B. Endothelin-1 stimulation of endothelial nitric oxide synthase in the pathogenesis of hepatopulmonary syndrome. *Am. J. Physiol.* **1999**, *277*, G944–G952. [CrossRef]
96. Tognetti, T.; Estevez, A.; Luchetti, C.G.; Sander, V.; Franchi, A.M.; Motta, A.B. Relationship between endothelin-1 and nitric oxide system in the corpus luteum regression. *Prostaglandins Leukot. Essent. Fat. Acids* **2003**, *69*, 359–364. [CrossRef]
97. Ferin, M. Stress and the reproductive cycle. *J. Clin. Endocrinol. Metab.* **1998**, *84*, 1768–1774. [CrossRef]
98. Pau, K.Y.F.; Orstead, K.M.; Hess, D.L.; Spies, H.G. Feedback effects ovarian steroids on the hypothalamic hypophyseal axis in the rabbit. *Biol. Reprod.* **1986**, *35*, 1009–1023. [CrossRef]
99. Schioth, H.B. The physiological role of melanocortin receptors. *Vitam. Horm.* **2001**, *63*, 195–232.
100. Guelfi, G.; Zerani, M.; Brecchia, G.; Parillo, F.; Dall'Aglio, C.; Maranesi, M.; Boiti, C. Direct actions of ACTH on ovarian function of pseudopregnant rabbits. *Mol. Cell Endocrinol.* **2011**, *339*, 63–71. [CrossRef]
101. Adashi, E.Y. The potential relevance of cytokines to ovarian physiology: The emerging role of resident ovarian cells of the white blood cell series. *Endocrin. Rev.* **1990**, *11*, 454–464. [CrossRef] [PubMed]

102. Bagavandoss, P.; Wiggins, R.C.; Kunkel, S.L.; Remick, D.G.; Keyes, P.L. Tumor necrosis factor production and accumulation of inflammatory cells in the corpus luteum of pseudopregnancy and pregnancy in rabbits. *Biol. Reprod.* **1990**, *42*, 367–376. [CrossRef] [PubMed]
103. Bagavandoss, P.; Kunkel, S.L.; Wiggins, R.C.; Keyes, P.L. Tumor necrosis factor-a (TNF-a) production and localization of macrophages and T lymphocytes in the rabbit corpus luteum. *Endocrinology* **1988**, *122*, 1185–1187. [CrossRef] [PubMed]
104. Takehara, Y.; Dharmarajan, A.M.; Kaufman, G.; Wallach, E.E. Effect of interleukin 1β on ovulation in the in vitro perfused rabbit ovary. *Endocrinology* **1994**, *134*, 1788–1793. [CrossRef]
105. Bréard, E.; Delarue, B.; Benhaïm, A.; Féral, C.; Leymarie, P. Inhibition by gonadotropins of interleukin-1 production by rabbit granulosa and theca cells: Effects on gonadotropin-induced progesterone production. *Eur. J. Endocrinol.* **1998**, *138*, 328–336. [CrossRef]
106. Gérard, N.; Caillaud, M.; Martoriati, A.; Goudet, G.; Lalmanach, A.C. The interleukin-1 system and female reproduction. *J. Endocrinol.* **2004**, *180*, 203–212. [CrossRef]
107. Narko, K.; Ritvos, O.; Ristmaki, A. Induction of cyclooxygenase-2 and prostaglandin F2α receptor expression by interleukin-1β in cultured human granulosa–luteal cells. *Endocrinology* **1997**, *138*, 3638–3644. [CrossRef]
108. Estevez, A.; Tognetti, T.; Rearte, B.; Sander, V.; Motta, A.B. Interleukin1-beta in the functional luteolysis. Relationship with the nitric oxide system. *Prostaglandins Leukot. Essent. Fat. Acids* **2002**, *67*, 411–417. [CrossRef]
109. Froment, P.; Gizard, F.; Defever, D.; Staels, B.; Dupont, J.; Monget, P. Peroxisome proliferator-activated receptors in reproductive tissues: From gametogenesis to parturition. *J. Endocrinol.* **2006**, *189*, 199–209. [CrossRef]
110. Komar, C.M. Peroxisome proliferator-activated receptors (PPARs) and ovarian function -implications for regulating steroidogenesis, differentiation, and tissue remodelling. *Reprod. Biol. Endocrinol.* **2005**, *3*, e41. [CrossRef]
111. Minge, C.E.; Robker, R.L.; Norman, R.J. PPAR gamma: Coordinating metabolic and immune contributions to female fertility. *PPAR Res.* **2008**, 243791. [CrossRef] [PubMed]
112. Braissant, O.; Foufelle, F.; Scotto, C.; Dauça, M.; Wahli, W. Differential expression of peroxisome proliferator-activated receptors (PPARs): Tissue distribution of PPAR-α, -β, and -γ in the adult rat. *Endocrinology* **1996**, *137*, 354–366. [CrossRef] [PubMed]
113. Cui, Y.; Miyoshi, K.; Claudio, E.; Siebenlist, U.K.; Gonzalez, F.J.; Flaws, J.; Wagner, K.U.; Hennighausen, L. Loss of the peroxisome proliferation-activated receptor γ (PPARγ) does not affect mammary development and propensity for tumor formation but leads to reduced fertility. *J. Biol. Chem.* **2002**, *277*, 17830–17835. [CrossRef] [PubMed]
114. Schoppee, P.D.; Garmey, J.C.; Veldhuis, J.D. Putative activation of the peroxisome proliferator-activated receptor γ impairs androgen and enhances progesterone biosynthesis in primary cultures of porcine theca cells. *Biol. Reprod.* **2002**, *66*, 190–198. [CrossRef]
115. Froment, P.; Fabre, S.; Dupont, J.; Pisselet, C.; Chesneau, D.; Staels, B.; Monget, P. Expression and functional role of peroxisome proliferator-activated receptor-γ in ovarian folliculogenesis in the sheep. *Biol. Reprod.* **2003**, *69*, 1665–1674. [CrossRef]
116. Löhrke, B.; Viergutz, T.; Shahi, S.K.; Pöhland, R.; Wollenhaupt, K.; Goldammer, T.; Walzel, H.; Kanitz, W. Detection and functional characterisation of the transcription factor peroxisome proliferator-activated receptor γ in lutein cells. *J. Endocrinol.* **1998**, *159*, 429–439. [CrossRef]
117. Sundvold, H.; Brzozowska, A.; Lien, S. Characterisation of bovine peroxisome proliferator-activated receptors γ1 and γ2: Genetic mapping and differential expression of the two isoforms. *Biochem. Biophys. Res. Commun.* **1997**, *239*, 857–861. [CrossRef]
118. Parillo, F.; Catone, G.; Gobbetti, A.; Zerani, M. Cell localization of ACTH, dopamine, and GnRH receptors and PPARγ in bovine corpora lutea during diestrus. *Acta Sci. Vet.* **2013**, *41*, e1129.
119. Maranesi, M.; Zerani, M.; Parillo, F.; Brecchia, G.; Gobbetti, A.; Boiti, C. Role of peroxisome proliferator-activated receptor gamma in corpora lutea of pseudopregnant rabbits. *Reprod. Domest. Anim.* **2011**, *46* (Suppl. 3), 126.
120. Lambe, K.G.; Tugwood, J.D. A human peroxisomeproliferator-activated receptor-γ is activated by inducers of adipogenesis, including thiazalidinedione drugs. *Eur. J. Biochem.* **1996**, *239*, 1–7. [CrossRef]
121. Mu, Y.M.; Yanase, T.; Nishi, Y.; Waseda, N.; Oda, T.; Tanaka, A.; Takayanagi, R.; Nawata, H. Insulin sensitizer, troglitazone, directly inhibits aromatase activity in human ovarian granulosa cells. *Biochem. Biophys. Res. Commun.* **2000**, *271*, 710–713. [CrossRef] [PubMed]
122. Doney, A.; Fischer, B.; Frew, D.; Cumming, A.; Flavell, D.M.; World, M.; Montgomery, H.E.; Boyle, D.; Morris, A.; Palmer, C.N. Haplotype analysis of the PPARγ Pro12Ala and C1431T variants reveals opposing associations with body weight. *BMC Genet.* **2002**, *3*, e21. [CrossRef] [PubMed]
123. Viergutz, T.; Lšhrke, B.; Poehland, R.; Becker, F.; Kanitz, W. Relationship between different stages of the corpus luteum and the expression of the peroxisome proliferator-activated receptor gamma protein in bovine large lutein cells. *J. Reprod. Fertil.* **2000**, *118*, 153–161. [CrossRef] [PubMed]
124. Willis, D.S.; White, J.; Brosens, S.; Franks, S. Effect of 15-deoxy-delta (12,14)-prostaglandin J2 (PGJ2) a peroxisome proliferator activating receptor γ (PPARγ) ligand on human ovarian steroidogenesis. *Endocrinology* **1999**, *140*, 1491.
125. Zerani, M.; Maranesi, M.; Brecchia, G.; Gobbetti, A.; Boiti, C.; Parillo, F. Evidence for a luteotropic role of peroxisome proliferator-activated receptor gamma: Expression and in vitro effects on enzymatic and hormonal activities in corpora lutea of pseudopregnant rabbits. *Biol. Reprod.* **2013**, *88*, 62. [CrossRef]

126. Parillo, F.; Maranesi, M.; Brecchia, G.; Gobbetti, A.; Boiti, C.; Zerani, M. In vivo chronic and in vitro acute effects of di(2-ethylhexyl) phthalate on pseudopregnant rabbit corpora lutea: Possible involvement of peroxisome proliferator-activated receptor gamma. *Biol. Reprod.* **2014**, *90*, 41. [CrossRef]
127. Yamamoto, K.; Vernier, P. The evolution of dopamine systems in chordates. *Front. Neuroanat.* **2011**, *5*, e21. [CrossRef]
128. Beaulieu, J.M.; Gainetdinov, R.R. The physiology, signaling, and pharmacology of dopamine receptors. *Pharm. Rev.* **2011**, *63*, 182–217. [CrossRef]
129. Mayerhofer, A.; Fritz, S.; Grunert, R.; Sanders, S.L.; Uffy, D.M.; Ojeda, S.R.; Stouffer, R.L. D1-receptor, darpp-32, and pp-1 in the primate corpus luteum and luteinized granulosa cells: Evidence for phosphorylation of darpp-32 by dopamine and human chorionic gonadotropin. *J. Clin. Endocrinol. Metab.* **2000**, *85*, 4750–4757. [CrossRef]
130. Mayerhofer, A.; Hemmings, H.C., Jr.; Snyder, G.L.; Greengard, P.; Boddien, S.; Berg, U.; Brucker, C. Functional dopamine-1 receptors and DARPP-32 are expressed in human ovary and granulosa luteal cells in vitro. *J. Clin. Endocrinol. Metab.* **1999**, *84*, 257–264. [CrossRef]
131. King, S.S.; Campbell, A.G.; Dille, E.A.; Roser, J.F.; Murphy, L.L.; Jones, K.L. Dopamine receptors in equine ovarian tissues. *Domest. Anim. Endocrinol.* **2005**, *28*, 405–415. [CrossRef] [PubMed]
132. Rey-Ares, V.; Lazarov, N.; Berg, D.; Berg, U.; Kunz, L.; Mayerhofer, A. Dopamine receptor repertoire of human granulosa cells. *Reprod. Biol. Endocrinol.* **2007**, *5*, e40. [CrossRef] [PubMed]
133. Parillo, F.; Maranesi, M.; Mignini, F.; Marinelli, L.; Di Stefano, A.; Boiti, C.; Zerani, M. Evidence for a dopamine intrinsic direct role in the regulation of the ovary reproductive function: In vitro study on rabbit corpora lutea. *PLoS ONE* **2014**, *9*, e104797. [CrossRef] [PubMed]
134. Gerendai, I.; Banczerowski, P.; Halász, B. Functional significance of the innervation of the gonads. *Endocrine* **2005**, *28*, 309–318. [CrossRef]
135. D'Albora, H.; Anesetti, G.; Lombide, P.; Dees, W.L.; Ojeda, S.R. Intrinsic neurons in the mammalian ovary. *Microsc. Res. Tech.* **2002**, *59*, 484–489. [CrossRef]
136. Skorupskaite, K.; George, J.T.; Anderson, R.A. The kisspeptin-GnRH pathway in human reproductive health and disease. *Hum. Reprod. Update* **2014**, *20*, 485–500. [CrossRef]
137. Wahab, F.; Atika, B.; Shahab, M.; Behr, R. Kisspeptin signalling in the physiology and pathophysiology of the urogenital system. *Nat. Rev. Urol.* **2016**, *13*, 21–32. [CrossRef]
138. Basini, G.; Grasselli, F.; Bussolati, S.; Ciccimarra, R.; Maranesi, M.; Bufalari, A.; Parillo, F.; Zerani, M. Presence and function of kisspeptin/KISS1R system in swine ovarian follicles. *Theriogenology* **2018**, *115*, 1–8. [CrossRef]
139. Peng, J.; Tang, M.; Zhang, B.P.; Zhang, P.; Zhong, T.; Zong, T.; Yang, B.; Kuang, H.B. Kisspeptin stimulates progesterone secretion via the Erk1/2 mitogen-activated protein kinase signaling pathway in rat luteal cells. *Fertil. Steril.* **2013**, *99*, 1436–1443. [CrossRef]
140. Laoharatchatathanin, T.; Terashima, R.; Yonezawa, T.; Kurusu, S.; Kawaminami, M. Augmentation of metastin/kisspeptin mRNA expression by the proestrous luteinizing hormone surge in granulosa cells of rats: Implications for luteinization. *Biol. Reprod.* **2015**, *93*, 15. [CrossRef]
141. Maranesi, M.; Petrucci, L.; Leonardi, L.; Bufalari, A.; Parillo, F.; Boiti, C.; Zerani, M. Kisspeptin/kisspeptin receptor system in pseudopregnant rabbit corpora lutea: Presence and function. *Sci. Rep.* **2019**, *9*, 5044. [CrossRef] [PubMed]
142. Pinilla, L.; Aguilar, E.; Dieguez, C.; Millar, R.P.; Tena-Sempere, M. Kisspeptins and reproduction: Physiological roles and regulatory mechanisms. *Physiol. Rev.* **2012**, *92*, 1235–1316. [CrossRef] [PubMed]
143. Franssen, D.; Tena-Sempere, M. The kisspeptin receptor: A key G-protein-coupled receptor in the control of the reproductive axis. *Best Pract. Res. Clin. Endocrinol. Metab.* **2018**, *32*, 107–123. [CrossRef] [PubMed]
144. Levi-Montalcini, R. The nerve growth factor 35 years later. *Science* **1987**, *237*, 1154–1162. [CrossRef] [PubMed]
145. Thoenen, H.; Barde, Y.A. Physiology of nerve growth factor. *Physiol. Rev.* **1980**, *60*, 1284–1335. [CrossRef]
146. Snider, W.D. Functions of the neurotrophins during nervous system development: What the knockouts are teaching us. *Cell* **1994**, *77*, 627–638. [CrossRef]
147. Garcia-Garcia, R.M.; Arias-Alvarez, M.; Sanchez-Rodriguez, A.; Lorenzo, P.L.; Rebollar, P.G. Role of nerve growth factor in the reproductive physiology of female rabbits: A review. *Theriogenology* **2020**, *15*, 321–328. [CrossRef]
148. Maranesi, M.; Petrucci, L.; Leonardi, L.; Piro, F.; Rebollar, P.G.; Millá, P.; Cocci, P.; Vullo, C.; Parillo, F.; Moura, A.; et al. New insights on a NGF-mediated pathway to induce ovulation in rabbits (*Oryctolagus cuniculus*). *Biol. Reprod.* **2018**, *98*, 634–643. [CrossRef]
149. Maranesi, M. Expression and potential mechanisms of action of b-NGF and its TRKA receptor in the reproductive system of rabbits. In Proceedings of the International Workshop: The β-Nerve Growth Factor in Rabbit Semen, Possible Role in Artificial Insemination, Perugia, Italy, 26 September 2014.
150. Silva, M.; Ulloa-Leal, C.; Valderrama, X.P.; Bogle, O.A.; Adams, G.P.; Ratto, M.H. Nerve growth factor from seminal plasma origin (spβ-NGF) increases CL vascularization and level of mRNA expression of steroidogenic enzymes during the early stage of Corpus Luteum development in llamas. *Theriogenology* **2017**, *103*, 69–75. [CrossRef]
151. Silva, M.; Ulloa-Leal, C.; Norambuena, C.; Fernández, A.; Adams, G.P.; Ratto, M.H. Ovulation-inducing factor (OIF/NGF) from seminal plasma origin enhances Corpus Luteum function in llamas regardless the preovulatory follicle diameter. *Anim. Reprod. Sci.* **2014**, *148*, 221–227. [CrossRef]

152. Tribulo, P.; Bogle, O.; Mapletoft, R.J.; Adams, G.P. Bioactivity of ovulation inducing factor (or nerve growth factor) in bovine seminal plasma and its effects onovarian function in cattle. *Theriogenology* **2015**, *83*, 1394–1401. [CrossRef] [PubMed]
153. Stewart, J.L.; Mercadante, V.R.G.; Dias, N.W.; Canisso, I.F.; Yau, P.; Imai, B.; Lima, F.S. Nerve Growth Factor-Beta, purified from bull seminal plasma, enhances corpus luteum formation and conceptus development in *Bos taurus* cows. *Theriogenology* **2018**, *106*, 30–38. [CrossRef] [PubMed]
154. Kershaw-Young, C.M.; Druart, X.; Vaughan, J.; Maxwell, W.M. β-Nerve growth factor is a major component of alpaca seminal plasma and induces ovulation in female alpacas. *Reprod. Fertil. Dev.* **2012**, *24*, 1093–1097. [CrossRef] [PubMed]
155. Stuart, C.C.; Vaughan, J.L.; Kershaw-Young, C.M.; Wilkinson, J.; Bathgate, R.; de Graaf, S.P. Effects of varying doses of β-nerve growth factor on the timing of ovulation, plasma progesterone concentration and corpus luteum size in female alpacas (*Vicugna pacos*). *Reprod. Fertil. Devel.* **2015**, *27*, 1181–1186. [CrossRef] [PubMed]

 *animals*

MDPI

Review

Strategies for Highly Efficient Rabbit Sperm Cryopreservation

Kazutoshi Nishijima [1,2,3,*], Shuji Kitajima [4], Fumikazu Matsuhisa [4], Manabu Niimi [5], Chen-chi Wang [6] and Jianglin Fan [5,7,*]

1 Center for Animal Resources and Collaborative Study, National Institutes of Natural Sciences, 38 Nishigonaka, Myodaiji, Okazaki 444-8585, Japan
2 National Institute for Physiological Sciences, National Institutes of Natural Sciences, 38 Nishigonaka, Myodaiji, Okazaki 444-8585, Japan
3 Department of Physiological Sciences, SOKENDAI (The Graduate University for Advanced Studies), Aichi, Okazaki 444-8585, Japan
4 Analytical Research Center for Experimental Sciences, Division of Biological Resources and Development, Saga University, 5-1-1 Nabeshima, Saga 849-8501, Japan; kitajims@cc.saga-u.ac.jp (S.K.); matsuf@cc.saga-u.ac.jp (F.M.)
5 Department of Molecular Pathology, Faculty of Medicine, Interdisciplinary Graduate School of Medical Sciences, University of Yamanashi, 1110 Shimokato, Chuo 409-3898, Japan; manabun@yamanashi.ac.jp
6 Animal Resources Section, Okinawa Institute of Science and Technology Graduate University, 1919-1 Tancha, Onna-son, Kunigami-gun, Okinawa 904-0495, Japan; ChenChi.Wang@oist.jp
7 School of Biotechnology and Health Sciences, Wuyi University, Jiangmen 529020, China
* Correspondence: kanish@nips.ac.jp (K.N.); jianglin@yamanashi.ac.jp (J.F.); Tel.: +81-564-557781 (K.N.); +81-55-2739519 (J.F.)

Simple Summary: The rabbit is a valuable animal for both the economy and biomedical sciences. Therefore, the preservation of many rabbit strains is vitally important. So far, sperm cryopreservation is one of the most efficient ways to preserve rabbit strains because it is easy to collect ejaculate repeatedly from a single male and perform artificial insemination to multiple females. Although this method is widely used, there are still some concerns regarding the cooling, freezing and thawing process of sperms, which markedly affects the quality of preserved sperms. In this article, we will review the progress made during the past years in terms of cryopreservation of rabbit sperms and discuss those factors that would possibly influence sperm damage including freezing extender, cryoprotectant, supplements, and procedures.

Abstract: The rabbit is a valuable animal for both the economy and biomedical sciences. Sperm cryopreservation is one of the most efficient ways to preserve rabbit strains because it is easy to collect ejaculate repeatedly from a single male and inseminate artificially into multiple females. During the cooling, freezing and thawing process of sperms, the plasma membrane, cytoplasm and genome structures could be damaged by osmotic stress, cold shock, intracellular ice crystal formation, and excessive production of reactive oxygen species. In this review, we will discuss the progress made during the past years regarding efforts to minimize the cell damage in rabbit sperms, including freezing extender, cryoprotectants, supplements, and procedures.

Keywords: rabbit; sperm quality; cryopreservation; animal model; assisted reproductive technology

Citation: Nishijima, K.; Kitajima, S.; Matsuhisa, F.; Niimi, M.; Wang, C.-c.; Fan, J. Strategies for Highly Efficient Rabbit Sperm Cryopreservation. *Animals* **2021**, *11*, 1220. https://doi.org/10.3390/ani11051220

Academic Editor: Rosa María García-García

Received: 15 March 2021
Accepted: 21 April 2021
Published: 23 April 2021

1. Introduction

Rabbits have been indispensable for human life because they are not only valuable for agriculture but also for biomedical research. Rabbits are widely used as a source of meat, hair and fur, and it is estimated that each year, around 300 million rabbits (and hares) are used in the world [1]. Because of their tame characters, rabbits are also raised as a pet. In addition, rabbits are the most-used animals for antibody production for biomedical research. Furthermore, rabbits are similar to humans in terms of cardiovascular physiology and lipid metabolism, and they play an important role in studying human diseases such as atherosclerosis and hypercholesterolemia [2,3]. Along with the development of genetic engineering, a number of gene-modified rabbits have been established as experimental models of human diseases. In addition to transgenic rabbits produced with the conventional pronuclear microinjection technique, knockout rabbits have been established using CRISPR/Cas9 genome editing technology [4]. These established genetically modified rabbits are rare and valuable and thus it is vitally important to breed and maintain rabbit strains for different purposes and preserve them as bio-resources [5].

There are two major ways to preserve rabbit strains. The common way to maintain a rabbit colony is carried out simply by repeat breeding. However, several difficulties with this method exist including space and cost. In particular, rabbit shows severe inbreeding depression [6–9], thus a considerable number of rabbits are required to keep a colony. For laboratory rabbits, they are usually housed in strictly controlled conditions in terms of temperature, humidity, illumination and microbiological examinations. Furthermore, living animals have a risk of annihilation or escape in the case of a disaster or accident.

The second method of maintaining the rabbit colony is the cryopreservation of gametes. Cryopreservation of gametes requires less space and cost than animal breeding. It is generally believed that properly cryopreserved zygotes and gametes can be preserved semi-eternally in a liquid nitrogen tank to keep their fertile and developmental ability. In the case of employing ovum or embryo preservation, ova or embryos are generally obtained with oviduct–uterus dissection from sacrificed females, and a skillful surgical operation is required for the embryo transfer. In contrast, ejected semen can be collected repeatedly without sacrificing males (Figure 1) and artificially inseminated into females can be conducted without specific skills, and thus, sperm cryopreservation would be the first choice for preservation of rabbit strains even though sperms preserved can bring paternal hereditary information only, and immediate offspring is always heterozygosity. However, when concerned with one specific transgene, homozygotes can be obtained in the second generation. Even in the case of livestock animals concerning pedigree related with polygenetic factors, the inbreeding can be avoided by mating live females and cryopreserved sperm with a generation gap.

As mentioned above, the successful preservation of rabbits depends on the efficiency and reliability of procedures in sperm cryopreservation. It is known that the process of sperm cryopreservation, including cooling, freezing and thawing, leads to cellular damage on membrane, cytoplasm and genome structures [10,11] caused by osmotic stress, cold shock, intracellular ice crystal formation, and excessive production of reactive oxygen species (ROS) [12] (Figure 2). During the cooling process, the sperm membrane is injured by cold shock which can be diminished by cooling rate [13] or materials stabilizing the membrane including egg yolk or skim milk [14]. The addition of cryoprotectants causes osmotic and toxic stress, which increases due to prolonged exposure during slow cooling [15]. In following freezing process, major problem is ice crystal formation which grow larger by recrystallization and injures cell [16]. The freezing rate and cryoprotectants application should be considered to diminish the problem. Recrystallization occurs during the thawing process because of entry through the recrystallization temperature zone. Since sperms suffer oxidative stress throughout the cryopreservation process [12], supplementation of antioxidants is considerable for the improvement of sperm quality.

Enormous efforts have been exerted to minimize these detrimental effects, and increase the efficiency and reliability of sperm cryopreservation in the rabbit. In this review, we will discuss recent findings and perspectives including extenders, cryoprotectants, supplements and procedures.

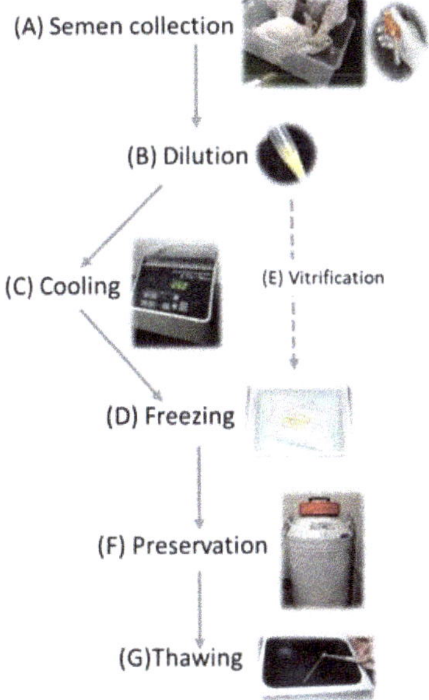

Figure 1. Schematic illustration of the essential process of sperm cryopreservation. (**A**) Semen is collected with an artificial vagina; (**B**) semen is diluted with a freezing extender; (**C**) sperm solution is cooled at slow rate (with a programming incubator); (**D**) sperm solution packed in freezing straws are frozen in vapor of liquid nitrogen; (**E**) vitrification is another option for sperm freezing; (**F**) sperms are cryopreserved in liquid nitrogen, and (**G**) sperms are thawed by immersion in a warm bath.

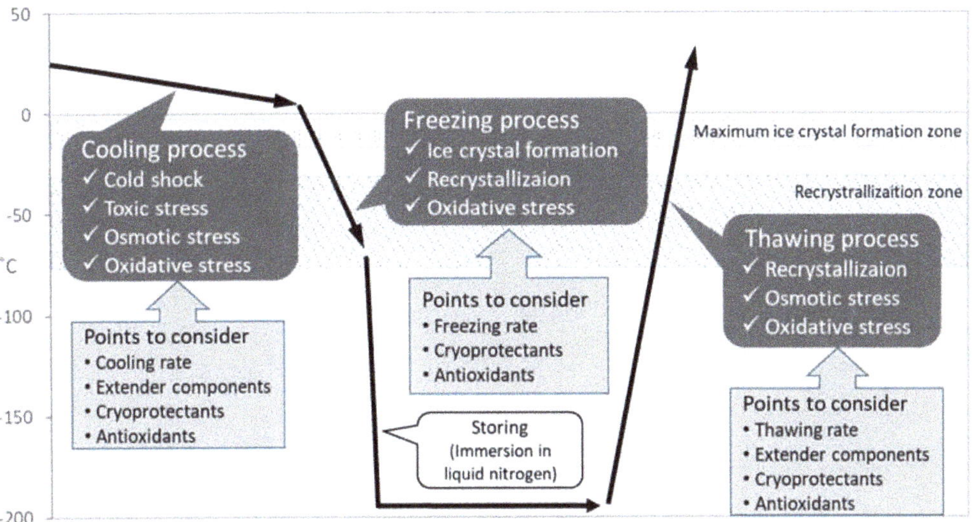

Figure 2. Schematic diagram of temperature changes during sperm cryopreservation process and associated problems.

2. Effects of Extender and Cryoprotectants

The sperm freezing extenders are commonly composed of a buffer (commonly Tris buffer for rabbit sperms), salt(s) and cryoprotectant(s) to avoid cell damage caused by inadequate pH, osmolality and cryogenic injury [13]. As a component in the freezing extender, egg yolk provides optimal results in rabbit sperm cryopreservation [17,18]. Nevertheless, the egg yolk contains both beneficial and detrimental components for sperms [19,20]. Additionally, there is a sanitary concern in using fresh biotic materials, like the egg yolk, which is subjected to quarantine inspection in the case of import/export. To avoid these risks, lecithin, also known as phosphatidylcholine, a component of egg yolk, is often used to prevent cold shock in sperm cryopreservation [21,22]. It is reported that non-animal originated soybean lecithin with minimal sanitary risks can be used as a substitute for egg yolk based on the motility and fertility of the frozen–thawed rabbit sperm [23]. Skim milk, another substitute, contains advantageous components in the freezing extender in various animal species [24–30] including rabbits [31]. However, using skim milk in the sperm freezing extender is less common than egg yolk in the rabbit.

The major problem in sperm cryopreservation is the mechanical invasion of sperm cells by ice crystals generated during the freezing process, which decreases the viability of sperm after thawing [32]. Dehydration of cellular water and percolation and immersion of cryoprotectants into the cell are usually performed to avoid the generation of the ice crystals. Both permeable (such as glycerol, dimethyl sulfoxide (DMSO), ethylene glycol, and amides) and non-permeable (saccharides, lipoprotein and Ficoll) are used as cryoprotectants. The permeant agent binds to intracellular free water leading to suppression of the generation of ice crystals. On the other hand, the non-permeant agents surrounding the cell increase extracellular osmotic pressure and enhance dehydration of the cell [33–35].

Lots of studies have been conducted regarding concentrations and combination effects of the cryoprotectants on rabbit sperm cryopreservation as encyclopedically reviewed by Mocé and Vicente [14]. Though glycerol and DMSO are the most common permeable cryoprotectant, it was suggested that glycerol is not suitable for rabbit sperm cryopreservation [14] possibly because of its low water permeability and high activation energy [36]. Additionally, it is known that high concentrations of DMSO show adverse effects on sperm quality in terms of motility and acrosome integrity [37,38]. Amides, namely lactamide and acetamide, are another candidate as a permeable cryoprotectant for rabbit sperm cryopreservation [39,40] and provide better results than glycerol or DMSO [41,42].

Lactose, sucrose, maltose, raffinose, treharose, and dextrans are used as non-permeable cryoprotectants in sperm cryopreservation, and have been shown to have an effect of stabilizing the plasma membrane during freeze–thaw process by interacting with membrane phospholipids [33–35] in addition to increasing the osmotic pressure.

3. Effects of Freezing and Thawing Procedure

Two principal techniques, slow freezing and vitrification, were employed for sperm freezing (Figure 1). During the freezing process of cells, extracellular water is frozen more quickly than intracellular water [42]. Generation of the ice corresponds to the decrement of water, which produces an imbalance of cellular osmotic pressure and intracellular water moving out of the cell. In slow freezing, generation of the intracellular ice crystal is suppressed since the cellular dehydration progresses slowly during a longer time of freezing. Concomitantly, concentrations of solute are deleterious to the cells and exposure time will be increased within a slow freezing procedure [42]. On the other hand, in vitrification, the cells are immediately cooled to −196 °C in liquid nitrogen and exposure time to high concentrated cryoprotectant is minimized. A countermeasure, including the application of high concentrated cryoprotectans, to suppress the ice crystal generation is required since dehydration is not always sufficient in vitrification.

Vitrification procedure has been explored in various mammalian sperm including humans [10,43]. It is known that sperms with larger heads are more susceptible to cold shock [44], which reduces sperm survivability [45]. The head of rabbit sperms is relatively large [32] and no efficient procedure in vitrification has been established. Rosato and Iaffaldano [46] compared the frozen–thawed sperm survival and fertility with/without various cryoprotectants between slow freezing and vitrification, and showed that the outcome from vitrification was far inferior to slow freezing. It is known that cells with a high water permeability show better tolerance for rapid freezing than those with a low water permeability [47], and rabbit sperm shows a low water permeability [36]. Therefore, conventional slow freezing is the prime choice for sperm cryopreservation at the current moment.

In the slow freezing procedure, sperms are cooled down to 5 °C before freezing to avoid the cold shock [45]. In the case of rabbits, Mocé et al. [48] reported that slow cooling to 5 °C improved neither fertility rate nor prolificacy in relation to cryopreservation. On the other hand, Maeda et al. [49] revealed that the viability of frozen–thawed rabbit sperm cooled at −0.1 or −0.2 °C/min (slower) were higher than at −0.8 °C/min. These data indicate that a faster cooling rate negatively affects sperm viability and does not improve reproductive performance after freezing and thawing. Obtaining more viable sperm results in efficiency improvement of cryopreservation, since the possibility of successful fertilization is increased. Additionally, in rabbits, time held at 5 °C affects the quality and reproductive performance of sperm. It was reported that longer holding time (90 min) at 5 °C increases the quality of frozen–thawed rabbit sperm and their fertilizing ability [50], and conversely, and shortened holding time (10 min) decreases it [51].

The ice crystal formation during the freezing process damage cells as described above and freezing protocol is also one of the concerns. In rabbit sperms, it was revealed that sperms frozen at slow (−15 °C/min) and fast (−60 °C/min) rates were lower in the quality and fertilizing ability than those frozen at medium (−40 °C/min) rate and in static liquid nitrogen vapor [52].

The thawing rate of frozen sperms is also known to affect the quality of sperms. Though frozen sperms are usually thawed at temperatures close to body temperature, it is generally recognized that high thawing rates provide better results [32,53]. It is possible that low thawing rates enhance recrystallization, a phenomenon that relatively many small ice crystals aggregate and form fewer larger ice crystals, which causes more severe damage to the cells [16]. Though thawing temperatures over 60 °C were adopted in some cases [54–57], exposing duration to a high temperature must be strictly controlled. Mocé et al. [58] compared the fertility rate and prolificacy of frozen rabbit sperm between thawed at 50 °C and 70 °C for 10–12 s, and concluded that thawing at 50 °C provided better results. In contrast, Chen and Foote [59] reported that the mortality of sperm thawed at 25 °C for 1 min was superior to those thawed at 45 °C for 30 s or 65 °C for 7 s following mechanical seeding at −6 °C. Therefore, appropriate thawing temperature can be affected by other conditions including cooling or freezing procedures, and further studies are demanded.

4. Cryopreservation Device

The choice of the freezing devices affects the quality of frozen–thawed sperm. Glass vials [31], plastic ampoules, polyvinylchloride tubing, [60] and pellet [61] addition to straws, which are widely employed in recent years, have been utilized for rabbit sperm cryopreservation. Thermal conductance to sperm depends on the shape, size and material of the device, which influences cooling, freezing and thawing rates. It was shown that rapid warming had a more dominant effect on survival than rapid cooling in mouse oocyte [62], and rapidly warming to the critical temperature range (−70 °C to −35 °C) at which intracellular ice is likely to form by recrystallization would improve cryopreservation efficiency of cell [63]. In the case of rabbit sperms, though the rapid freezing method (vitrification) has not been commonly utilized, the development of a new method and device for rapid thawing could lead to a great improvement of sperm cryopreservation efficiency.

5. Effects of Supplements

As mentioned above, lots of studies have been performed to address components of the sperm freezing extenders; however, results were not always satisfying. Some additional supplements can improve the efficiency in sperm cryopreservation (Table 1).

The antioxidant is one of the most expected supplements for sperm cryopreservation by eliminating the excessive production of ROS [12]. Zhu and colleague investigated the effects of supplementation of amin E analogue [64], cysteine [65], glutamine [66], trehalose [67] and melatonin [68], and revealed that supplementation of these antioxidative agents in Tris-citrate-glucose extender decreased ROS levels and improved the quality of frozen–thawed rabbit sperms. Fadl et al. confirmed 1.0 mM melatonin improved the motility, viability, membrane and acrosome integrities, and DNA integrity of frozen–thawed rabbit sperms in different extender (INRA-82) [69] Additionally, curcumin and curcumin nanoparticles were confirmed to improve the post-thawed quality of rabbit sperms via redox signaling and reduce the apoptosis process [70]. On the other hand, Maya-Soriano et al. reported that supplementation of bovine serum albumin, retinol and retinyl in the sperm freezing extender has no beneficial effect on the viability, mortality, progressivity, and acrosome and morphological integrity of frozen–thawed rabbit sperms at the concentrations they tested [71]. In this sense, other researchers have revealed the importance of the antioxidant concentration to achieve beneficial effects on rabbit sperm quality after thawing. Thus, while 4 mM of glutathione (GSH) improved [72] the viability, mortality, progressivity and acrosome integrity of frozen–thawed rabbit sperms, 0.5 mM of GSH did not provide similar results [73]. More studies are necessary for the adjustment of the freezing media in this species.

There are also some supplements with controversy surrounding their effects on rabbit sperm cryopreservation. Bovine serum albumin (BSA) is known to have a dual effect to protect sperm from osmotic stress by increasing membrane resistance [12] and oxidative stress by trapping free radicals [74]. Therefore, supplementation of BSA has been shown to improve the quality of frozen–thawed sperms in some species [75–78]. However, in rabbits, controversial results regarding the usage of BSA have been reported. Thus, while Maya-Soriano et al. [71] did not find any beneficial effect on sperm viability, mortality, progressivity, and acrosome and morphological integrity, Rosato and Iaffaldano [46] showed that frozen–thawed rabbit sperm display better mortality and DNA integrity when BSA is combined with sucrose or trehalose. These facts suggest that concentration and/or combination with other contents of the extender are important for exerting the beneficial effects of BSA. Moreover, gelatin is known to have protective effects including reduction of the sperm sedimentation and maintenance of the pH homogeneity in cooled semen [79] which has been confirmed in rabbits as well [80,81]. However, Cortell et al. [82] reported that gelatin addition did not improve the motility and viability nor fertility and prolificacy of frozen–thawed sperms in the rabbit. In later years, it was reported adding 2% gelatin enhanced the freezability and fertility of frozen–thawed rabbit sperms [83].

Antifreeze proteins (AFPs) are known to stabilize cell membranes and inhibit ice crystal growth and ice recrystallization [84,85]. With such functions, supplementation of AFP type III in the sperm freezing extender can improve the quality of frozen–thawed rabbit sperms [86]. However, AFPs are not easy to extract from natural resources such as fungi, bacteria, plants, insects and fish that are adapted to cold environments [87] and AFPs derived from other organisms can be detrimental to sperms or inseminated female animals. Tekin and Daşkın [88] utilized polyvinyl alcohol (PVA) instead of AFPs for ice recrystallization inhibition and showed that supplementation enhances motility, viability, acrosome integrity and mitochondrial activity in frozen–thawed rabbit sperms.

Table 1. Effects of supplements on quality of thawed rabbit sperm frozen by slow freezing method.

		Effective Concetration	Sperm Quality	EXtender Additives	Dilution Rate (Sperm:Extender)	Referrence
Antioxdants						
	Trolox (vitamine E analogue)	200 µM	Improved	4% DMSO 20% egg yolk	1:1	[64]
	Cysteine	5, 7.5 mM	Improved			[65]
	Glutamine	20 mM	Improved			[66]
	Trehalone	100 mM	Improved			[67]
	Melatonin	0.1 mM	Improved			[68]
		1.0 mM	Improved	INRA−82 (0.15% skim milk)	1:1	[69]
	Oxidised glutathione	0.5 mM	No effect	3.5 M DMSO 0.1 M sucrose	1:1	[73]
	Reduceed glutathione	0.5 mM	No effect			
		4 mM	Improved			[72]
	Curcumin nanoparticles	1.5 µg/mL/0.3	Improved	7% glycerol 20% egg yolk	1:4	[70]
	Retinol	50, 100, 200 µM	No effect	Gent B® * (egg yolk, glycerol)	1:2	[71]
	Retinyl	0.282, 2.82 µg/mL	Deteriorated			
Bovine serum albumin		5, 30, 60 mg/mL	No effect			
		0.5%	Improved	10% DMSO 0.1 M sucrose/0.1 M treharose	1:1	[46]
Gelatin		2%	No effect	3.5 M DMSO 0.1 M sucrose.	1:1	[82]
		2%	Improved	2% glycerol 10% egg yolk	1:5	[83]
Antifreeze protein III		0.1, 1 mg/mL	Improved	20% egg yolk	1:5	[86]
Polyvinyl alcohol		0.001, 0.01, 0.1, 1, 2 %	Improved	5 [a], 4 [b], 3 [c], 3 [d], 2 [e]% glycerol 5 % egg yolk	N/A	[88]
Sericin		0.1%	Improved	16% Me_2SO 2% sucrose	1:2	[89]
Cholesterol-loaded cyclodextrin		1 mg/40 × 10^6 sperm	Improved	6% acetamide 20% egg yolk	1:5	[90]

*: Minitüb, Tiefenbach, Germany. Concetrations of glycerol when supplemented with a: 0.001, b: 0.01, c: 0.1, d: 1, e: 2% of polyvinl alchool.

By stabilization of cell membrane, supplementation of a silk protein, sericin [89] and cholesterol-loaded cyclodextrin [90] improved motility and quality of frozen–thawed rabbit sperm. However, since these supplements are known to inhibit acrosome reaction [89,91], further studies on enhancing its fertility are required for practical use.

6. Another Preservation Strategy

The freeze-drying technique is an alternative technology for the long-term preservation of sperms [92]. Liu et al. [93] showed that freeze-dried rabbit sperms maintain the ability for full-term development in spite of immobilization, membrane breaking, and tail fragmentation. It is important to store freeze-dried sperms at low temperatures for stable, long-term preservation, and properly stored freeze-dried sperms maintain their fertility for years [92]. It is necessary to operate intracytoplasmic sperm injection (ICSI) [94] and embryo transfer, which require a skillful technique and particular device for using the freeze-dried sperms. One of the advantages of sperm preservation in rabbits is the applicability of artificial insemination as mentioned above, and preservation of freeze-dried sperms would be just a spare option in rabbit sperm preservation.

7. Conclusions and Perspectives

As stated above, numerous studies on improving rabbit sperm cryopreservation have been conducted from various aspects including freezing procedure, type, concentration and combination of cryoprotectants. In spite of this, a standard procedure for rabbit sperm cryopreservation has not been well established due to various and irreproducible results from each study.

One of the reasons for the irreproducibility in rabbit sperm cryopreservation may be derived from the differences in sperm conditions. It is known that freezabilities of rabbit sperms differ among individuals [95] or breeds [96]. However, rabbit breeds used in some reproductive studies have not strictly and well defined rather than other livestock animals such as cattle, horses or pigs, and some authors even did not provide enough information about the rabbit breed examined. There are several reports which indicate associations between the sperm freezability and abundance of particular components in seminal plasma including proteins and fatty acids in some species [97–102]. The individual and breed difference in sperm freezability can be explained by such seminal plasma traits. It would be possible to improve the sperm freezability and resolve the individual or breed differences by complemental supplementations according to the seminal plasma trait. In the rabbit, it was revealed that genotype, i.e., breed, affects the abundance of some seminal plasma proteins [103], which are associated with sperm quality [104]. Furthermore, there is no information about the association between sperm freezability and seminal plasma traits in rabbits, and further studies are needed. Again, lack of information about the rabbit breed can disturb the improvement of rabbit sperm cryopreservation efficiency.

Another possible reason for the irreproducibility is disunified evaluation criteria of the quality of the sperms among the reports. Some studies report both the quality of the frozen–thawed sperms and their fertility and prolificacy, and others do only one of them. Examination of fertility and prolificacy involves artificial insemination procedures which affect the results of the study. On the other hand, fertility and prolificacy cannot be estimated by the sperm quality alone, even though obtaining more motile sperms is generally advantageous for efficient reproduction [105–107]. Additionally, the rate of rapid and progressive motile sperm would be important for successful artificial insemination, since the inseminated sperm need to reach the ova via a long reproductive tract [108–110].

In any case, to achieve a consensus on the efficient method for rabbit sperm cryopreservation, extensive investigations are required under unified evaluation criteria and conditions except for factors like extenders, cryoprotectants and procedures to be examined. On the other hand, it seemed that the supplements like antioxidants generally just add their effects without interfering with other components in the extender. Therefore, most of the supplements exert expected effects on the quality of frozen–thawed rabbit sperm even in various conditions including the extender (Table 1). The dose and combination of the supplements can be a key subject for highly efficient rabbit sperm cryopreservation in future studies.

As Dr. Robert G. Edwards was awarded The Nobel Prize in Physiology or Medicine for the development of in vitro fertilization (IVF) in 2010, assisted reproductive technology (ART) including IVF is an indispensable medical procedure in this modern age. The rabbit is also known as a prime reproductive model for human health, because of (1) exact staging of early embryonic developmental and maternal pregnancy stages, (2) large-sized blastocysts amenable to micromanipulation, (3) cell-lineage-specific analyses, (4) gastrulation stages representative of mammalian development, and (5) placental morphology and function similar to the human [111]. Therefore, the development of reproductive technology in rabbits, which leads to the improvement of medical procedures in humans, is very important and desirable.

Author Contributions: Writing—original draft preparation, K.N.; review and editing, J.F. and M.N.; supervision, S.K., F.M. and C.-c.W. All authors have read and agreed to the published version of the manuscript.

Funding: This work was supported in part by JSPS KAKENHI (JP17K08783 and JP15H04718), the National Natural Science Foundation of China (No. 81941001 and 81770457) and the JSPS-CAS Bilateral Joint Research Program (JPJSBP 120187204).

Acknowledgments: The authors express their sincere respect and thanks to the emeritus Teruo Watanabe, Saga University and Masatoshi Morimoto of Fukuoka Jo Gakuin Nursing University for their valuable advice in preparing the article.

Conflicts of Interest: The authors declare no conflict of interest.

References

1. Food and Agriculture Organization of the United Nations. FAO Database. 2019. Available online: http://www.faostat.fao.org (accessed on 14 March 2021).
2. Fan, J.; Kitajima, S.; Watanabe, T.; Xu, J.; Zhang, J.; Liu, E.; Chen, Y.E. Rabbit models for the study of human atherosclerosis: From pathophysiological mechanisms to translational medicine. *Pharmacol. Ther.* **2015**, *146*, 104–119. [CrossRef] [PubMed]
3. Shiomi, M. The History of the WHHL Rabbit, an Animal Model of Familial Hypercholesterolemia (I)—Contribution to the Elucidation of the Pathophysiology of Human Hypercholesterolemia and Coronary Heart Disease. *J. Atheroscler. Thromb.* **2020**, *27*, 105–118. [CrossRef] [PubMed]
4. Matsuhisa, F.; Kitajima, S.; Nishijima, K.; Akiyoshi, T.; Morimoto, M.; Fan, J. Transgenic Rabbit Models: Now and the Future. *Appl. Sci.* **2020**, *10*, 7416. [CrossRef]
5. Mazur, P.; Leibo, S.P.; Seidel, G.E., Jr. Cryopreservation of the germplasm of animals used in biological and medical research: Importance, impact, status, and future directions. *Biol. Reprod.* **2007**, *78*, 2–12. [CrossRef] [PubMed]
6. Casellas, J.; Vidal-Roqueta, D.; Flores, E.; Casellas-Vidal, D.; Llach-Vila, M.; Salgas-Fina, R.; Casellas-Molas, P. Epistasis for founder specific inbreeding depression in rabbits. *J. Hered.* **2011**, *102*, 157–164. [CrossRef]
7. Chai, C.K. Effects of inbreeding in rabbits. Inbred lines, discrete characters, breeding performance, and mortality. *J. Hered.* **1969**, *60*, 64–70. [CrossRef]
8. Chai, C.K.; Degenhardt, K.H. Developmental anomaties in inbred rabbits. *J. Hered.* **1962**, *53*, 174–182. [CrossRef]
9. Ragab, M.; Sánchez, J.P.; Baselga, M. Effective population size and inbreeding depression on litter size in rabbits. A case study. *J. Anim. Breed. Genet.* **2015**, *132*, 68–73. [CrossRef] [PubMed]
10. Isachenko, E.; Isachenko, V.; Katkov, I.I.; Dessole, S.; Nawroth, F. Vitrification of mammalian spermatozoa in the absence of cryoprotectants: From past practical difficulties to present success. *Reprod. Biomed. Online* **2003**, *6*, 191–200. [CrossRef]
11. Watson, P.F. The causes of reduced fertility with cryopreserved semen. *Anim. Reprod. Sci.* **2000**, *60–61*, 481–492. [CrossRef]
12. Amidi, F.; Pazhohan, A.; Shabani Nashtaei, M.; Khodarahmian, M.; Nekoonam, S. The role of antioxidants in sperm freezing: A review. *Cell Tissue Bank* **2016**, *17*, 745–756. [CrossRef] [PubMed]
13. Parrish, J.J.; Foote, R. Fertility of cooled and frozen rabbit sperm measured by competitive fertilization. *Biol. Reprod.* **1986**, *35*, 253–257. [CrossRef]
14. Mocé, E.; Vicente, J.S. Rabbit sperm cryopreservation: A review. *Anim. Reprod. Sci.* **2009**, *110*, 1–24. [CrossRef] [PubMed]
15. Mazur, P. The role of intracellular freezing in the death of cellscooled at supraoptimal rates. *Cryobiology* **1977**, *14*, 251–272. [CrossRef]
16. Koshimoto, C.; Mazur, P. Effects of warming rate, temperature, and antifreeze proteins on the survival of mouse spermatozoa frozen at an optimal rate. *Cryobiology* **2002**, *45*, 49–59. [CrossRef]
17. Salamon, S.; Maxwell, W.M. Storage of ram semen. *Anim. Reprod. Sci.* **2000**, *62*, 77–111. [CrossRef]
18. Arriola, J.; Foote, R.H. Accessory sperm as an indication of fertilizing ability of rabbit spermatozoa frozen in egg yolk-acetamide with detergent. *J. Androl.* **2001**, *22*, 458–643. [CrossRef]
19. Aires, V.A.; Hinsch, K.D.; Mueller-Schloesser, F.; Bogner, K.; Mueller-Schloesser, S.; Hinsch, E. In vitro and in vivo comparison of egg yolk-based and soybean lecithin-based extenders for cryopreservation of bovine semen. *Theriogenology* **2003**, *60*, 269–279. [CrossRef]
20. Watson, P.F.; Martin, I.C. Artificial insemination of sheep: The effect of semen diluents containing egg yolk on the fertility of ram semen. *Theriogenology* **1976**, *6*, 559–564. [CrossRef]
21. Amirat, L.; Tainturier, D.; Jeanneau, L.; Thorin, C.; Gerard, O.; Courtens, J.L.; Anton, M. Bull semen in vitro fertility after cryopreservation using egg yolk LDL: A comparison with Optidyl, a commercial egg yolk extender. *Theriogenology* **2004**, *61*, 895–907. [CrossRef]
22. Moussa, M.; Matinet, V.; Trimeche, A.; Tainturier, D.; Anton, M. Low density lipoproteins extracted from hen egg yolk by an easy method: Cryoprotective effect on frozen-thawed bull semen. *Theriogenology* **2002**, *57*, 1695–1706. [CrossRef]
23. Nishijima, K.; Kitajima, S.; Koshimoto, C.; Morimoto, M.; Watanabe, T.; Fan, J.; Matsuda, Y. Motility and fertility of rabbit sperm cryopreserved using soybean lecithin as an alternative to egg yolk. *Theriogenology* **2015**, *84*, 1172–1175. [CrossRef]
24. Abe, Y.; Lee, D.S.; Sano, H.; Akiyama, K.; Yanagimoto-Ueta, Y.; Asano, T.; Suwa, Y.; Suzuki, H. Artificial insemination with canine spermatozoa frozen in a skim milk/glucose-based extender. *J. Reprod. Dev.* **2008**, *54*, 290–294. [CrossRef] [PubMed]
25. Dorado, J.; Rodríguez, I.; Hidalgo, M. Cryopreservation of goat spermatozoa: Comparison of two freezing extenders based on post-thaw sperm quality and fertility rates after artificial insemination. *Theriogenology* **2007**, *68*, 168–177. [CrossRef] [PubMed]

26. Fiser, P.S.; Fairfull, R.W. Combined effects of glycerol concentration, cooling velocity, and osmolality of skim milk diluents on cryopreservation of ram spermatozoa. *Theriogenology* **1986**, *25*, 473–484. [CrossRef]
27. Nakagata, N. Cryopreservation of mouse spermatozoa. *Mamm. Genome* **2000**, *11*, 572–576. [CrossRef] [PubMed]
28. Papa, P.M.; Papa, F.O.; Oliveira, L.A.; Guasti, P.N.; Castilho, C.; Giometti, I.C. Different extenders in the cryopreservation of bovine epididymal spermatozoa. *Anim. Reprod. Sci.* **2015**, *161*, 58–63. [CrossRef]
29. Scherzer, J.; Fayrer-Hosken, R.A.; Aceves, M.; Hurley, D.J.; Ray, L.E.; Jones, L.; Heusner, G.L. Freezing equine semen: The effect of combinations of semen extenders and glycerol on post-thaw motility. *Aust. Vet. J.* **2009**, *87*, 275–279. [CrossRef]
30. Varisli, O.; Scott, H.; Agca, C.; Agca, Y. The effects of cooling rates and type of freezing extenders on cryosurvival of rat sperm. *Cryobiology* **2013**, *67*, 109–116. [CrossRef]
31. O'Shea, T.; Wales, R.G. Further studies of the deep freezing of rabbit spermatozoa in reconstituted skim milk powder. *Aust. J. Biol. Sci.* **1969**, *22*, 709–719. [CrossRef]
32. Kumar, A.; Prasad, J.K.; Srivastava, N.; Ghosh, S.K. Strategies to minimize various stress-related freeze-thaw damages during conventional cryopreservation of mammalian spermatozoa. *Biopreserv. Biobank* **2019**, *17*, 603–612. [CrossRef]
33. Aisen, E.G.; Medina, V.H.; Venturino, A. Cryopreservation and post-thawed fertility of ram semen frozen in different trehalose concentrations. *Theriogenology* **2002**, *57*, 1801–1808. [CrossRef]
34. Anchordoguy, T.J.; Rudolph, A.S.; Carpenter, J.F.; Crowe, J.H. Modes of interaction of cryoprotectants with membrane phospholipids during freezing. *Cryobiology* **1987**, *24*, 324–331. [CrossRef]
35. Sieme, H.; Harriëtte Oldenhof, H.; Wolkers, W.F. Mode of action of cryoprotectants for sperm preservation. *Anim. Reprod. Sci.* **2016**, *169*, 2–5. [CrossRef] [PubMed]
36. Curry, M.R.; Redding, B.J.; Watson, P.F. Determination of water permeability coefficient and its activation energy for rabbit spermatozoa. *Cryobiology* **1995**, *32*, 175–181. [CrossRef] [PubMed]
37. Rasul, Z.; Ahmed, N.; Anzar, M. Antagonist effect of DMSO on the cryoprotection ability of glycerol during cryopreservation of buffalo sperm. *Theriogenology* **2007**, *68*, 813–819. [CrossRef]
38. Si, W.; Hildebrandt, T.B.; Reid, C.; Krieg, R.; Ji, W.; Fassbender, M.; Hermes, R. The successful double cryopreservation of rabbit (Oryctolagus cuniculus) semen in large volume using the directional freezing technique with reduced concentration of cryoprotectant. *Theriogenology* **2006**, *65*, 788–798. [CrossRef] [PubMed]
39. Hanada, A.; Nagase, H. Cryoprotective effects of some amides on rabbit spermatozoa. *J. Reprod. Fertil.* **1980**, *60*, 247–252. [CrossRef]
40. Kashiwazaki, N.; Okuda, Y.; Seita, Y.; Hisamatsu, S.; Sonoki, S.; Shino, M.; Masaoka, T.; Inomata, T. Comparison of glycerol, lactamide, acetamide and dimethylsulfoxide as cryoprotectants of Japanese white rabbit spermatozoa. *J. Reprod. Dev.* **2006**, *52*, 511–556. [CrossRef]
41. Okuda, Y.; Seita, Y.; Hisamatsu, S.; Sonoki, S.; Shino, M.; Masaoka, T.; Inomata, T.; Kamijo, S.; Kashiwazaki, N. Fertility of spermatozoa cryopreserved with 2% acetamide or glycerol through artificial insemination in the Japanese white rabbit. *Exp. Anim.* **2007**, *56*, 29–34. [CrossRef]
42. Simione, F.P., Jr. Key issues relating to the genetic stability and preservation of cells and cell banks. *J. Parent. Sci. Technol.* **1992**, *46*, 226–232. [PubMed]
43. Isachenko, E.; Mallmann, P.; Rahimi, G.; Risopatrón, J.; Schulz, M.; Isachenko, V.; Sánchez, R. Vitrification technique-new possibilities for male gamete low temperature storage. In *Current Frontiers in Cryobiology*; Katkov, I.I., Ed.; IntechOpen: Rijeka, Croatia, 2012; pp. 41–76. [CrossRef]
44. Isachenko, E.; Isachenko, V.; Sanchez, R.; Katkov, I.I.; Kreienberg, R. Cryopreservation of spermatozoa. Old routine and new perspectives. In *Principles and Practice of Fertility Preservation*; Kim, D., Ed.; Cambridge University Press: Cambridge, UK, 2011; pp. 176–198. [CrossRef]
45. Mahadevan, M.; Trounson, A.O. Effect of cooling, freezing and thawing rates and storage conditions on preservation of human spermatozoa. *Andrologia* **1984**, *16*, 52–60. [CrossRef] [PubMed]
46. Rosato, M.; Iaffaldano, N. Cryopreservation of rabbit semen: Comparing the effects of different cryoprotectants, cryoprotectant-free vitrification, and the use of albumin plus osmoprotectants on sperm survival and fertility after standard vapor freezing and vitrification. *Theriogenology* **2013**, *79*, 508–516. [CrossRef] [PubMed]
47. Nei, T.; Araki, T.; Matsusaka, T. Freezing injury to aerated and non-aerated cultures of Escherichia coli. In *Freezing and Drying of Microorganisms*; Nei, T., Ed.; University of Tokyo Press: Tokyo, Japan, 1969; pp. 3–16. [CrossRef]
48. Mocé, E.; Lavara, R.; Vicente, J.S. Effect of cooling rate to 5 °C, straw size and farm on fertilizing ability of cryopreserved rabbit sperm. *Reprod. Domest. Anim.* **2010**, *45*, e1–e7. [CrossRef] [PubMed]
49. Maeda, T.; Liu, E.; Nishijima, K.; Tanaka, M.; Yamaguchi, S.; Morimoto, M.; Watanabe, T.; Fan, J.; Kitajima, S. Effect of the primary cooling rate on the motility and fertility of the frozen-thawed rabbit spermatozoa. *World Rabbit Sci.* **2012**, *20*, 65–70. [CrossRef]
50. Di Iorio, M.; Colonna, M.A.; Miranda, M.; Principe, P.; Schiavitto, M.; Cerolini, S.; Manchisi, A.; Iaffaldano, N. Initial cooling time before freezing affects post-thaw quality and reproductive performance of rabbit semen. *Anim. Sci. J.* **2018**, *89*, 1240–1244. [CrossRef]
51. Mocé, E.; Blanch, E.; Talaván, A.; de Castro, M.P.V. Reducing the time rabbit sperm are held at 5 °C negatively affects their fertilizing ability after cryopreservation. *Theriogenology* **2014**, *82*, 1049–1053. [CrossRef]

52. Mocé, E.; Blanch, E.; Talaván, A.; Viudes de Castro, M.P. Effect of different freezing velocities on the quality and fertilising ability of cryopreserved rabbit spermatozoa. *Reprod. Fertil. Dev.* **2015**, *27*, 846–851. [CrossRef]
53. Diskin, M.G. Review: Semen handling, time of insemination and insemination technique in cattle. *Animal* **2018**, *12*, s75–s84. [CrossRef]
54. Chandler, J.E.; Adkinson, R.W.; Nebel, R.L. Thawing optimums for bovine spermatozoa processed by three methods and packaged in Continental and French straws. *J. Dairy Sci.* **1984**, *67*, 398–404. [CrossRef]
55. Correa, J.R.; Pace, M.M.; Zavos, P.M. Relationships among frozen-thawed sperm characteristics accessed via the routine semen analysis, sperm functional tests and fertility of bulls in an artificial insemination program. *Theriogenology* **1997**, *48*, 721–731. [CrossRef]
56. Eriksson, B.M.; Rodriguez-Martinez, H. Effect of freezing and thawing rates on the post-thaw viability of boar spermatozoa frozen in FlatPacks and Maxi-straws. *Anim. Reprod. Sci.* **2000**, *63*, 205–220. [CrossRef]
57. Senger, P.L. Handling frozen bovine semenfactors which influence viability and fertility. *Theriogenology* **1980**, *13*, 51–62. [CrossRef]
58. Mocé, E.; Vicente, J.S.; Lavara, R. Effect of freezing-thawing protocols on the performance of semen from three rabbit lines after artificial insemination. *Theriogenology* **2003**, *60*, 115–123. [CrossRef]
59. Chen, Y.; Foote, R.H. Survival of rabbit spermatozoa frozen and thawed at different rates with and without seeding. *Anim. Reprod. Sci.* **1994**, *35*, 131–143. [CrossRef]
60. Stranzinger, G.F.; Maurer, R.R.; Paufler, S.K. Fertility of frozen rabbit semen. *J. Reprod. Fertil.* **1971**, *2*, 111–113. [CrossRef]
61. Awad, M.M.; Kishk, W.; Hassenin, A.M.; El-Alamy, M.A. Freezing rabbit semen in pellet form using cold surface of paraffin wax compared to straws. In Proceedings of the 7th World Rabbit Congress, Valencia, Spain, 4–7 July 2000; Volume 8, pp. 89–95, ISSN 2308-1910.
62. Seki, S.; Mazur, P. Effect of warming rate on the survival of vitrified mouse oocytes and on the recrystallization of intracellular ice. *Biol. Reprod.* **2008**, *79*, 727–737. [CrossRef] [PubMed]
63. Seki, S.; Basaki, K.; Komatsu, Y.; Fukuda, Y.; Yano, M.; Matsuo, Y.; Obata, T.; Matsuda, Y.; Nishijima, K. Vitrification of one-cell mouse embryos in cryotubes. *Cryobiology* **2018**, *81*, 132–137. [CrossRef] [PubMed]
64. Zhu, Z.; Fan, X.; Lv, Y.; Zhang, N.; Fan, C.; Zhang, P.; Zeng, W. Vitamin E analogue improves rabbit sperm quality during the process of cryopreservation through its antioxidative action. *PLoS ONE* **2015**, *10*, e0145383. [CrossRef]
65. Zhu, Z.; Ren, Z.; Fan, X.; Pan, Y.; Lv, S.; Pan, C.; Lei, A.; Zeng, W. Cysteine protects rabbit spermatozoa against reactive oxygen species-induced damages. *PLoS ONE* **2017**, *12*, e0181110. [CrossRef] [PubMed]
66. Zhu, Z.; Fan, X.; Lv, Y.; Lin, Y.; Wu, D.; Zeng, W. Glutamine protects rabbit spermatozoa against oxidative stress via glutathione synthesis during cryopreservation. *Reprod. Fertil. Dev.* **2017**, *29*, 2183–2194. [CrossRef]
67. Zhu, Z.; Fan, X.; Pan, Y.; Lu, Y.; Zeng, W. Trehalose improves rabbit sperm quality during cryopreservation. *Cryobiology* **2017**, *75*, 45–51. [CrossRef] [PubMed]
68. Zhu, Z.; Li, R.; Lv, Y.; Zeng, W. Melatonin protects rabbit spermatozoa from cryo-damage via decreasing oxidative stress. *Cryobiology* **2019**, *88*, 1–8. [CrossRef] [PubMed]
69. Fadl, A.M.; Ghallab, A.R.M.; Abou-Ahmed, M.M.; Moawad, A.R. Melatonin can improve viability and functional integrity of cooled and frozen/thawed rabbit spermatozoa. *Reprod. Domest. Anim.* **2021**, *56*, 103–111. [CrossRef]
70. Abdelnour, S.A.; Hassan, M.A.E.; Mohammed, A.K.; Alhimaidi, A.R.; Al-Gabri, N.; Al-Khaldi, K.O.; Swelum, A.A. The effect of adding different levels of curcumin and its nanoparticles to extender on post-thaw quality of cryopreserved rabbit sperm. *Animals* **2020**, *10*, 1508. [CrossRef]
71. Maya-Soriano, M.J.; Taberner, E.; Sabés-Alsina, M.; Piles, M.; Lopez-Bejar, M. Absence of beneficial effects on rabbit sperm cell cryopreservation by several antioxidant agents. *Zygote* **2015**, *23*, 1–10. [CrossRef]
72. Ahmad, E.; Naseer, Z.; Aksoy, M. Supplementation of extender with reduced gluatathione (GSH) preserve rabbit sperm quality after cryopreservation. In Proceedings of the 11th World Rabbit Congress, Qingdao, China, 15–18 June 2016; pp. 171–174, ISSN 2308-1910.
73. Marco-Jimenez, F.; Lavara, R.; Vicente, J.S.; Viudes-de-Castro, M.P. Cryopreservation of rabbit spermatozoa with freezing media supplemented with reduced and oxidised glutathione. *Cryo Lett.* **2006**, *27*, 261–268. [PubMed]
74. Roche, M.; Rondeau, P.; Singh, N.R.; Tarnus, E.; Bourdon, E. The antioxidant properties of serum albumin. *FEBS Lett.* **2008**, *582*, 1783–1787. [CrossRef]
75. Ashrafi, I.; Kohram, H.; Ardabili, F.F. Antioxidative effects of melatonin on kinetics, microscopic and oxidative parameters of cryopreserved bull spermatozoa. *Anim. Reprod. Sci.* **2013**, *139*, 25–30. [CrossRef]
76. Matsuoka, T.; Imai, H.; Kohno, H.; Fukui, Y. Effects of bovine serum albumin and trehalose in semen diluents for improvement of frozen-thawed ram spermatozoa. *J. Reprod. Dev.* **2006**, *52*, 675–683. [CrossRef]
77. Naijian, H.R.; Kohram, H.; ZareShahneh, A.; Sharafi, M. Effects of various concentrations of BSA on microscopic and oxidative parameters of Mahabadi goat semen following the freeze–thaw process. *Small Rumin. Res.* **2013**, *113*, 371–375. [CrossRef]
78. Uysal, O.; Bucak, O.M. Effects of Oxidized Glutathione, Bovine Serum Albumin, Cysteine and Lycopene on the Quality of Frozen-Thawed Ram Semen. *Acta Vet. Brno* **2007**, *76*, 383–390. [CrossRef]
79. Bandeira, A.M.P.; Matos, J.E.; Maria, A.N.; Carneiro, P.C.F.; Purdy, P.H.; Bucak, O.M.; Azevedo, H.C. The effects of gelatin supplementation prior to cooling on ram semen quality and fertility. *Anim. Reprod.* **2018**, *15*, 23–28. [CrossRef]

80. Nagy, S.; Sinkovics, G.; Kovács, A. Viability and acrosome integrity of rabbit spermatozoa processed in a gelatin-supplemented extender. *Anim. Reprod. Sci.* **2002**, *70*, 283–286. [CrossRef]
81. Ragab, A.A.; El-Sherbieny, M.A.; El-Siefy, E.M.; Adel-Khalek, A.E. Effect of gelatin supplementation on the quality and fertility of rabbit spermatozoa preserved at room or refrigerator temperature degrees. *J. Anim. Poult. Prod.* **2011**, *2*, 579–588. [CrossRef]
82. Cortell, C.; de Castro, M.P.V. Effect of gelatin addition to freezing extender on rabbit semen parameters and reproductive performance. In Proceedings of the 9th World Rabbit Congress, Verona, Italy, 10–13 June 2008; pp. 327–332.
83. El-Sherbieny, M.A.; Kalaba, Z.M.; El-Siefy, E.M.E.; Ragab, A.A. Freezing and Fertilizing Capacity of Frozen Rabbit Semen Extended with Gelatin Addition. *Asian J. Anim. Sci.* **2012**, *6*, 291–299. [CrossRef]
84. Carpenter, J.F.; Hansen, T.N. Antifreeze protein modulates cell survival during cryopreservation: Mediation through influence on ice crystal growth. *Proc. Natl. Acad. Sci. USA* **1992**, *89*, 8953–8957. [CrossRef]
85. Crevel, R.W.; Fedyk, J.K.; Spurgeon, M.J. Antifreeze proteins: Characteristics, occurrence and human exposure. *Food Chem. Toxicol.* **2002**, *40*, 899–903. [CrossRef]
86. Nishijima, K.; Tanaka, M.; Sakai, Y.; Koshimoto, C.; Morimoto, M.; Watanabe, T.; Fan, J.; Kitajima, S. Effects of type III antifreeze protein on sperm and embryo cryopreservation in rabbit. *Cryobiology* **2014**, *69*, 22–25. [CrossRef]
87. Rubinsky, B.; Arav, A.; Fletcher, G.L. Hypothermic protection a fundamental property of "antifreeze" proteins. *Biochem. Biophys. Res. Commun.* **1991**, *180*, 566–571. [CrossRef]
88. Tekin, K.; Daşkın, A. Effect of polyvinyl alcohol on survival and function of angora buck spermatozoa following cryopreservation. *Cryobiology* **2019**, *89*, 60–67. [CrossRef]
89. Raza, S.; Uçan, U.; Aksoy, M.; Erdoğan, G.; Ceylan, A.; Serin, I. Silk protein sericin pretreatment enhances osmotic tolerance and post-thaw sperm quality but reduces the ability of sperm cells to undergo in vitro induced acrosome reaction in rabbit. *Cryobiology* **2019**, *90*, 1–7. [CrossRef] [PubMed]
90. Nishijima, K.; Yamaguchi, S.; Tanaka, M.; Sakai, Y.; Koshimoto, C.; Morimoto, M.; Watanabe, T.; Fan, J.; Kitajima, S. Effects of cholesterol-loaded cyclodextrins on the rate and the quality of motility in frozen and thawed rabbit sperm. *Exp. Anim.* **2014**, *63*, 149–154. [CrossRef] [PubMed]
91. Aksoy, M.; Akman, O.; Lehimcioğlu, N.C.; Erdem, H. Cholesterol-loaded cyclodextrin enhances osmotic tolerance and inhibits the acrosome reaction in rabbit spermatozoa. *Anim. Reprod. Sci.* **2010**, *120*, 166–172. [CrossRef] [PubMed]
92. Gil, L.; Olaciregui, M.; Luño, V.; Malo, C.; González, N.; Martínez, F. Current status of freeze-drying technology to preserve domestic animals sperm. *Reprod. Domest. Anim.* **2014**, *49*, 72–81. [CrossRef]
93. Liu, J.L.; Kusakabe, H.; Chang, C.C.; Suzuki, H.; Schmidt, D.W.; Julian, M.; Pfeffer, R.; Bormann, C.L.; Tian, X.C.; Yanagimachi, R.; et al. Freeze-dried sperm fertilization leads to full-term development in rabbits. *Biol. Reprod.* **2004**, *70*, 1776–1781. [CrossRef]
94. Li, Q.Y.; Hou, J.; Chen, Y.F.; An, X.R. Full-term development of rabbit embryos produced by ICSI with sperm frozen in liquid nitrogen without cryoprotectants. *Reprod. Domest. Anim.* **2010**, *45*, 717–722. [CrossRef] [PubMed]
95. Mocé, E.; Lavara, R.; Vicente, J.S. Influence of the donor male on the fertility of frozen-thawed rabbit sperm after artificial insemination of females of different genotypes. *Reprod. Domest. Anim.* **2005**, *40*, 516–521. [CrossRef]
96. Kulíková, B.; Oravcová, M.; Baláži, A.; Supuka, P.; Chrenek, P. Factors affecting storage of Slovak native rabbit semen in the gene bank. *Zygote* **2017**, *25*, 592–600. [CrossRef]
97. Martínez-Soto, J.C.; Landeras, J.; Gadea, J. Spermatozoa and seminal plasma fatty acids as predictors of cryopreservation success. *Andrology* **2013**, *1*, 365–375. [CrossRef]
98. Rickard, J.P.; Leahy, T.; Soleilhavoup, C.; Tsikis, G.; Labas, V.; Harichaux, G.; Lynch, G.W.; Druart, X.; de Graaf, S.P. The identification of proteomic markers of sperm freezing resilience in ram seminal plasma. *J. Proteom.* **2015**, *126*, 303–311. [CrossRef]
99. Vilagran, I.; Castillo, J.; Bonet, S.; Sancho, S.; Yeste, M.; Estanyol, J.M.; Oliva, R. Acrosin-binding protein (ACRBP) and triosephosphate isomerase (TPI) are good markers to predict boar sperm freezing capacity . *Theriogenology* **2013**, *80*, 443–450. [CrossRef]
100. Vilagran, I.; Yeste, M.; Sancho, S.; Castillo, J.; Oliva, R.; Bonet, S. Comparative analysis of boar seminal plasma proteome from different freezability ejaculates and identification of Fibronectin 1 as sperm freezability marker. *Andrology* **2015**, *3*, 345–356. [CrossRef] [PubMed]
101. Wang, P.; Wang, Y.F.; Wang, H.; Wang, C.W.; Zan, L.S.; Hu, J.H.; Li, Q.W.; Jia, Y.H.; Ma, G.J. HSP90 expression correlation with the freezing resistance of bull sperm. *Zygote* **2014**, *22*, 239–245. [CrossRef] [PubMed]
102. Yeste, M.; Estrada, E.; Casas, I.; Bonet, S.; Rodríguez-Gil, J.E. Good and bad freezability boar ejaculates differ in the integrity of nucleoprotein structure after freeze-thawing but not in ROS levels. *Theriogenology* **2013**, *79*, 929–939. [CrossRef]
103. Casares-Crespo, L.; Fernández-Serrano, P.; Vicente, J.S.; Marco-Jiménez, F.; Viudes-de-Castro, M.P. Rabbit seminal plasma proteome: The importance of the genetic origin. *Anim. Reprod. Sci.* **2018**, *189*, 30–42. [CrossRef] [PubMed]
104. Bezerra, M.J.B.; Arruda-Alencar, J.M.; Martins, J.A.M.; Viana, A.G.A.; Viana Neto, A.M.; Rêgo, J.P.A.; Oliveira, R.V.; Lobo, M.; Moreira, A.C.O.; Moreira, R.A.; et al. Major seminal plasma proteome of rabbits and associations with sperm quality. *Theriogenology* **2019**, *128*, 156–166. [CrossRef] [PubMed]
105. Baker, R.D.; Dziuk, P.J.; Norton, H.W. Effect of volume of semen, number of sperm and drugs on transport of sperm in artificially inseminated gilts. *J. Anim. Sci.* **1968**, *27*, 88–93. [CrossRef]
106. Flowers, W.L. Increasing fertilization rate of boars: Influence of number and quality of spermatozoa inseminated. *J. Anim. Sci.* **2002**, *80*, E47–E53. [CrossRef]

107. Langford, G.A.; Marcus, G.J. Influence of sperm number and seminal plasma on fertility of progestagen-treated sheep in confinement. *J. Reprod. Fertil.* **1982**, *65*, 325–329. [CrossRef]
108. Berker, B.; Şükür, Y.E.; Kahraman, K.; Atabekoğlu, C.S.; Sönmezer, M.; Özmen, B.; Ateş, C. Absence of rapid and linear progressive motile spermatozoa "grade A" in semen specimens: Does it change intrauterine insemination outcomes? *Urology* **2012**, *80*, 1262–1266. [CrossRef] [PubMed]
109. Bollendorf, A.; Check, J.H.; Lurie, D. Evaluation of the effect of the absence of sperm with rapid and linear progressive motility on subsequent pregnancy rates following intrauterine insemination or in vitro fertilization. *J. Androl.* **1996**, *17*, 550–557. [CrossRef] [PubMed]
110. Nagata, M.P.B.; Endo, K.; Ogata, K.; Yamanaka, K.; Egashira, J.; Katafuchi, N.; Yamanouchi, T.; Matsuda, H.; Goto, Y.; Sakatani, M.; et al. Live births from artificial insemination of microfluidic-sorted bovine spermatozoa characterized by trajectories correlated with fertility. *Proc. Natl. Acad. Sci. USA* **2018**, *115*, E3087–E3096. [CrossRef]
111. Fischer, B.; Chavatte-Palmer, P.; Viebahn, C.; Navarrete Santos, A.; Duranthon, V. Rabbit as a reproductive model for human health. *Reproduction* **2012**, *144*, 1–10. [CrossRef] [PubMed]

MDPI

Article

Modulation of Cathepsin S (*CTSS*) Regulates the Secretion of Progesterone and Estradiol, Proliferation, and Apoptosis of Ovarian Granulosa Cells in Rabbits

Guohua Song [1,†], Yixuan Jiang [1,†], Yaling Wang [1], Mingkun Song [1], Xuanmin Niu [2], Huifen Xu [1,*] and Ming Li [1,*]

1 College of Animal Science and Technology, Henan Agricultural University, Zhengzhou 450046, China; m13782525956@163.com (G.S.); J19203090@163.com (Y.J.); 15637410238@163.com (Y.W.); m18625951565@163.com (M.S.)
2 Yangguang Rabbit Technology Co., Ltd., Jiyuan 454650, China; damoguyue@163.com
* Correspondence: huifen221@126.com (H.X.); 13803849306@163.com (M.L.); Tel.: +86-0371-5699-0218 (M.L.)
† Guohua Song and Yixuan Jiang contributed equally to this paper.

Simple Summary: In goat and sheep, *CTSS* is reported to be important for the development and maturation of oocytes by regulating cell proliferation and apoptosis. The purpose of this study was to investigate the role of *CTSS* in regulating cell apoptosis and hormone secretion in rabbit granulosa cells. Our results suggested that the *CTSS* gene can promote the proliferation of granulosa cells and reduce its apoptosis in vitro, while overexpression of *CTSS* promoted the secretion of progesterone and estrogen in rabbit granulosa cells. Therefore, manipulation of *CTSS* may improve development of oocytes, and thus provide an approach for better manipulation of rabbit reproductive performance.

Abstract: Cathepsin S (*CTSS*) is a member of cysteine protease family. Although many studies have demonstrated the vital role of *CTSS* in many physiological and pathological processes including tumor growth, angiogenesis and metastasis, the function of *CTSS* in the development of rabbit granulosa cells (GCS) remains unknown. To address this question, we isolated rabbit GCS and explored the regulatory function of the *CTSS* gene in cell proliferation and apoptosis. *CTSS* overexpression significantly promoted the secretion of progesterone (P4) and estrogen (E2) by increasing the expression of *STAR* and *CYP19A1* ($p < 0.05$). We also found that overexpression of *CTSS* increased GCS proliferation by up-regulating the expression of proliferation related gene (*PCNA*) and anti-apoptotic gene (*BCL2*). Cell apoptosis was markedly decreased by *CTSS* activation ($p < 0.05$). In contrast, *CTSS* knockdown significantly decreased the secretion of P4 and E2 and the proliferation of rabbit GCS, while increasing the apoptosis of rabbit GCS. Taken together, our results highlight the important role of *CTSS* in regulating hormone secretion, cell proliferation, and apoptosis in rabbit GCS. These results might provide a basis for better understanding the molecular mechanism of rabbit reproduction.

Keywords: CTSS; granulosa cells; proliferation; apoptosis; hormone secretion

Citation: Song, G.; Jiang, Y.; Wang, Y.; Song, M.; Niu, X.; Xu, H.; Li, M. Modulation of Cathepsin S (*CTSS*) Regulates the Secretion of Progesterone and Estradiol, Proliferation, and Apoptosis of Ovarian Granulosa Cells in Rabbits. *Animals* **2021**, *11*, 1770. https://doi.org/10.3390/ani11061770

Academic Editors: Rosa María García-García and Maria Arias Alvarez

Received: 28 May 2021
Accepted: 10 June 2021
Published: 13 June 2021

1. Introduction

Rabbits are multi-purpose domestic animals, which can be used as pet animals or biomedical model animals for scientific research, but in people's lives, rabbits are mainly used for meat and fur production [1–3]. Rabbit meat has high nutritional value and is widely accepted by many people for its low content in fat, cholesterol, and high content in protein [4,5]. However, as the demand for rabbit meat increases rapidly, global rabbit meat production is still relatively low [6]. Accelerating the reproductive performance is an important strategy to increase rabbit meat production, thus it is vital to investigate the regulatory mechanism of factors affecting rabbit reproduction.

Ovarian granulosa cells (GCS) are an important component of follicles and play a vital role in follicle growth and development. Each follicle contains an oocyte, numerous

GCS and theca cells [7]. GCS deliver regulatory signals to oocytes through zone projections (TZPs), and also provide nutrition and metabolite for oocyte [8]. Many studies reported that apoptosis of GCS affect cellular connection between GCS and oocytes [9]. It causes ovarian follicular atresia [10], which occurs in more than 99% of developmental follicles in mammals [11,12]. Another important role of GCS is the secretion of progesterone and estradiol, both of which play an important role in animal breeding and reproduction. It was reported that estradiol can promote the expression of FSH receptors in GCS [13], and promote the formation of rat antral follicles [14]. In bovine, low concentration of progesterone enhanced the follicular development, and the diameter of follicular increased [15,16]. Moreover, after follicle ovulation, GCS differentiated into luteal granulosa cells to form the corpus luteum. The main function of corpus luteum is to secrete progesterone to maintain pregnancy. Therefore, both GCS progesterone secretion disorder and GCS apoptosis will result in a decrease in rabbit reproduction performance.

Cathepsin S (*CTSS*), which is located on chromosome 13 in rabbits, is one member of 11 cysteine proteases, and plays an important role in extracellular matrix degradation and remodeling, antigen presentation, inflammation, and angiogenesis [17]. In malignant tumors, *CTSS* induces tumor proliferation, invasion and metastasis through various mechanisms [18–20]. It was reported that *CTSS* might be a potential predictor of chemotherapy response because up-regulation of *CTSS* is associated with tumor progression and poor prognosis [21]. Studies in vitro also showed that *CTSS* promoted adipocyte differentiation by degrading fibrinolytic proteins [22]. Studies have also found that the polymorphisms of the *CTSS* gene are associated with obesity-related traits [23], and *CTSS* circulating levels are associated with triglycerides synthesis and accumulation [24,25]. *CTSS* can also regulate blood sugar by reducing glucose output [26].

The regulatory function of *CTSS* in cell proliferation has been widely investigated, while its function in the development and maturation of rabbits' oocytes has not been reported. Our hypothesis was that *CTSS* plays an important role in follicle development and ovulation through regulating cell proliferation and hormone secretion. To verify this hypothesis, the present study aimed to investigate the functional role of the *CTSS* gene in cell proliferation and apoptosis, as well as progesterone and estradiol secretion in rabbit GCS. These results provide evidence with which to further uncover the regulatory mechanism which mediated the ovulation process and reproductive performance of rabbits.

2. Materials and Methods

2.1. Ethic Statement

The present study was designed and performed according to the guidelines of Institutional Animals Care and Use Committee College (IACUC) of College of Animal Science and Technology of Henan Agricultural University, China (Permit Number: 11-0085; Data: 06-2011).

2.2. Tissue Sample Collection

In this experiment, three female New Zealand white rabbits (180 days old) were selected from the animal house of Henan Agricultural University. The rabbits were anesthetized and slaughtered for ovary collection. The collected ovaries were placed in PBS (containing 1% penicillin/streptomycin, 37 °C) for rabbit granulosa cells isolation.

2.3. Isolation and Culture of Rabbits GCS

The rabbits' GCS were isolated and cultured according to the previously published papers with few modifications. Briefly, the collected ovaries were washed three times with PBS (phosphate buffer saline) supplemented with 1% penicillin/streptomycin at 37 °C. Follicles were needled with a 1 mL syringe in a basal medium supplemented with DMEM/F12 medium (Gibco, MD, USA), 15% FBS (Gibco, CA, USA), and 1% penicillin/streptomycin (Hyclone, Logan, UT, USA) [27]. The GCS were centrifuged (5 min, 1000 rpm/min) and incubated with basal medium at 37 °C in 5% CO_2; cell medium was changed every 24 h [28]. The GCS isolated from the 3 rabbits were isolated individually and pooled together for experimental treatments. Before experiments, GCS were seeded in 6-well plate at a density of 3×10^5 cells per cm^2. After 24 h incubation (approximately 90% confluence), GCS were treated with siRNA and recombinant adenovirus for 24 h. All treatments were performed with 3 independent biological replicates.

2.4. Immunostaining

FSHR (follicle stimulating hormone receptor) protein is usually used as a marker protein to identify GCS. The cells were seeded on culture plates plated with cell-climbing slices. After reaching 70% confluence, cells were fixed with 4% paraformaldehyde and then permeated with 0.1% Triton X-100. For immunohistochemical staining, the cells were incubated with anti-FSHR (MAB65591, 1: 200, R&D Systems, Minneapolis, MN, USA) for 1 h at 37 °C, and then incubated with goat anti-rabbit IgG (bs-0296G-FITC, 1:400, Bioss, Beijing, China) at 37 °C for 45 min. Finally, the cells were taken out and mounted with DAPI (4′,6-Diamidino-2-phenylindole, Dihydrochloride) medium (Solarbio, Beijing, China). The cells were then observed and imaged under a fluorescence microscope (Nikon Eclipse C1, Tokyo, Japan).

2.5. HE (Hematoxylin-Eosin) Staining

The morphology examination of GCS was measured by HE staining. GCS were seeded in 6-well culture plates plated with cell-climbing slices. Then, the cells were fixed with 95% ethanol for 20 min and stained with hematoxylin dye solution for 3 min. After dehydration with gradient alcohol, the cells were dyed with eosin dye solution for 5 min. Finally, the cells were mounted with neutral gum and observed by microscope (Nikon Eclipse E100, Tokyo, Japan).

2.6. SiRNA (Small Interfering RNA) Interference

GCS were seeded into 6-well plates and transfected with siRNA-*CTSS* and siRNA-NC by using Lipofectamine 3000 (Invitrogen, Carlsbad, CA, USA) according to the manufacturer's instructions. The specific siRNA sequence targeting *CTSS* was designed and synthesized by GenePharma (Shanghai, China). siRNA-*CTSS*, sense: 5′-GGAAGAAAGCCUACG GCAATT-3′; antisense:5′-UUGCCGUAGGCUUUCUUCCTT-3′. siRNA-NC, sense: 5′-UUCUCCGAACGUGUCACGUTT-3′, antisense: 5′-ACGUGACACGUUCGGAGAATT-3′.

2.7. Recombinant Adenoviruses Generation

The production of recombinant adenovirus has been previously described [29]. In short, *CTSS* was amplified and subcloned into the shuttle vector pAdTrack-CMV. Then the linearized product of pAdTrack-CMV-*CTSS* was transformed into Escherichia coli BJ5183 competent cells containing the backbone vector pAdEasy-1 to obtain the positive recombinants of pAd-*CTSS*. Then, linearized pAd-*CTSS* fragment was transfected into 293 cells for adenovirus packaging and amplification. GCS were infected with Ad-*CTSS* at MOI (multiplicity of infection) value of 40.

2.8. RNA Extraction and RT-qPCR

Total RNA of rabbit GCS were extracted by using Trizol reagent (Invitrogen, CA, USA) according to the instructions of the manufacturer. Total RNA was quantified with a Nanodrop ND-2000 spectrophotometer (Thermo Fisher Scientific, Waltham, MA, USA). cDNA was synthesized by using Prime Script RT reagent Kit (Takara, Tokyo, Japan) following the instructions. The SYBR Premix Ex Taq II kit (Takara, Tokyo, Japan) was used for RT-qPCR. Primers used for RT-qPCR were listed in Table 1. *β-actin* was selected as reference gene.

Table 1. Name, accession number, sequences, amplicon length of primer pairs of RT-qPCR.

Gene/Accession No.	Primers	Sequence (5′ to 3′)	bp
PCNA	F.432	TGCACGTATATGCCGAGACC	240
XM_017341762.1	R.671	GTAGGAGAAAGCGGAGTGGC	
Notch 2	F.3798	CAACCGCCAGTGTGTTCAAG	232
XM_017345939.1	R.4029	CTTCCGCTTTCGTTTTGCCA	
Cyclin D	F.44	GCAGCCCTTTCAATGCTGAC	244
XM_017348091.1	R.697	CTTTGGACGCTCTGACCAGT	
STAR1	F.477	GATTGGGAAGGACACGGTCA	179
XM_017350353.1	R.655	CACCCCTGATGACGCCTTT	
CYP19A1	F.463	CCTGGGCTTGTTCAGATGGT	171
NM_001170921.2	R.633	ACTTTCGTCCATGGGGATGC	
CYP11A1	F.1259	GCCGGGAACCTAGCTTCTTT	155
XM_008253734.2	R.1413	GATGCTCATCTCCACCTCGG	
BCL2	F.492	GACTGAGTACCTGAACCGGC	166
XM_008261439.2	R.65	GAGGGTGATGCAAGCTCCTA	
Bax	F.11	CCGGGGAGCAGTCCAGA	167
XM_002723696.3	R.177	CAGCTTCTTGGTGGACTCGT	
Caspase-3	F.1	ATGGAGAACAACGAAACCTCC	191
NM_001082117.1	R.191	CGGGACGACATTCCAGTGTT	
β-actin	F.1140	ATGCAGAAGGAGATCACCGC	148
NM_001101683.1	R.191	ACTCCTGCTTGCTGATCCAC	
CTSS	F.128	CTACGGCAAGCAATACAA	187
XM_002715580.3	R.296	TCTCAGGGAACTCATCAAA	

Annealing temperature for all primers listed in this table is 60°C: F, forward primer; R, reverse primer. *PCNA*, proliferating cell nuclear antigen; *STAR1*, steroidogenic acute regulatory protein 1; *CYP19A1*, cytochrome P450 family 19 subfamily A member 1; *CYP11A1*, cytochrome P450 family 11 subfamily 11 subfamily A member 1; *BCL2*, B-cell lymphoma-2.

2.9. Measurement of Progesterone and Estradiol Secretion

After 48 h of treatment with Ad-*CTSS* and siRNA-*CTSS*, cell culture medium was collected for determination of progesterone and estradiol concentration. Secretion levels of progesterone and estradiol were evaluated with enzyme linked immunosorbent assay (ELISA) according to the instructions of the manufacturer. The kits were all obtained from Nanjing Ruixin Biology (Quanzhou, China). The absorbance was measured using a microplate reader (Bio-Rad) at 450 nm.

2.10. Cell Viability Assay

GCS were seeded onto 96-well plates at 1×10^4/well and treated with recombinant adenovirus and siRNA for 0 h, 12 h, 24 h, 36 h and 48 h, separately. Cell proliferation was detected with Cell Counting Kit-8 (CCK8) (US Everbright® Inc., Nanjing, China). According to the instructions of the manufacturer, the cells were added to 10 μL of CCK8 solution and incubated at 37 °C for 2 h. Then, the optical density (OD) value at a wavelength of 450 nm was detected by a microplate reader (Bio-Rad, Hercules, CA, USA).

2.11. Cell Apoptosis Analysis

After treatment with recombinant adenovirus and siRNA for 48 h, GCS apoptosis was detected by Annexin V-Alexa Fluor 647/7-AAD Apoptosis Kit (4Abiotech., Beijing, China). According to the instructions of the manufacturer, GCS were digested with trypsin (without EDTA) and resuspended with cold PBS, and then mixed with 5 μL Annexin V/Alexa Fluor 647 at room temperature for 5 min. Finally, the cells were treated with 10 μL 7-AAD and 400 μL PBS. The apoptosis rate of the GCS was measured with a CytoFLEX flow cytometer (Beckman CytoFlex, CA, USA).

2.12. Statistical Analysis

All data analyses were conducted using SPSS 22.0 software (SPSS, Chicago, IL, USA). All the treatments were performed in three independent biological replicates and three technical replicates. Statistical analyses were performed with one-way ANOVA to compare the effects of *CTSS* overexpression and knockdown relative to negative control. The fixed effect was treatment and the random effect was replicated in the statistic model. Relative mRNA expression of RT-qPCR was calculated by using $2^{-\Delta\Delta Ct}$ method, where Ct is the cycle threshold. p value < 0.05 was indicative of a statistically significant difference, and data are presented as means ± SE.

3. Results

3.1. Isolation and Identification of Rabbit GCS

Rabbit GCS were identified by HE staining and immunofluorescence by using FSHR antibody, our HE staining results showed that the nucleus of GCS is blue and the cytoplasm is red, and the pseudopodia between the cells was closely connected (Figure 1A). As shown in Figure 1B, cell nuclei were stained with DAPI and presented to be blue; FSHR protein was green and located in the cytoplasm. The positive rate of FSHR was 99%, indicating that the isolated GCS were 99%.

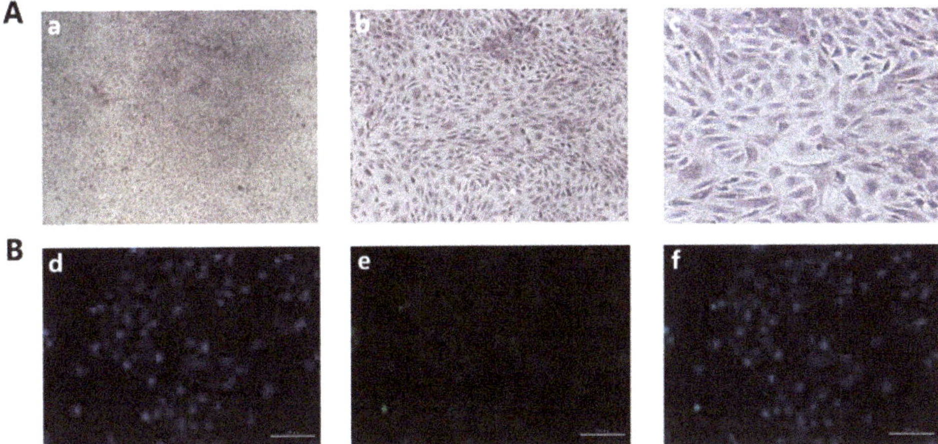

Figure 1. Isolation and identification of rabbit GCS: (**A**): GCS were stained with Hematoxylin and eosin (H&E) staining, (**a–c**): Original magnification, 40×, 100×, 200×. (**B**): The identification of rabbit GCS was measured by immunofluorescence staining with fluorescein isothiocyanate (FITC)-goat anti-rabbit IgG. (**d**): Cell nucleus were stained by DAPI (blue), (**e**): FSHR were stained with fluorescein isothiocyanate (FITC)-goat anti-rabbit IgG (green), (**f**): Merge of (**d**) and (**e**). (Scale bar = 100 μm).

3.2. Efficiency of CTSS Overexpression and SiRNA Interference

RT-qPCR was performed to determine the efficiency of Ad-*CTSS* and siRNA-*CTSS* in rabbit GCS. Our results showed that *CTSS* expression was remarkably increased by Ad-*CTSS* compared with Ad-GFP (Figure 2A, $p < 0.05$), and its expression was significantly decreased by siRNA-*CTSS* relative to siRNA-NC (Figure 2B, $p < 0.05$).

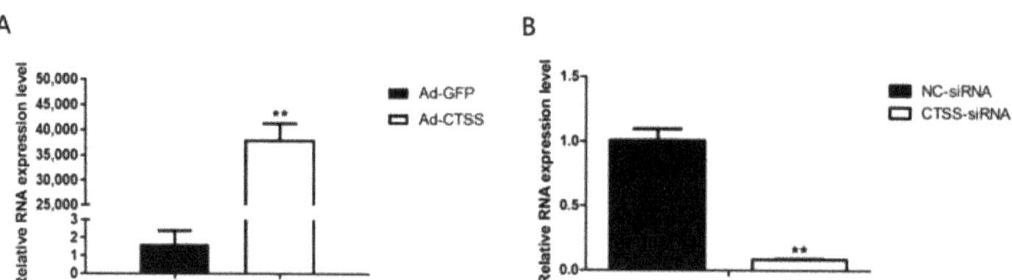

Figure 2. The efficiency of Ad-*CTSS* and siRNA-*CTSS* in rabbit GCS: (**A**): mRNA expression changes of *CTSS* after infection with Ad-GFP. (**B**): mRNA expression changes of *CTSS* after transfection with siRNA-*CTSS*. Values are presented as means ± SEM for three biological replicates. ** $p < 0.01$.

3.3. *CTSS* Promotes the Secretion of Progesterone and Estradiol in Rabbit GCS

To explore the role of *CTSS* on rabbit reproduction, we measured the effects of *CTSS* overexpression and knockdown on hormone secretion in rabbit GCS. As shown in Figure 3A,D, genes responsible for steroid hormone synthesis (Steroidogenic acute regulatory protein, *STAR*) and estrogen synthesis (Cytochrome p450 family 19 subfamily A member 1, *CYP19A1*) were remarkably increased and decreased by *CTSS* overexpression and knockdown, respectively ($p < 0.05$). Consistently, levels of progesterone and estradiol in cell medium were significantly increased by *CTSS* overexpression (Figure 3B,C). While *CTSS* knockdown only decreased progesterone secretion (Figure 3E, $p < 0.05$), estradiol secretion was not changed by siRNA-*CTSS* (Figure 3F, $p > 0.05$).

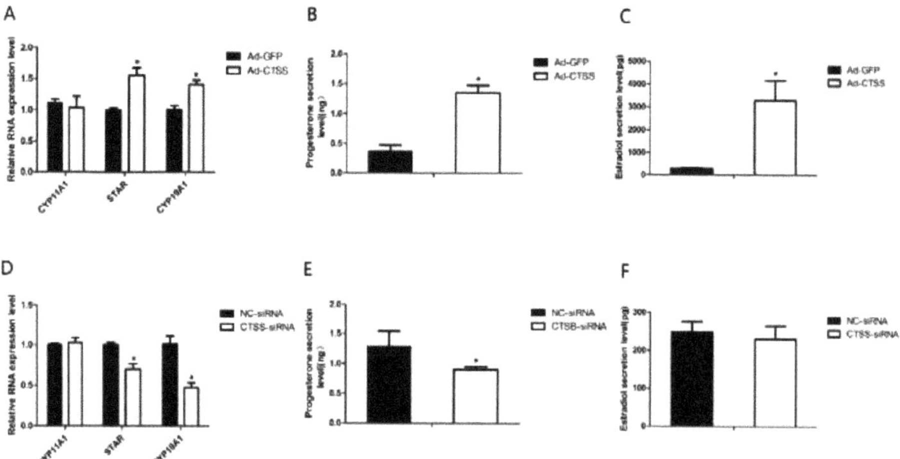

Figure 3. *CTSS* overexpression and interference on progesterone and estradiol secretion in rabbit GCS: Effects of *CTSS* overexpression and knockdown on (**A**,**D**) mRNA expression changes of genes related to progesterone and estradiol secretion; (**B**,**E**) progesterone secretion and (**C**,**F**) estradiol secretion. Values are presented as means ± SEM for 3 biological replicates. * $p < 0.05$.

3.4. *CTSS* Promotes Cell Proliferation Activity in Rabbit GCS

After treatment with Ad-*CTSS* and siRNA-*CTSS* for 0 h, 12 h, 24 h, 36 h and 48 h, the proliferation activity of GCS was measured. As shown in Figure 4A, genes responsible for cell proliferation (*PCNA*, proliferating cell nuclear antigen) were significantly increased by *CTSS* overexpression ($p < 0.05$), and cell proliferation activity of the Ad-*CTSS* infected group was remarkably higher than the Ad-GFP infected group (Figure 4B, $p < 0.05$). *CTSS* knockdown decreased *Notch2* expression and cell proliferation activity in rabbit GCS (Figure 4C,D, $p < 0.05$).

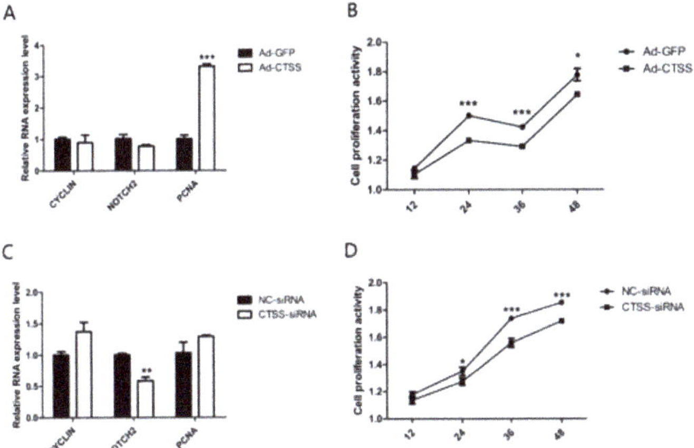

Figure 4. *CTSS* overexpression and interference on cell proliferation in rabbit GCS: Effects of *CTSS* overexpression and knockdown on (**A**,**C**): mRNA expression changes of cell proliferation related genes and (**B**,**D**): cell proliferation activity. Values are presented as means ± SEM for 3 biological replicates. * $p < 0.05$; ** $p < 0.01$; *** $p < 0.001$.

3.5. Activation of CTSS Deceased Cell Apoptosis in Rabbit GCS

As shown in Figure 5A, compared with the Ad-GFP infected group, mRNA level of *BCL2* (one of the main anti-apoptotic gene) was significantly increased by *CTSS* overexpression ($p < 0.05$). The other important pro-apoptotic gene, *BAX,* was decreased significantly by *CTSS* knockdown (Figure 5E, $p < 0.05$). Flow cytometry analysis revealed that *CTSS* overexpression inhibited the apoptosis of rabbit GCS significantly (Figure 5B–D, $p < 0.01$). As shown in Figure 5F–H, *CTSS* knockdown led to a remarkable increase in apoptotic cells in rabbit GCS ($p < 0.05$).

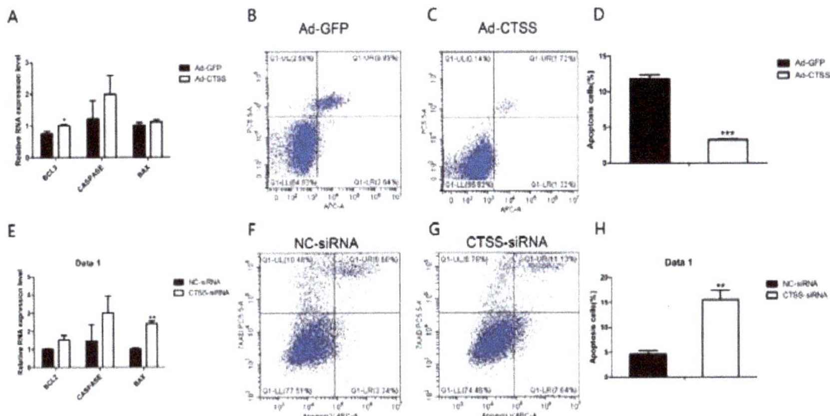

Figure 5. Effect of *CTSS* overexpression and interference on cell apoptosis in rabbit GCS: Relative mRNA expression changes of marker genes related to cell apoptosis after *CTSS* overexpression (**A**) and knockdown (**E**). Cell apoptosis rate was measured by using flow cytometry after *CTSS* overexpression (**B–D**) and knockdown (**F–H**). * $p < 0.05$; ** $p < 0.01$; *** $p < 0.001$.

4. Discussions

Reproductive performance is an important index for animal husbandry. GCS play an important role in oocyte development and affect female breeding performance. A number of studies have shown that small molecules including Vitamin D3 [30], FSH [31], Bisphenol A [32] and genes including *CITED4* [33], *CXCL12* [34], and *STAT4* [35] regulate cell proliferation, apoptosis and hormone secretion in mammals, thus affecting estrus of female animals and oocyte maturation. However, as one species of the high fertility husbandry animals, the effect of *CTSS* gene on rabbit reproduction has been rarely studied.

Cathepsin S (*CTSS*), one kind of lysosomal protease, is a major participant in proteolysis [36], and participates in many physiological processes. Studies have shown that *CTSS* gene played an important role in lipid metabolism in tumor tissue. To further investigate the role of *CTSS* in rabbit reproduction, we isolated rabbit GCS. GCS exist in the follicle and play an important role in the development of the follicle, as it can provide nutrients to oocytes during follicular development [37]. Studies have reported many methods for in vitro isolation and culture of GCS in pigs [38], cattle [39], and mice [40]. Isolation of rabbit GCS in the present study was performed according to the published methods with little modifications [41], providing material for further study of the complex mechanism of rabbit reproduction.

Progesterone and estradiol are mainly secreted by GCS and play an important role in the breeding process of mammals. They can promote estrus of female rabbits and early fetal pregnancy. To investigate whether *CTSS* affects progesterone and estradiol secretion in GCS, we overexpressed and interfered the expression of *CTSS* gene in rabbit GCS. Our results demonstrated that *CTSS* gene can regulate the secretion of progesterone hormone, and that this effect might be mediated by *STAR* and *CYP19A1*. As summarized in Figure 6, activation of *CTSS* resulted in up-regulation of progesterone hormone-related genes (*STAR* and *CYP19A1*) and promoted the secretion of progesterone. In mammals, progesterone plays a key role in follicular development and early fetal pregnancy [40]. Progesterone promotes follicular development, and low levels of progesterone increase follicular diameter [16]. However, the insufficiency of progesterone secretion will cause the fetus to be unable to survive or, even worse, will result in the abortion phenomenon [42]. In this experiment, our results demonstrated that *CTSS* play an important role in regulating the secretion of progesterone and estradiol in GCS.

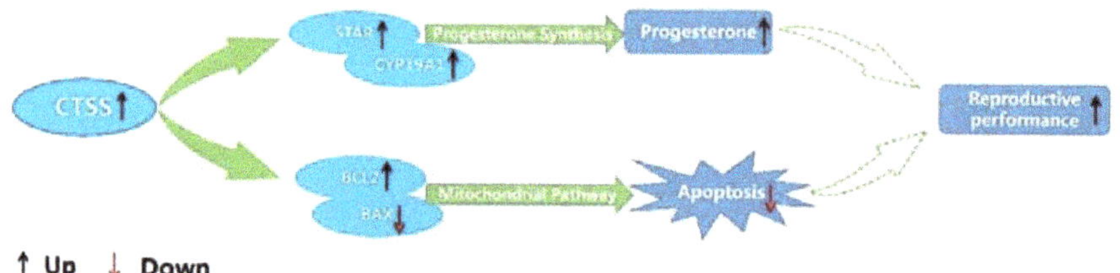

Figure 6. Molecular mechanism of *CTSS* regulating cell apoptosis and progesterone secretion in rabbit GCS.

By providing nutrition and metabolite to oocytes, GCS play an important role in oocyte development. GCS also transmit signals and information for oocytes, and make oocytes mature. Apoptosis of GCS affects follicle development and leads to follicular atresia. In this study, we found that *CTSS* gene could affect the proliferation and apoptosis of GCS, and that the overexpression of *CTSS* gene in GCS increased the expression of proliferation gene (*PCNA*) and anti-apoptotic gene (*BCL2*), and promoted the proliferation of GCS while decreasing its apoptosis (Figure 6). Consistently, *CTSS* knockdown decreased the proliferation rate while increasing the apoptosis of rabbit GCS. These findings are in accordance with recent research reporting that *CTSS* interference inhibited the proliferation while inducing the apoptosis in mhCC97-H cells [43,44], and recent studies have also shown that *CTSS* plays a crucial role in the invasion and apoptosis of cancer cells [45,46]. Taken together, our results have demonstrated that *CTSS* can promote the proliferation while reducing the apoptosis of rabbit GCS.

5. Conclusions

In summary, our data suggest that *CTSS* regulates the proliferation and apoptosis of granulosa cells by regulating the expression of key genes involved in these processes. In addition, *CTSS* is also involved in the process of progesterone and estradiol secretion in rabbit GCS. Overall, the present study provided evidence that *CTSS* may play an important role in regulating follicular development by regulating cell apoptosis and hormone secretion in rabbits. However, whether manipulation of *CTSS* expression in vivo can improve the reproductive performance of rabbits needs further research.

Author Contributions: Conceptualization, H.X., M.L. and G.S.; funding acquisition, M.L.; supervision, X.N.; formal analysis, Y.W.; validation, Y.J. and M.S.; writing—original draft, G.S. and Y.J.; writing—review and editing, H.X. and M.L. All authors have read and agreed to the published version of the manuscript.

Funding: The research was jointly supported by the National Key Research and Development Program of China (2018YFD0502203) and the Special Fund for Henan Agriculture Research System (S2013-08-G01).

Institutional Review Board Statement: The study was conducted according to the guidelines of Declaration of Helsinki, and approved by the Ethics Committee of College of Animal Science and Technology of Henan Agricultural University, China (Permit Number: 11-0085; Data: 06-2011).

Informed Consent Statement: Not applicable.

Data Availability Statement: Not applicable.

Conflicts of Interest: The authors declare no competing financial interest.

References

1. Mage, R.G.; Esteves, P.J.; Rader, C. Rabbit models of human diseases for diagnostics and therapeutics development. *Dev. Ccomp. Immunol.* **2019**, *92*, 99–104. [CrossRef] [PubMed]
2. Esteves, P.J.; Abrantes, J.; Baldauf, H.-M.; Ben Mohamed, L.; Chen, Y.; Christensen, N.; González-Gallego, J.; Giacani, L.; Hu, J.; Kaplan, G.; et al. The wide utility of rabbits as models of human diseases. *Exp. Mol. Med.* **2018**, *50*, 1–10. [CrossRef] [PubMed]
3. Fontanesi, L. Rabbit Genetic Resources Can Provide Several Animal Models to Explain at the Genetic Level the Diversity of Morphological and Physiological Relevant Traits. *Appl. Sci.* **2021**, *11*, 373. [CrossRef]
4. Dalle Zotte, A. Rabbit farming for meat purposes. *Anim. Front.* **2014**, *4*, 62–67. [CrossRef]
5. Li, S.; Zeng, W.; Li, R.; Hoffman, L.C.; He, Z.; Sun, Q.; Li, H. Rabbit meat production and processing in China. *Meat Sci.* **2018**, *145*, 320–328. [CrossRef] [PubMed]
6. Trocino, A.; Cotozzolo, E.; Zomeño, C.; Petracci, M.; Xiccato, G.; Castellini, C. Rabbit production and science: The world and Italian scenarios from 1998 to 2018. *Ital. J. Anim. Sci.* **2019**, *18*, 1361–1371. [CrossRef]
7. Chang, H.-M.; Qiao, J.; Leung, P.C.K. Oocyte-somatic cell interactions in the human ovary-novel role of bone morphogenetic proteins and growth differentiation factors. *Hum. Reprod. Update* **2016**, *23*, 1–18. [CrossRef]
8. Li, R.; Albertini, D.F. The road to maturation: Somatic cell interaction and self-organization of the mammalian oocyte. *Nat. Rev. Mol. Cell Biol.* **2013**, *14*, 141–152. [CrossRef]
9. Emanuelli, I.P.; Costa, C.B.; Rafagnin Marinho, L.S.; Seneda, M.M.; Meirelles, F.V. Cumulus-oocyte interactions and programmed cell death in bovine embryos produced in vitro. *Theriogenology* **2019**, *126*, 81–87. [CrossRef]
10. Tilly, J.L.; Kowalski, K.I.; Johnson, A.L.; Hsueh, A.J. Involvement of apoptosis in ovarian follicular atresia and postovulatory regression. *Endocrinology* **1991**, *129*, 2799–2801. [CrossRef]
11. Hughes, F.M.; Gorospe, W.C. Biochemical identification of apoptosis (programmed cell death) in granulosa cells: Evidence for a potential mechanism underlying follicular atresia. *Endocrinology* **1991**, *129*, 2415–2422. [CrossRef]
12. Manabe, N.; Goto, Y.; Matsuda-Minehata, F.; Inoue, N.; Maeda, A.; Sakamaki, K.; Miyano, T. Regulation mechanism of selective atresia in porcine follicles: Regulation of granulosa cell apoptosis during atresia. *J. Reprod. Dev.* **2004**, *50*, 493–514. [CrossRef]

13. Cai, L.; Sun, A.; Li, H.; Tsinkgou, A.; Yu, J.; Ying, S.; Chen, Z.; Shi, Z. Molecular mechanisms of enhancing porcine granulosa cell proliferation and function by treatment in vitro with anti-inhibin alpha subunit antibody. *Reprod. Biol. Endocrin.* **2015**, *13*, 26. [CrossRef] [PubMed]
14. Gore-Langton, R.E.; Daniel, S.A. Follicle-stimulating hormone and estradiol regulate antrum-like reorganization of granulosa cells in rat preantral follicle cultures. *Biol. Reprod.* **1990**, *43*, 65–72. [CrossRef] [PubMed]
15. Cerri, R.L.A.; Chebel, R.C.; Rivera, F.; Narciso, C.D.; Oliveira, R.A.; Thatcher, W.W.; Santos, J.E.P. Concentration of progesterone during the development of the ovulatory follicle: I. Ovarian and embryonic responses. *J. Dairy Sci.* **2011**, *94*, 3342–3351. [CrossRef] [PubMed]
16. Cerri RL, A.; Chebel, R.C.; Rivera, F.; Narciso, C.D.; Oliveira, R.A.; Thatcher, W.W.; Santos, J.E.P. Effect of progesterone concentrations, follicle diameter, timing of artificial insemination, and ovulatory stimulus on pregnancy rate to synchronized artificial insemination in postpubertal Nellore heifers. *Theriogenology* **2014**, *81*, 446–453. [CrossRef]
17. Zhao, P.; Lieu, T.M.; Barlow, N.; Metcalf, M.; Bunnett, N.W. Cathepsin S causes inflammatory pain via biased agonism of PAR2 and TRPV4. *J. Biol. Chem.* **2014**, *289*, 35858. [CrossRef]
18. Karimkhanloo, H.; Keenan, S.N.; Sun, E.W.; Wattchow, D.A.; Keating, D.J.; Montgomery, M.K.; Watt, M.J. Circulating cathepsin S improves glycemic control in mice. *J. Endocrinol.* **2021**, *248*, 167–179. [CrossRef]
19. Sevenich, L.; Bowman, R.L.; Mason, S.D.; Quail, D.F.; Rapaport, F.; Elie, B.T.; Brogi, E.; Brastianos, P.K.; Hahn, W.C.; Holsinger, L.J. Analysis of tumour- and stroma-supplied proteolytic networks reveals a brain-metastasis-promoting role forcathepsin S. *Nat. Cell Biol.* **2014**, *16*, 876–888. [CrossRef]
20. Joyce, J.A.; Baruch, A.; Chehade, K.; Meyer-Morse, N.; Giraudo, E.; Tsai, F.Y.; Greenbaum, D.C.; Hager, J.H.; Bogyo, M.; Hanahan, D. Cathepsin cysteine proteases are effectors of invasive growth and angiogenesis during multistage tumorigenesis. *Cancer Cell.* **2004**, *5*, 443–453. [CrossRef]
21. Harbeck, N.; Alt, U.; Berger, U.; Krüger, K.; Thomssen, C.F.J.; Höfler, H.; Kates, R.E.; Schmitt, M. Prognostic Impact of Proteolytic Factors (Urokinase-Type Plasminogen Activator, Plasminogen Activator Inhibitor 1, and Cathepsins B, D, and L) in Primary Breast Cancer Reflects Effects of Adjuvant Systemic Therapy. *Clin. Cancer Res.* **2001**, *7*, 2757. [CrossRef] [PubMed]
22. Taleb, S.; Cancello, R.; Clément, K.; Lacasa, D. Cathepsins promotes human preadipocyte differentiation: Possible involvement of fibronectin degradation. *Endocrinology* **2006**, *147*, 4950–4959. [CrossRef] [PubMed]
23. Hooton, H.; Ängquist, L.; Holst, C.; Hager, J.; Rousseau, F.; Hansen, R.D.; Tjønneland, A.; Roswall, N.; Overvad, K.; Jakobsen, M.U.; et al. Dietary Factors Impact on the Association between *CTSS* Variants and Obesity Related Traits. *PLoS ONE* **2012**, *7*, e40394. [CrossRef] [PubMed]
24. Taleb, S.; Lacasa, D.; Bastard, J.P.; Poitou, C.; Cancello, R.; Pelloux, V.; Viguerie, N.; Benis, A.; Zucker, J.D.; Bouillot, J.L. Cathepsin S, a novel biomarker of adiposity: Relevance to atherogenesis. *Faseb. J.* **2005**, *19*, 1540–1542. [CrossRef] [PubMed]
25. Nadia, N.; Christine, R.; Soraya, F.; Marie-Eve, L.; Christine, P.; Mayoura, K.; Delphine, E.; Steve, S.; Salwa, R.; Jean-Philippe, B. Cathepsins in Human Obesity: Changes in Energy Balance Predominantly Affect Cathepsin S in Adipose Tissue and in Circulation. *J. Clin. Endocrinol. Metab.* **2010**, *95*, 1861–1868. [CrossRef]
26. Xu, H.F.; Luo, J.; Zhao, W.S.; Yang, Y.C.; Tian, H.B.; Shi, H.B.; Bionaz, M. Overexpression of SREBP1 (sterol regulatory element binding protein 1) promotes de novo fatty acid synthesis and triacylglycerol accumulation in goat mammary epithelial cells. *J. Dairy Sci.* **2016**, *99*, 783–795. [CrossRef] [PubMed]
27. Lovasco, L.A.; Seymour, K.A.; Zafra, K.; O'Brien, C.W.; Schorl, C.; Freiman, R.N. Accelerated ovarian aging in the absence of the transcription regulator TAF4B in mice. *Biol. Reprod.* **2010**, *82*, 23–34. [CrossRef]
28. Jiajie, T.; Yanzhou, Y.; Albert Cheung, H.-H.; Chen, Z.-J.; Chan, W.-Y. Conserved miR-10 family represses proliferation and induces apoptosis in ovarian granulosa cells. *Sci. Rep.* **2017**, *7*, 41304. [CrossRef]
29. Bakhshalizadeh, S.; Amidi, F.; Alleyassin, A.; Soleimani, M.; Shirazi, R.; Shabani Nashtaei, M. Modulation of steroidogenesis by vitamin D3 in granulosa cells of the mouse model of polycystic ovarian syndrome. *Syst. Biol. Reprod. Med.* **2017**, *63*, 150–161. [CrossRef]
30. Tang, X.; Ma, L.; Guo, S.; Liang, M.; Jiang, Z. High doses of FSH induce autophagy in bovine ovarian granulosa cells via the AKT/mTOR pathway. *Reprod. Domest. Anim.* **2021**, *56*, 324–332. [CrossRef]
31. Shi, J.; Liu, C.; Chen, M.; Yan, J.; Wang, C.; Zuo, Z.; He, C. The interference effects of bisphenol A on the synthesis of steroid hormones in human ovarian granulosa cells. *Environ. Toxicol.* **2021**, *36*, 665–674. [CrossRef] [PubMed]
32. Yao, X.; Ei-Samahy, M.A.; Xiao, S.; Wang, Z.; Meng, F.; Li, X.; Bao, Y.; Zhang, Y.; Wang, Z.; Fan, Y.; et al. CITED4 mediates proliferation, apoptosis and steroidogenesis of Hu sheep granulosa cells in vitro. *Reproduction* **2021**, *161*, 255–267. [CrossRef] [PubMed]
33. Jin, L.; Ren, L.; Lu, J.; Wen, X.; Zhuang, S.; Geng, T.; Zhang, Y. CXCL12 and its receptors regulate granulosa cell apoptosis in PCOS rats and human KGN tumor cells. *Reproduction* **2021**, *161*, 145–157. [CrossRef] [PubMed]
34. Jin, L.; Ren, L.; Lu, J.; Wen, X.; Zhuang, S.; Geng, T.; Zhang, Y. STAT4 targets KISS1 to promote the apoptosis of ovarian granulosa cells. *J. Ovarian Res.* **2020**, *13*, 135. [CrossRef]
35. Riese, R.J.; Mitchell, R.N.; Villadangos, J.A.; Shi, G.P.; Palmer, J.T.; Karp, E.R.; De Sanctis, G.T.; Ploegh, H.L.; Chapman, H.A. Cathepsin S activity regulates antigen presentation and immunity. *J. Clin. Investig.* **1998**, *101*, 2351–2363. [CrossRef]
36. Kim, S.; Jin, H.; Seo, H.-R.; Lee, H.J.; Lee, Y.-S. Regulating BRCA1 protein stability by cathepsin S-mediated ubiquitin degradation. *Cell Death Differ.* **2019**, *26*, 812–825. [CrossRef]

37. Zhang, J.Q.; Gao, B.W.; Guo, H.X.; Ren, Q.L.; Wang, X.W.; Chen, J.F.; Wang, J.; Zhang, Z.J.; Ma, Q.; Xing, B.S. miR-181a promotes porcine granulosa cell apoptosis by targeting TGFBR1 via the activin signaling pathway. *Mol. Cell Endocrinol.* **2020**, *499*, 110603. [CrossRef]
38. Morrell, B.C.; Perego, M.C.; Maylem ER, S.; Zhang, L.N.; Schutz, L.F.; Spicer, L.J. Regulation of the transcription factor E2F1 mRNA in ovarian granulosa cells of cattle. *J. Anim. Sci.* **2020**, *98*, skz376. [CrossRef]
39. Yao, G.; Yin, M.; Lian, J.; Tian, H.; Liu, L.; Li, X.; Sun, F. MicroRNA-224 is involved in transforming growth factor-beta-mediated mouse granulosa cell proliferation and granulosa cell function by targeting Smad4. *Mol. Endocrinol.* **2010**, *24*, 540–551. [CrossRef]
40. Aisemberg, J.; Vercelli, C.A.; Bariani, M.V.; Billi, S.C.; Wolfson, M.L.; Franchi, A.M. Progesterone is essential for protecting against LPS-induced pregnancy loss. LIF as a potential mediator of the anti-inflammatory effect of progesterone. *PLoS ONE* **2013**, *8*, e56161. [CrossRef]
41. Sirotkin, A.V.; Petrak, J.; Alwasel, S.; Harrath, A.H. Apoptosis signal-regulating kinase (ASK1) and transcription factor tumor suppressor protein TP53 suppress rabbit ovarian granulosa cell functions. *Anim. Reprod. Sci.* **2019**, *204*, 140–151. [CrossRef]
42. Ke, R.W. Endocrine Basis for Recurrent Pregnancy Loss. *Obstet. Gynecol. Clin. N. Am.* **2014**, *41*, 103–112. [CrossRef]
43. Wang, X.; Xiong, L.; Yu, G.; Li, D.; Peng, T.; Luo, D.; Xu, J. Cathepsin S silencing induces apoptosis of human hepatocellular carcinoma cells. *Am. J. Transl. Res.* **2015**, *7*, 100–110. [CrossRef]
44. Fan, Q.; Wang, X.; Zhang, H.; Li, C.; Fan, J.; Xu, J. Silencing cathepsin S gene expression inhibits growth, invasion and angiogenesis of human hepatocellular carcinoma in vitro. *Biochem. Biophys. Res. Commun.* **2012**, *425*, 703–710. [CrossRef] [PubMed]
45. Ärnlöv, J. Cathepsin S as a biomarker: Where are we now and what are the future challenges? *Biomark. Med.* **2012**, *6*, 9–11. [CrossRef] [PubMed]
46. Small, D.M.; Burden, R.E.; Jaworski, J.; Hegarty, S.M.; Spence, S.; Burrows, J.F.; McFarlane, C.; Kissenpfennig, A.; McCarthy, H.O.; Johnston, J.A.; et al. Cathepsin S from both tumor and tumor-associated cells promote cancer growth and neovascularization. *Int. J. Cancer* **2013**, *133*, 2102–2112. [CrossRef] [PubMed]

Article

Semen Modulates Inflammation and Angiogenesis in the Reproductive Tract of Female Rabbits

Jaume Gardela [1,2], Amaia Jauregi-Miguel [3], Cristina A. Martinez [1], Heriberto Rodríguez-Martinez [1], Manel López-Béjar [2,4] and Manuel Álvarez-Rodríguez [1,2,*]

[1] Department of Biomedical and Clinical Sciences (BKV), Division of Children's and Women Health (BKH), Obstetrics and Gynecology, Linköping University, 58185 Linköping, Sweden; jaume.gardela@uab.cat (J.G.); cristina.martinez-serrano@liu.se (C.A.M.); heriberto.rodriguez-martinez@liu.se (H.R.-M.)

[2] Department of Animal Health and Anatomy, Veterinary Faculty, Universitat Autònoma de Barcelona, 08193 Bellaterra, Spain; manel.Lopez.Bejar@uab.cat

[3] Division of Molecular Medicine and Virology (MMV), Linköping University, 58185 Linköping, Sweden; amaya.jauregi.miguel@liu.se

[4] College of Veterinary Medicine, Western University of Health Sciences, Pomona, CA 91766, USA

* Correspondence: manuel.alvarez-rodriguez@liu.se

Received: 19 October 2020; Accepted: 18 November 2020; Published: 25 November 2020

Simple Summary: In mammals, the expression of regulatory genes is modified by the interaction between semen and the female reproductive tract. This study intends to unveil how mating or insemination with sperm-free seminal plasma, as well as the presence of preimplantation embryos, affects inflammation and angiogenesis in different segments of the reproductive tract of female rabbits. Gene expression of anti-inflammatory cytokines and angiogenesis mediators was analyzed in segmented tracts (cervix to infundibulum) in response to mating and sperm-free seminal plasma infusion. Moreover, the gene expression at different times post-mating was also analyzed. Results showed that gene expression changes were mainly localized in the uterus in the natural mating group, describing a clear temporal variation, while limited to the oviduct in the sperm-free seminal plasma group. These changes suggest an early response in the uterus and late modulation in the oviduct, distinctly demonstrating that semen and seminal plasma, through their interaction with the female reproductive tract, can differentially modulate the expression of anti-inflammatory and angiogenesis mediators.

Abstract: The maternal environment modulates immune responses to facilitate embryo development and ensure pregnancy. Unraveling this modulation could improve the livestock breeding systems. Here it is hypothesized that the exposure of the female rabbit reproductive tract to semen, as well as to early embryos, modulates inflammation and angiogenesis among different tissue segments. qPCR analysis of the gene expression changes of the anti-inflammatory interleukin-10 (IL10) and transforming growth factor beta family (TGFβ1–3) and the angiogenesis mediator vascular endothelial growth factor (VEGF-A) were examined in response to mating or insemination with sperm-free seminal plasma (SP). Reproductive tract segment (cervix to infundibulum) samples were obtained in Experiment 1, 20 h after gonadotropin-releasing hormone (GnRH) stimulation (control), natural mating (NM) or vaginal infusion with sperm-free SP (SP-AI). Additionally, segmented samples were also obtained at 10, 24, 36, 68 or 72 h after GnRH-stimulation and natural mating (Experiment 2). The results of gene expression, analyzed by quantitative PCR, showed that NM effects were mainly localized in the uterine tissues, depicting clear temporal variation, while SP-AI effects were restricted to the oviduct. Changes in anti-inflammatory and angiogenesis mediators indicate an early response in the uterus and a late modulation in the oviduct either induced by semen or preimplantation embryos. This knowledge could be used in the implementation of physiological strategies in breeding systems to face the new challenges on rabbit productivity and sustainability.

Keywords: gene expression; endometrium; oviduct; spermatozoa; seminal plasma; inflammation; angiogenesis; rabbit

1. Introduction

Considered either laboratory animals, pets, invasive species or livestock, the rabbit (*Oryctolagus cuniculus*) is one of the most versatile and multipurpose animal species bred by humans [1–3]. Subjected today to intense production for meat and fur, does are affected by requirements on fertility and lifespan [4]. Contrary to other livestock species, rabbits require the generation of genital-somatosensory signals during mating to generate the preovulatory peak of gonadotropin-releasing hormone (GnRH) [5,6]. The consequent release of luteinizing hormone from the anterior pituitary induces ovulation [5]. Despite requiring hormonal stimulation if artificial insemination (AI) is used [7], this technique is extensively employed, giving similar or better results than natural mating [8] when using fresh or cooled semen [9]. The success of AI with diluted semen in several species suggest that seminal plasma (SP) components, excluding spermatozoa, are not required for pregnancy [10]. However, in some species, such as rodents or pigs, reproductive success and pregnancy can be jeopardized when the SP signaling is disrupted [10]. Several studies demonstrated that SP has multiple effects on the female reproductive tract essential for conceptus and pregnancy, improving reproductive performance [10].

In different species of mammals, the direct interaction between the female reproductive tract and semen, as well as the sensorial stimulation during the act of mating [11], induce molecular and cellular modifications into the female reproductive tract [12–14], modulating the immune system and optimizing the reproductive outcomes [15]. The immune response generated as a result of mating leads to a suitable environment for embryo survival, implantation success, optimal fetal and placental development and the overall reproductive process [16].

The first description of female post-mating inflammation responses was made by observing rabbits in 1952 [17]. The observations of McDonald and collaborators showed a leucocytic influx in the uterus in response to SP but not by sterile spermatozoa administration [17]. Later studies demonstrated that both SP and spermatozoa triggered a uterine leucocytic response in rabbits [18,19]. Contrary to the vagina, cervix and uterus, the oviduct seems to respond to insemination without a reduction of the spermatozoa population at the time of ovulation [20], providing the optimal conditions for spermatozoa survival until fertilization takes place [7]. The interaction between spermatozoa and oviductal epithelial cells could regulate the immunological environment of the oviduct-inducing anti-inflammatory cytokines [21], like transforming growth factor beta 1 (TGFβ1) or interleukin-10 (IL10), protecting spermatozoa from immune responses [13].

The transforming growth factor beta (TGFβ) family regulates many cellular activities, including cell growth, proliferation, differentiation, tissue remodeling, extracellular matrix formation, control of cell surface molecules, immunoregulation, angiogenesis and apoptosis [22]. The TGFβ belong to a large superfamily of cytokines composed of three 25 kDa homodimeric proteins (TGFβ1–3) whose biological effects are mediated by three transforming growth factor beta receptors (TGFβR1–3) [22,23]. The IL10, first described as cytokine synthesis inhibitory factor or CSIF [24], is a cytokine that inhibits inflammatory responses preventing the secretion of inflammatory cytokines and regulating the differentiation and proliferation of immune cells [25].

The vascular endothelial growth factor (VEGF, also referred to as VEGF-A), a heparin-binding homodimeric glycoprotein, is a mitogen for endothelial cells and a potent inducer of angiogenesis [26] that also promotes vascular permeability [27]. VEGF protein family includes other factors like VEGF-B, VEGF-C, VEGF-D, VEGF-E, VEGF-F and placental growth factor (P1GF) [28,29]. VEGF-A is expressed in the endometrium of several species [30–32], being proposed in the rabbit as a local signal between the implantation embryo and vascular structures in the receptive endometrium [33].

This study aimed to determine whether natural mating and/or sperm-free SP infusion modulates the expression of anti-inflammatory cytokines and angiogenesis mediator genes (*IL10*, *TGFβ1–3* and *VEGF-A*) in the maternal environment of the doe at 20 h post-exposure. Moreover, we determine the expression of these mediators when early embryo development occurs along the reproductive tract of the doe up to 72 h post-mating. Additionally, we analyzed the spatial changes of these genes along the tubular reproductive tract of the doe.

2. Materials and Methods

2.1. Ethics Statement

Rabbits were handled according to the Spanish Royal Decree 1201/2005 (BOE, 2005: 252:34367-91) and the Directive 2010/63/EU of the European Parliament and of the Council of 22 September 2010 on the protection of animals used for scientific purposes (2010; 276:33–79). The Committee of Ethics and Animal Welfare of the Universitat Autónoma de Barcelona, Spain, approved this study (Expedient #517).

2.2. Animals and Experimental Design

Animals were housed in the nucleus colony at the farm of the Institut de Recerca i Tecnologia Agroalimentaries (IRTA-Torre Marimon, Caldes de Montbui, Barcelona, Spain) under a controlled photoperiod and temperature [34]. Single cage equipped with plastic footrests, a feeder (restricted to 180 g/day of an all-mash pellet) and nipple-drinkers (ad libitum access to water) were used for each rabbit.

Six New Zealand White (NZW) adult bucks (from seven to thirteen months old) were included in the study. At 4.5 months of age, the bucks were started to be trained with a homemade polyvinyl chloride artificial vagina at 50 °C to ensure 41–42 °C at the time of semen collection. One ejaculate was collected per male, discarding ejaculates that contained urine and calcium carbonate deposits on visual inspection.

Two separate experiments were performed in this study (Figure 1). In Experiment 1, 9 NZW adult does were randomly allocated into three experimental groups. Sequential segments of the right side of female reproductive tracts were retrieved after 20 h post-induction of the ovulation with 0.03 mg GnRH (Fertagyl®, Esteve Veterinaria, Barcelona, Spain) intramuscularly (im) (control of the ovulation; control, $n = 3$), 20 h post-induction of the ovulation with 0.03 mg GnRH im and sperm-free SP vaginal infusion (SP-AI, $n = 3$) and 20 h post-induction of the ovulation with 0.03 mg GnRH im and natural mating (NM, $n = 3$). In Experiment 2, 15 NZW adult does were sequentially euthanized at 10, 24, 36, 68 or 72 h post-induction of the ovulation with 0.03 mg GnRH im and natural mating ($n = 3$/collection time). The 10 h group was established as the reference group to compare the ovulatory moment (10 h post-mating) with some different time-point steps of the preimplantation embryo development: 2–4 cell embryo (24 h), 8-cell embryo (36 h), early morula (68 h) and morula (72 h).

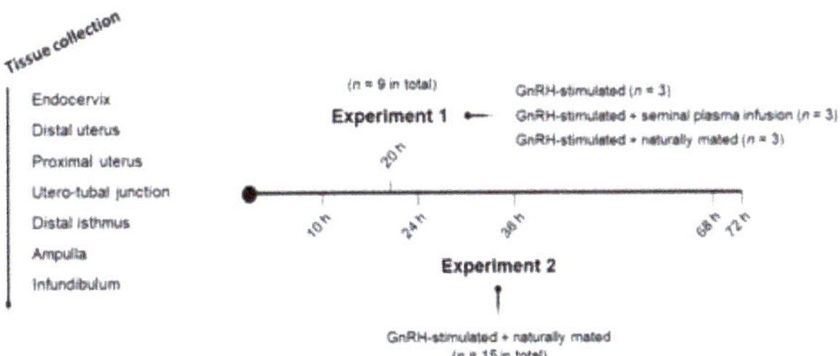

Figure 1. Representation of the experimental design and tissue sections obtained from does. Sequential tissue segments derived from does were endocervix, distal uterus, proximal uterus, utero-tubal junction, distal isthmus, ampulla and infundibulum. Intramuscular injection of 0.03 mg gonadotropin-releasing hormone (GnRH) was used to induce ovulation in all groups of both experiments.

2.3. Mating and Semen Collection

Two randomly selected bucks were used to mate the does included in the mating group of Experiment 1 and Experiment 2 after the hormonal induction of the ovulation with 0.03 mg GnRH im. Additionally, semen was collected from the same rabbit bucks through an artificial vagina, as described above. The sperm-free SP was obtained after centrifugation at 2000× *g* for 10 min and checked for the absence of spermatozoa. The harvested sperm-free SP was immediately used for sperm-free SP vaginal infusions of Experiment 1.

2.4. Tissue Sample Collection

For each experimental condition, seven consecutive tissue sections from the female reproductive tracts (endocervix, distal uterus, proximal uterus, utero-tubal junction, ampulla, isthmus and infundibulum; Figure 1) were obtained after the euthanasia of the does [34]. The tissue segments were retrieved and stored in RNAlater solution (Thermo Fisher Scientific, Waltham, MA, USA) at −80 °C.

Before tissue segmentation of does included in experiment 2, the oviduct was isolated and flushed (phosphate-buffer saline supplemented with 5% fetal calf serum and 1% antibiotic-antimycotic solution) to collect the embryos, which were examined by number and developmental stage. The number of ovulated ovarian follicles (5.08 ± 2.06 follicles, mean ± standard deviation (SD)) and embryos (4.67 ± 3.14, mean ± SD) were counted for each side of the reproductive organs, as previously published elsewhere [34]. Briefly, at 24 h (2 and 4-cell), 36 h (8-cell stage), 68 h (early morula) and 72 h (morula), the embryos were collected by the method described above.

2.5. Quantitative PCR Analyses

Briefly, total RNA was extracted from the tissue segments following a TRIzol-based protocol [34]. The RNA concentration of the extracts was determined from the absorbance of 260 nm with Thermo Scientific NanoDrop™ 2000 (Fisher Scientific, Gothenburg, Sweden). The quality of the RNA was determined by the Agilent 2100 Bioanalyzer (Agilent Technologies, Palo Alto, CA, USA), using the samples with an RNA integrity number higher than 8. High-Capacity RNA-to-cDNA™ Kit (Applied Biosystems™, Foster City, CA, USA) was used to synthesize the first-strand cDNA for quantitative polymerase chain reaction (qPCR) analyses (CFX96™; Bio-Rad Laboratories, Inc; Hercules, CA, USA). Following our previous qPCR protocol [34], the gene relative expression levels were quantified using the Pfaffl method [35]. The primer sequences, product sizes and efficiencies are shown in Table 1. For the *β-ACTIN* gene, commercial gene-specific PCR primers for rabbit samples were used (PrimePCR™SYBR® Green Assay: ACTB, Rabbit; Bio-Rad Laboratories, Inc., Hercules, CA, USA). Product sizes were confirmed by loading the amplicons in an agarose gel using a gel imaging system (ChemiDoc XRS+ System, BioRad Laboratories, Inc., Hercules, CA, USA).

Table 1. Primers used for the quantitative PCR analyses.

Gene	Primer Sequence (5′–3′)	Product Size (bp)	Efficiency (%)
β-ACTIN	F: commercial, not available R: commercial, not available	120	88.6
IL10	F: TTGTTAACCGAGTCCCTGCT R: TTTTCACAGGGGAGAAATCG	209	107.5
TGFβ1	F: CACCATTCACAGCATGAACC R: TTCTCTGTGGAGCTGAAGCA	129	99.3
TGFβ2	F: CCGGAGGTGATCTCCATCTA R: AGGTACCCACAGAGCACCTG	236	94.7
TGFβ3	F: CACGAACCCAAGGGCTACTA R: GGTCCTCCCGACGTAGTACA	186	97.5
VEGF-A	F: GGAGACAATAAACCCCACGA R: CTGCATGGTGACGTTGAACT	219	87.0

β-ACTIN: β-actin, *IL10*: interleukin 10, *TGFβ1*: transforming growth factor β 1, *TGFβ2*: transforming growth factor β 2, *TGFβ3*: transforming growth factor β 3, *VEGF-A*: vascular endothelial growth factor A, F: forward, R: reverse, A: adenine, C: cytosine, G: guanine, T: thymine, bp: base pair.

2.6. Statistical Analyses

CFX Maestro™ 1.1 software version 4.1.2433.1219 (Bio-Rad Laboratories, Inc; Hercules, CA, USA) was used to export all data. Normal distribution and homoscedasticity of the data were analyzed using the Shapiro–Wilk normality test and Levene's test. Non-normal data distribution was restored using Log(x) transformation prior to analysis. R version 3.6.1. [36] was used to conduct the statistical analyses with *nlme* [37] to perform linear mixed-effects (LME) models and *multcomp* [38] to perform pairwise comparisons adjusted by Tukey's test. The threshold for significance was set at $p < 0.05$. Data are presented as median (minimum, maximum) unless otherwise stated.

A first LME model included the treatments of Experiment 1 (control, SP-AI, NM) as fixed effects and the females as the random part of the model. Pairwise comparisons were adjusted by Tukey's test. A second LME model included the different collection times of Experiment 2 (10, 24, 36, 68 and 72 h post-mating) as fixed effects and the females as the random part of the model. Post hoc comparisons were performed using Tukey's multiple comparisons test.

The differential expression changes in qPCR results among tissues in both Experiments were re-analyzed separately. The utero-tubal junction was used as an arbitrary anatomical compartment reference among all tissues examined to compare the gene expression changes per gene [34], issued both by control, NM or SP-AI (Experiment 1) or by different times post-mating (Experiment 2). The LME model was followed by Tukey's multiple comparison test to analyze the differences among each anatomical region of the female reproductive tract. Data are presented as median (minimum, maximum) unless otherwise stated. Data on the differential expression among tissues are presented as Row Z-scores.

3. Results

3.1. Natural Mating and Sperm-Free Seminal Plasma Infusion Differentially Modulated Anti-Inflammatory Cytokines and VEGF-A Genes

Differences in *IL10*, *TGFβ1*, *TGFβ2*, *TGFβ3* and *VEGF-A* expression among groups included in Experiment 1 are displayed in Figure 2 (and Figure S1). The NM treatment downregulated the *IL10* expression in the utero-tubal junction ($p < 0.05$); upregulated the *TGFβ1* and *TGFβ3* expression in the distal ($p < 0.01$) and proximal uterus ($p < 0.001$); upregulated the *TGFβ1* and *TGFβ2* expression in the infundibulum ($p < 0.01$); and upregulated the *VEGF-A* expression in the distal uterus ($p < 0.05$). The SP-AI treatment upregulated the *TGFβ1* expression in the proximal uterus ($p < 0.01$), upregulated the *TGFβ1* and *TGFβ3* expression in the infundibulum ($p < 0.01$), and upregulated the *VEFG-A* expression in the infundibulum ($p < 0.01$).

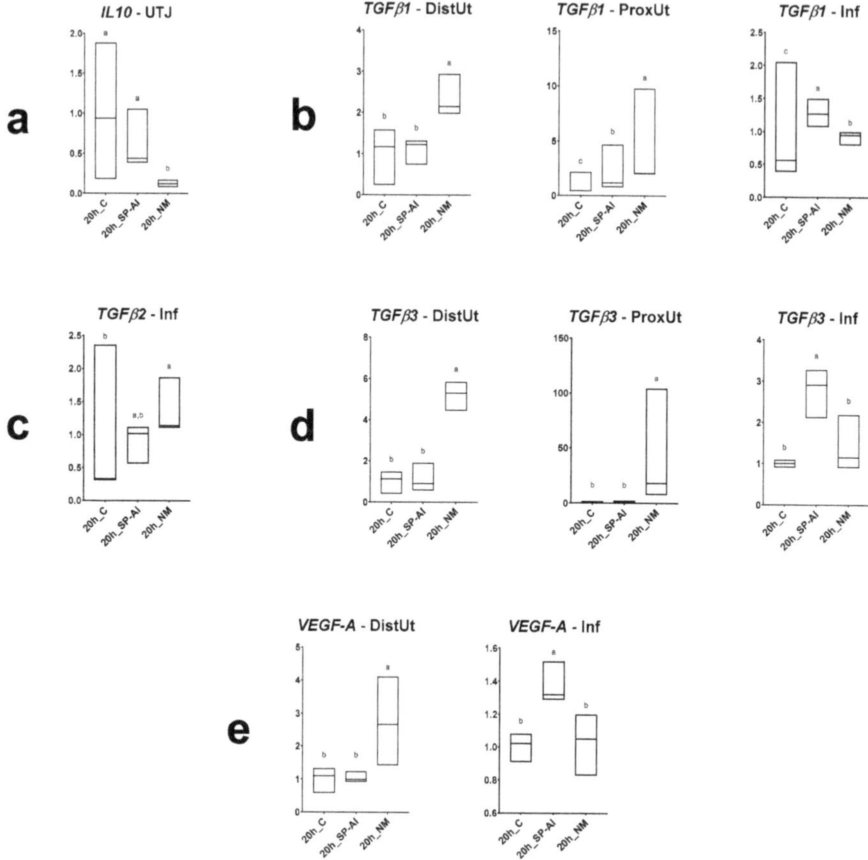

Figure 2. Results of Experiment 1. Gene expression of (**a**) *IL10*, (**b**) *TGFβ1*, (**c**) *TGFβ2*, (**d**) *TGFβ3* and (**e**) *VEGF-A* in differentially expressed segments ($p < 0.05$) (DistUt: distal uterus, ProxUt: proximal uterus, UTJ: utero-tubal junction, Amp: ampulla and Inf: infundibulum) of the does internal reproductive tract after 20 h of the induction of the ovulation with 0.03 mg gonadotropin-releasing hormone (GnRH) intramuscularly (20 h_C, $n = 3$), 20 h post-GnRH-stimulation and seminal plasma vaginal infusion (20h_SP-AI, $n = 3$) and 20 h post-GnRH-stimulation and natural mating (20 h_NM, $n = 3$). Fold changes relative to the control of the ovulation group are shown. Different letters (a,b) represent statistical differences between groups ($p < 0.05$). Median (minimum, maximum).

3.2. Temporal Gene Expression of Anti-Inflammatory cytokines and VEGF-A at 10 h up to 72 h Post-Mating

Differences in *IL10*, *TGFβ1*, *TGFβ2*, *TGFβ3* and *VEGF-A* expression in does sampled at different times post-mating are displayed in Figure 3 (and Figure S2), with 10 h post-mating as the control group. The *IL10* expression was downregulated at 36, 68 and 72 h post-mating in the endocervix and distal uterus ($p < 0.001$); downregulated at 24, 36, 68 and 72 h post-mating in the utero-tubal junction and infundibulum ($p < 0.001$); and upregulated at 68 and 72 h post-mating in the distal isthmus ($p < 0.05$). The *TGFβ1* expression was downregulated at 36, 68 and 72 h post-mating in the distal uterus ($p < 0.05$); and downregulated at 24, 36, 68 and 72 h post-mating in the infundibulum ($p < 0.001$). The *TGFβ2* expression was downregulated at 24, 36, 68 and 72 h post-mating in the distal uterus and infundibulum ($p < 0.001$); downregulated at 24 h post-mating in the proximal uterus ($p < 0.01$); and downregulated at 24 and 72 h post-mating in the utero-tubal junction ($p < 0.05$). The *TGFβ3* expression was downregulated at 36 h post-mating in the endocervix ($p < 0.05$); downregulated at 36, 68 and 72 h post-mating in the distal isthmus ($p < 0.05$); and downregulated at 24, 36, 68 and 72 h post-mating in the infundibulum ($p < 0.001$). The *VEGF-A* expression was upregulated at 24 h post-mating in the utero-tubal junction ($p < 0.05$); downregulated at 24 and 72 h post-mating in the ampulla ($p < 0.05$); and downregulated at 24, 36, 68 and 72 h post-mating in the infundibulum ($p < 0.001$).

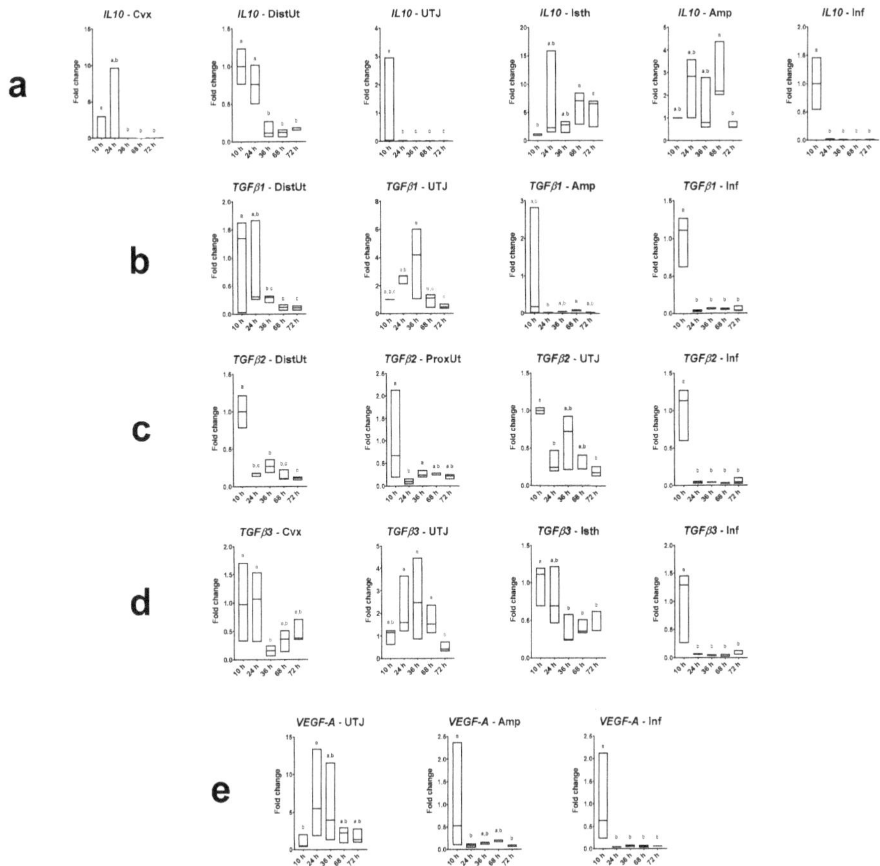

Figure 3. Results of Experiment 2. Gene expression of (**a**) *IL10*, (**b**) *TGFβ1*, (**c**) *TGFβ2*, (**d**) *TGFβ3* and (**e**) *VEGF-A* was statistically significant in rabbit endocervix (Cvx), distal uterus (DistUt), proximal uterus (ProxUt), utero-tubal junction (UTJ), distal isthmus (Isth), ampulla (Amp) and infundibulum (Inf) after 10, 24, 36, 68 or 72 h post-induction of the ovulation (intramuscular injection of 0.03 mg gonadotropin-releasing hormone) and natural mating (n = 3/collection time). Fold changes relative to 10 h post-mating group are shown. Different letters (a–c) represent statistical differences between groups ($p < 0.05$). Median (minimum, maximum).

3.3. Spatial Gene Expression Triggered by Natural Mating and Sperm-Free Seminal Plasma Infusion

The tissue analysis expression showed significant differences in *IL10*, *TGFβ1–3* and *VEGF-A* expression among tissues in the different groups included in Experiment 1 (Figure 4a and Figure S3). The infundibulum presented the highest *IL10* expression in the SP-AI and control groups ($p < 0.05$), whereas the cervix presented the highest *IL10* expression in the NM group ($p < 0.05$). In the case of *TGFβ1* expression, the endocervix presented the highest expression in the NM group ($p < 0.05$). The distal isthmus presented the highest *TGFβ2* expression in all groups included in Experiment 1

($p < 0.05$). Similarly, the oviductal tissues presented the highest *TGFβ3* expression in the SP-AI and control groups ($p < 0.05$). However, the proximal uterus presented the highest *TGFβ3* expression in the NM group ($p < 0.01$). The oviductal tissues presented the highest *VEGF-A* expression in the SP-AI and control groups ($p < 0.05$), whereas the uterine tissues presented the highest *VEGF-A* expression in the NM group ($p < 0.05$).

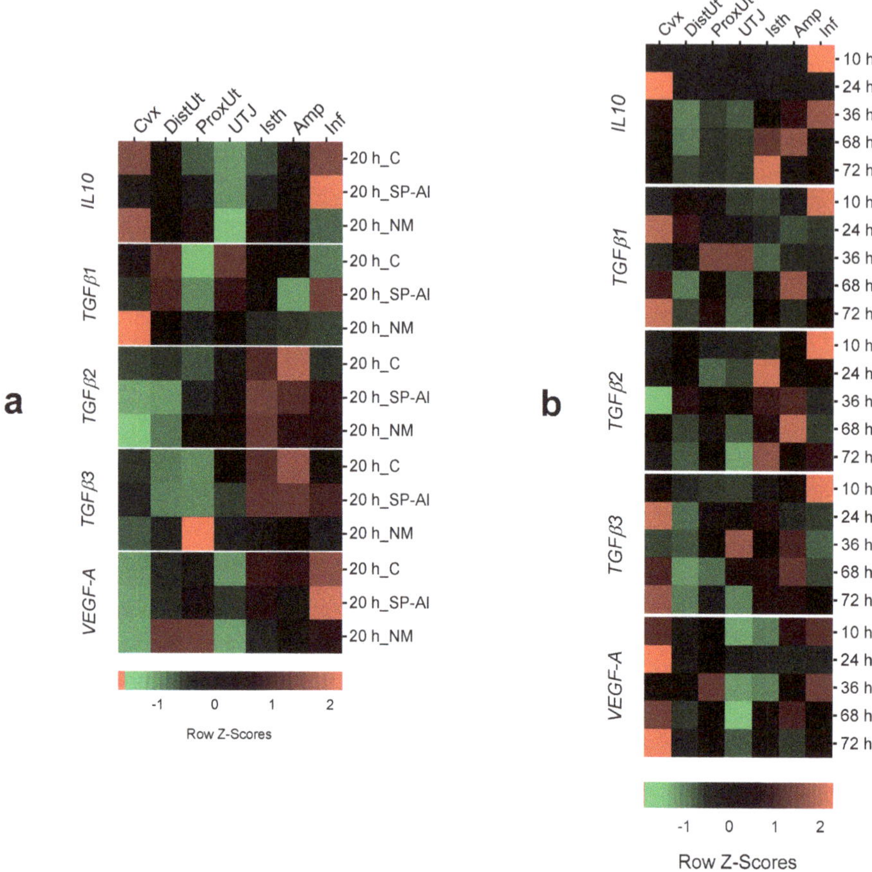

Figure 4. Heatmap composition of the gene expression changes among tissues in the different groups included in the study. Changes in *IL10*, *TGFβ1*, *TGFβ2*, *TFGβ3* and *VEGF-A* expression among tissues (endocervix, Cvx; distal uterus, DistUt; proximal uterus, ProxUt; utero-tubal junction, UTJ; distal isthmus, Isth; ampulla, Amp; and infundibulum, Inf) (**a**) at 20 h after the induction of ovulation with 0.03 mg of gonadotropin-releasing hormone (GnRH) intramuscularly, as the control of ovulation (20 h_C); 20 h post-GnRH stimulation and sperm-free seminal plasma vaginal infusion (20 h_SP-AI); and 20 h post-GnRH stimulation and natural mating (20 h_NM); (**b**) at 10, 24, 36, 68 and 72 h post-GnRH-stimulation and natural mating. Row Z-scores of the mean fold change relative to the reference group (UTJ) are shown. Red indicates upregulation, and green indicates downregulation.

3.4. Spatial Gene Expression at 10 h up to 72 h Post-Mating

The tissue analysis expression showed significant differences in *IL10*, *TGFβ1–3* and *VEGF-A* expression among tissues in the different groups included in Experiment 2 (Figure 4b and Figure S4). At 10 h post-mating (time of ovulation), the infundibulum presented the highest *IL10* expression

($p < 0.01$). Conversely, at 24 h post-mating, the endocervix presented the highest *IL10* expression ($p < 0.05$). At 36, 68 and 72 h post-mating, the oviductal segments presented higher *IL10* expression compared to the uterine tissues ($p < 0.05$).

At 10 h post-mating, the infundibulum presented the highest *TGFβ1–3* expressions ($p < 0.05$). The *TGFβ1* expression was higher in the uterine tissues compared to the oviduct segments ($p < 0.05$), whereas the *TGFβ2* expression was higher in the oviduct segments compared to the uterine tissues ($p < 0.05$). Both uterine and oviductal segments presented the highest *TGFβ3* expressions ($p < 0.05$). The uterine tissues presented higher *VEGF-A* expressions ($p < 0.05$) compared to oviduct segments. However, at 10 and 36 h post-mating, the infundibulum presents the highest *VEGF-A* expression ($p < 0.05$).

4. Discussion

The results showed a differential expression triggered by sperm-free SP infusion and natural mating on anti-inflammatory cytokines and angiogenesis-related genes, as well as the differential expression produced at different times post-mating in the rabbit, an induced ovulatory species. Additionally, the study provides the differential expression on anti-inflammatory cytokines and angiogenesis-related genes along the doe reproductive tract triggered by natural mating and sperm-free SP infusion, as well as by preimplantation embryo at different times post-mating.

The SP interacts with the female reproductive tract-inducing female immune adaptation processes required to tolerate male antigens suppressing inflammation and immune rejection responses [10]. This maternal immune tolerance is essential for ongoing pregnancy success [39]. Additionally, the SP induces molecular and cellular modifications in the uterus and in the highest reproductive tract promoting embryo development and implantation competence [10]. In this regard, different soluble components present in the SP play key roles in these processes, such as the TGFβs, which can drive immune cells to tolerogenic phenotypes [40]. This study revealed differences between sperm-free SP and natural mating (that also contains SP) in the *IL10*, *TGFβ1*, *TGFB3* and *VEGF-A* expression. The data suggest that natural mating, at 20 h post-exposure, downregulated the *IL10* expression in the utero-tubal junction, probably related to the spermatozoa selection or clearance in this region. Conversely, the *TGFβ1*, *TGFβ3* and *VEGF-A* expressions were upregulated in the uterine tissues in the natural mating group, whereas upregulated in the infundibulum in the sperm-free SP infusion group.

Even though the SP appears to not reach the upper segments of the oviduct after mating, it seems that the signaling originated in lower segments travel to upper segments in response to SP [10]. The range of effects produced by SP in different species includes clearance of microorganisms in the uterus after mating, sperm selection, induction of ovulation, the formation of corpus luteum and supporting the development of the preimplantation embryo, among others [10]. Our observations indicate that the sperm-free SP infusion increased the expression of *TGFβ1*, *TGFβ3* and *VEGF-A* in the infundibulum, whereas the expression of these genes in the natural mating group, also containing SP, remained unaltered compared to the control group. Moreover, the natural mating group increased *VEGF-A* expression in the distal uterus, but not in the oviduct, suggesting that the rabbit uterus reacts to post-mating inflammation [15], likely to process seminal material and recovery for uterine tissue homeostasis after mating. Our data suggest that the responsiveness of the doe reproductive tract differs between the sperm-free SP and sperm-containing SP treatments. The signaling originated by the sperm-free SP infusion was able to travel to upper segments of the oviduct, as previously demonstrated in several species [10], whereas the signal generated by the natural mating group remained in the uterine tissues. This pattern was also observed in the temporal tracking experiment of the study, suggesting that the rabbit uterus may react to the early stages of embryo development before its arrival from the oviduct, but the mechanism involved is yet not fully understood.

The TGFβs isoforms present a diverse range of functions in reproductive tissues, including gonad development, gamete production, embryo implantation, fetal and placental development, development of secondary sex organs and immune tolerance of gametes and conceptus antigens [22]. The effects of

TGFβs are mediated by antigen-presenting cells such as macrophages and dendritic cells, creating and maintaining an immunotolerant environment for the conceptus antigens if an anti-inflammatory dominance is established [22]. In this regard, our results revealed an upregulation of *TGFβs* expression at 20 h post-mating in the endometrium, suggesting that mating induces suppression of anti-sperm immune response [39]. Similar results were found at 10 h post-mating in oviductal segments, probably related to the ovulation that takes place 10 h after coitus or GnRH stimulation [7] or the suppression of anti-sperm immune response induced by the transport of spermatozoa through the oviduct before fertilization [41,42]. However, the upregulation of *TGFβs* expression in the oviduct was not present in the rest of the times after mating compared to 10 h post-mating, when the early embryo stages of development are present along the oviduct [43]. Thus, the effect produced by semen on the *TGFβs* expression changes seemed to be stronger than the effect of the embryo transport through the oviduct at different stages of development, presumably due to the different number of cells in contact with the female reproductive tract (300 million spermatozoa [44] vs. 4–12 embryos [45]). Although the effects of spermatozoa and the early embryo stages of development on *TGFβs* expression seems to differ in magnitude, tissue analyses revealed a *TGFβ2* and *TGFβ3* expression pattern in the oviduct, increasing the expression of those genes. Moreover, the *TGFβ1* expression pattern in the uterus in the presence of spermatozoa and the early stages of embryo development may indicate that both antigenic compounds induced an anti-inflammatory response creating and maintaining an immunotolerance environment in the oviduct and uterus, essential for the pregnancy success [39]. Similarly, this immunotolerant environment to spermatozoa may be observed by the upregulation of the *IL10* expression in the endocervix and distal uterus at 10 and 24 h post-mating.

The tissue and temporal analyses revealed an interesting pattern along the oviduct during the preimplantation embryo transport. When fertilization occurs in the ampulla [46], embryos are delayed at the ampulla–isthmus junction until 48 h post-mating and then pass through the isthmus portion after 70 h post-mating to finally reach the utero-tubal junction and enter the uterus [43]. Our observations showed an upregulation of *TGFβs* expression in the infundibulum at the time of induced ovulation (10 h post-mating), similarly, as the pattern observed in the *IL10* expression at the same time, might relate with the inflammation process associated with ovulation [47]. At 24 h post-mating, the endocervix displayed the highest expression of *IL10, TGFβ1, TGFβ3* and *VEGF-A*, whereas the *TGFβ2* expression was higher in the isthmus compared to the endocervix. Although fertilization takes place at 12–13 h post-mating [46], this upregulation of anti-inflammatory cytokines and angiogenesis mediators in the endocervix may be related to preserving spermatozoa viability and fertilization competence [13], in accordance with previous research that proposed the cervix as the first anatomical barrier for spermatozoa reservoir in rabbits [20,48,49].

The coordination between the preimplantation embryos and the differentiation of the uterus to a receptive state must be closely synchronized for successful embryo implantation [50]. Following the stromal and epithelial cell changes, the endometrial vascular bed undergoes a marked expansion before implantation [50,51]. This vascular expansion, in size and number, is markedly enhanced during implantation and placental structure development [51]. In the rabbit, the vascular supply during pregnancy is produced by the addition of new vessels and the expansion of the existing ones [52]. Our results revealed that *VEGF-A* expression was upregulated at 10 h post-mating in the infundibulum and ampulla, probably related to the ongoing processes like ovulation [7] and the imminent fertilization [46]. The *VEGF-A* expression displayed higher relative expression values at 20 h post-mating in the endometrium compared to the oviduct, perhaps related to preparing the receptive endometrium for the embryo implantation. However, this effect was not observable at different times post-mating, suggesting that additional factors, other than the analyzed in our experimental setup, are involved in the differentiation of the uterus to a receptive state before embryos reach the uterus.

5. Conclusions

Our results confirmed a differential modulation of the expression of anti-inflammatory and angiogenesis-related-genes by sperm-free SP and natural mating. Moreover, the gene expression analysis up to 72 h after induction of the ovulation might suggest a concerted effect of the presence of sperm and preimplantation embryos in the differential expression of these genes along the reproductive tract of the doe. Our findings warrant further research both to discriminate the effects of mating, spermatozoa, SP and preimplantation embryos on the reproductive tract of the doe and to fully describe the mechanism involved in such differential gene expression. Thus, mimicking the gene modulations induced by natural mating and sperm-free SP using physiological strategies in the current rabbit breeding systems may enhance the efficiency, productivity and sustainability of such systems.

Supplementary Materials: The following are available online at http://www.mdpi.com/2076-2615/10/12/2207/s1, Figure S1: Changes in (a) *IL10*, (b) *TGFβ1*, (c) *TGFβ2*, (d) *TGFβ1* and (e) *VEGF-A* expression among different treatments at 20 h post-treatment: 20 h post-induction of the ovulation, control (20 h_C); 20 h post-seminal plasma infusion, 20 h_SP; and 20 h post-natural mating, 20 h_NM. Tissue anatomical regions of the rabbit female reproductive tract (endocervix, Cvx; distal uterus, DistUt; proximal uterus, ProxUt; utero-tubal junction, UTJ; distal isthmus, Isth; ampulla, Amp; and infundibulum, Inf).-fold changes relative to the reference group (20 h_C) are shown. Different letters (a–c) represent statistical differences between tissues ($p < 0.05$). Median (minimum, maximum); Figure S2: Changes in (a) *IL10*, (b) *TGFβ1*, (c) *TGFβ2*, (d) *TGFβ1* and (e) *VEGF-A* expression at different times (from 10 to 72 h) post-mating: 10, 24, 36, 68 and 72 h post-natural mating. Tissue anatomical regions of the female rabbit reproductive tract (endocervix, Cvx; distal uterus, DistUt; proximal uterus, ProxUt; utero-tubal junction, UTJ; distal isthmus, Isth; ampulla, Amp; and infundibulum, Inf).-fold changes relative to the reference group (10 h post-mating) are shown. Different letters (a–c) represent statistical differences between tissues ($p < 0.05$). Median [minimum, maximum]; Figure S3: Changes in (a) *IL10*, (b) *TGFβ1*, (c) *TGFβ2*, (d) *TGFβ1* and (e) *VEGF-A* expression among different tissues at 20 h post-treatment. 20 h post-induction of the ovulation, control (20 h_C); 20 h post-seminal plasma infusion, 20 h_SP; and 20 h post-natural mating, 20 h_NM. Tissue anatomical regions of the female rabbit reproductive tract (endocervix, Cvx; distal uterus, DistUt; proximal uterus, ProxUt; utero-tubal junction, UTJ; distal isthmus, Isth; ampulla, Amp; and infundibulum, Inf).-fold changes relative to the reference group (UTJ) are shown. Different letters (a–d) represent statistical differences between tissues ($p < 0.05$). Median [minimum, maximum]; Figure S4: Changes in (a) *IL10*, (b) *TGFβ1*, (c) *TGFβ2*, (d) *TGFβ1* and (e) *VEGF-A* expression among different tissues in the period 10–72 h (10, 24, 36, 68 and 72 h post-natural mating). Tissue anatomical regions of the female rabbit reproductive tract (endocervix, Cvx; distal uterus, DistUt; proximal uterus, ProxUt; utero-tubal junction, UTJ; distal isthmus, Isth; ampulla, Amp; and infundibulum, Inf).-fold changes relative to the reference group (UTJ) are shown. Different letters (a–d) represent statistical differences between tissues ($p < 0.05$). Median [minimum, maximum].

Author Contributions: Conceptualization, H.R.-M. and M.Á.-R.; methodology, M.L.-B.; software, J.G., A.J.-M. and M.Á.-R.; validation, A.J.-M.; formal analysis, J.G., A.J.-M. and M.Á.-R.; investigation, J.G. and C.A.M.; resources, H.R.-M., M.L.-B. and M.Á.-R.; data curation, J.G.; writing—original draft preparation, J.G.; writing—review and editing, A.J.-M., C.A.M., H.R.-M., M.L.-B. and M.Á.-R.; visualization, J.G.; supervision, M.L.M and M.Á.-R.; project administration, M.Á.-R.; funding acquisition, H.R.-M., M.L.-B. and M.Á.-R. All authors have read and agreed to the published version of the manuscript.

Funding: This research was funded by the Research Council FORMAS, Stockholm (Project 2017-00946 and Project 2019-00288), The Swedish Research Council (Vetenskaprådet, VR; project 2015-05919) and Juan de la Cierva Incorporación Postdoctoral Research Program (MICINN; IJDC-2015-24380). J.G. is supported by the Generalitat de Catalunya, Agency for Management of University and Research Grants co-financed with the Eropean Social Found (grants for the recruitment of new research staff 2018 FI_B 00236).

Acknowledgments: We thank Annette Molbaek and Asa Schippert, from the Genomics Core Facility at LiU for expert assistance when running the bioanalyzer. We appreciate the kind support provided by Míriam Piles and Oscar Perucho and the technical staff from Torre Marimon—Institut de Recerca i Tecnologia Agroalimentàries (IRTA, Caldes de Montbui, Barcelona, Spain) and Annaïs Carbajal, Sergi Olvera-Maneu and Mateo Ruiz-Conca for their kind assistance in sample handling.

Conflicts of Interest: The authors declare no conflict of interest. The funders had no role in the design of the study; in the collection, analyses or interpretation of data; in the writing of the manuscript or in the decision to publish the results.

References

1. United States Department of Agriculture. Annual Report Animal Usage by Fiscal Year. Available online: https://www.aphis.usda.gov/animal_welfare/downloads/reports/Annual-Report-Animal-Usage-by-FY2017.pdf (accessed on 29 May 2020).
2. Roy-Dufresne, E.; Lurgi, M.; Brown, S.C.; Wells, K.; Cooke, B.; Mutze, G.; Peacock, D.; Cassey, P.; Berman, D.; Brook, B.W.; et al. The Australian National Rabbit Database: 50 yr of population monitoring of an invasive species. *Ecology* **2019**, *100*, e2750. [CrossRef] [PubMed]
3. Petracci, M.; Soglia, F.; Leroy, F. Rabbit meat in need of a hat-trick: From tradition to innovation (and back). *Meat Sci.* **2018**, *146*, 93–100. [CrossRef] [PubMed]
4. Lorenzo, P.L.; García-García, R.M.; Árias-Álvarez, M.; Rebollar, P.G. Reproductive and nutritional management on ovarian response and embryo quality on rabbit does. *Reprod. Domest. Anim.* **2014**, *49*, 49–55. [CrossRef] [PubMed]
5. Bakker, J.; Baum, M.J. Neuroendocrine regulation of GnRH release in induced ovulators. *Front. Neuroendocrinol.* **2000**, *21*, 220–262. [CrossRef] [PubMed]
6. Ratto, M.H.; Berland, M.; Silva, M.E.; Adams, G.P. New insights of the role of B-NGF in the ovulation mechanism of induced ovulating species. *Reproduction* **2019**, *157*, R199–R207. [CrossRef] [PubMed]
7. Dukelow, W.R.; Williams, W.L. Survival of Capacitated Spermatozoa in the Oviduct of the Rabbit. *J. Reprod. Fert.* **1967**, *14*, 477–479. [CrossRef] [PubMed]
8. Morrell, J.M. Artificial insemination in rabbits. *Br. Vet. J.* **1995**, *151*, 477–488. [CrossRef]
9. Mocé, E.; Vicente, J.S. Rabbit sperm cryopreservation: A review. *Anim. Reprod. Sci.* **2009**, *110*, 1–24. [CrossRef]
10. Schjenken, J.E.; Robertson, S.A. Seminal Fluid Signalling in the Female Reproductive Tract: Implications for Reproductive Success and Offspring Health. In *Advances in Experimental Medicine and Biology*; Springer: Cham, Switzerland, 2015; Volume 868, pp. 127–158, ISBN 978-3-319-18880-5.
11. Parada-Bustamante, A.; Oróstica, M.L.; Reuquen, P.; Zuñiga, L.M.; Cardenas, H.; Orihuela, P.A. The role of mating in oviduct biology. *Mol. Reprod. Dev.* **2016**, *83*, 875–883. [CrossRef]
12. Ghersevich, S.; Massa, E.; Zumoffen, C. Oviductal secretion and gamete interaction. *Reproduction* **2015**, *149*, R1–R14. [CrossRef]
13. Suarez, S.S. Interactions of Gametes with the Female Reproductive Tract. *Cell Tissue Res* **2016**, *363*, 185–194. [CrossRef] [PubMed]
14. Alvarez-Rodriguez, M.; Atikuzzaman, M.; Venhoranta, H.; Wright, D.; Rodriguez-Martinez, H. Expression of Immune Regulatory Genes in the Porcine Internal Genital Tract Is Differentially Triggered by Spermatozoa and Seminal Plasma. *Int. J. Mol. Sci.* **2019**, *20*, 513. [CrossRef] [PubMed]
15. Katila, T. Post-mating Inflammatory Responses of the Uterus. *Reprod. Domest. Anim.* **2012**, *47*, 31–41. [CrossRef] [PubMed]
16. Hansen, P.J. The Immunology of Early Pregnancy in Farm Animals. *Reprod. Domest. Anim.* **2011**, *46*, 18–30. [CrossRef] [PubMed]
17. McDonald, L.E.; Black, W.G.; Casida, L.E. The Response of the Rabbit Uterus to Instillation of Semen at Different Phases of the Estrous Cycle. *Am. J. Vet. Res.* **1952**, *13*, 419–424. [PubMed]
18. Phillips, D.M.; Mahler, S. Phagocytosis of spermatozoa by the rabbit vagina. *Anat. Rec.* **1977**, *189*, 61–71. [CrossRef] [PubMed]
19. Phillips, D.M.; Mahler, S. Leukocyte emigration and migration in the vagina following mating in the rabbit. *Anat. Rec.* **1977**, *189*, 45–59. [CrossRef]
20. El-Banna, A.A.; Hafez, E.S.E. Sperm transport and distribution in rabbit and cattle female tract. *Fertil. Steril.* **1970**, *21*, 534–540. [CrossRef]
21. Yousef, M.S.; Marey, M.A.; Hambruch, N.; Hayakawa, H.; Shimizu, T.; Hussien, H.A.; Abdel-Razek, A.R.K.; Pfarrer, C.; Miyamoto, A. Sperm binding to oviduct epithelial cells enhances TGFB1 and IL10 expressions in epithelial cells aswell as neutrophils in vitro: Prostaglandin E2 as a main regulator of anti-inflammatory response in the bovine oviduct. *PLoS ONE* **2016**, *11*, 1–19. [CrossRef]
22. Ingman, W.V.; Robertson, S.A. Defining the actions of transforming growth factor beta in reproduction. *BioEssays* **2002**, *24*, 904–914. [CrossRef]
23. Massagué, J. TGF-β signal transduction. *Annu. Rev. Biochem.* **1998**, *67*, 753–791. [CrossRef] [PubMed]

24. Fiorentino, D.F.; Bond, M.W.; Mosmann, T.R. Two types of mouse T helper cell. IV. Th2 clones secrete a factor that inhibits cytokine production by Th1 clones. *J. Exp. Med.* **1989**, *170*, 2081–2095. [CrossRef] [PubMed]
25. Moore, K.W.; de Waal Malefyt, R.; Coffman, R.L.; O'Garra, A. Interleukin-10 and the interleukin-10 receptor. *Annu. Rev. Immunol.* **2001**, *19*, 683–765. [CrossRef] [PubMed]
26. Ferrara, N.; Houck, K.; Jakeman, L.; Leung, D.W. Molecular and biological properties of the vascular endothelial growth factor family of proteins. *Endocr. Rev.* **1992**, *13*, 18–32. [CrossRef]
27. Senger, D.R.; Perruzzi, C.A.; Feder, J.; Dvorak, H.F. A Highly Conserved Vascular Permeability Factor Secreted by a Variety of Human and Rodent Tumor Cell Lines. *Cancer Res.* **1986**, *46*, 5629–5632.
28. Apte, R.S.; Chen, D.S.; Ferrara, N. VEGF in Signaling and Disease: Beyond Discovery and Development. *Cell* **2019**, *176*, 1248–1264. [CrossRef]
29. Melincovici, C.S.; Boşca, A.B.; Şuşman, S.; Mărginean, M.; Mihu, C.; Istrate, M.; Moldovan, I.M.; Roman, A.L.; Mihu, C.M. Vascular endothelial growth factor (VEGF)—Key factor in normal and pathological angiogenesis. *Rom. J. Morphol. Embryol.* **2018**, *59*, 455–467.
30. Rabbani, M.; Rogers, P. Role of vascular endothelial growth factor in endometrial vascular events before implantation in rats. *Reproduction* **2001**, *122*, 85–90. [CrossRef]
31. Welter, H.; Wollenhaupt, K.; Tiemann, U.; Einspanier, R. Regulation of the VEGF-System in the Endometrium during Steroid-Replacement and Early Pregnancy of Pigs. *Exp. Clin. Endocrinol. Diabetes* **2003**, *111*, 33–40. [CrossRef]
32. Chennazhi, K.; Nayak, N. Regulation of Angiogenesis in the Primate Endometrium: Vascular Endothelial Growth Factor. *Semin. Reprod. Med.* **2009**, *27*, 080–089. [CrossRef]
33. Das, S.K.; Chakraborty, I.; Wang, J.; Dey, S.K.; Hoffman, L.H. Expression of Vascular Endothelial Growth Factor (VEGF) and VEGF-Receptor Messenger Ribonucleic Acids in the Peri-Implantation Rabbit Uterus. *Biol. Reprod.* **1997**, *56*, 1390–1399. [CrossRef] [PubMed]
34. Gardela, J.; Jauregi-Miguel, A.; Martinez, C.A.; Rodriguez-Martinez, H.; Lopez-Bejar, M.; Alvarez-Rodriguez, M. Semen Modulates the Expression of NGF, ABHD2, VCAN, and CTEN in the Reproductive Tract of Female Rabbits. *Genes* **2020**, *11*, 758. [CrossRef] [PubMed]
35. Pfaffl, M.W. A new mathematical model for relative quantification in real-time RT-PCR. *Nucleic Acids Res.* **2001**, *29*, e45. [CrossRef] [PubMed]
36. R Core Team. *R: A Language and Environment for Statistical Computing*; R Foundation for Statistical Computing: Vienna, Austria, 2013. Available online: https://www.r-project.org/ (accessed on 10 August 2020).
37. Pinheiro, J.; Bates, D.; DebRoy, S.; Sarjar, D.; Team, R.C. nlme: Linear and Nonlinear Mixed Effects Models. R package version 3.1-145. Available online: https://cran.r-project.org/package=nlme (accessed on 10 August 2020).
38. Hothorn, T.; Bretz, F.; Westfall, P. Simultaneous Inference in General Parametric Models. *Biom. J.* **2008**, *50*, 346–363. [CrossRef]
39. Robertson, S.A.; Sharkey, D.J. The role of semen in induction of maternal immune tolerance to pregnancy. *Semin. Immunol.* **2001**, *13*, 243–254. [CrossRef]
40. Robertson, S.A.; Ingman, W.V.; O'Leary, S.; Sharkey, D.J.; Tremellen, K.P. Transforming growth factor β—A mediator of immune deviation in seminal plasma. *J. Reprod. Immunol.* **2002**, *57*, 109–128. [CrossRef]
41. Overstreet, J.W.; Cooper, G.W. Sperm Transport in the Reproductive Tract of the Female Rabbit: I. The Rapid Transit Phase of Transport. *Biol. Reprod.* **1978**, *19*, 101–114. [CrossRef]
42. Overstreet, J.W.; Cooper, G.W. Sperm Transport in the Reproductive Tract of the Female Rabbit: II. The Sustained Phase of Transport. *Biol. Reprod.* **1978**, *19*, 115–132. [CrossRef]
43. Greenwald, G.S. A study of the transport of ova through the rabbit oviduct. *Fertil. Steril.* **1961**, *12*, 80–95. [CrossRef]
44. Roca, J.; Martínez, S.; Orengo, J.; Parrilla, I.; Vázquez, J.M.; Martínez, E.A. Influence of constant long days on ejaculate parameters of rabbits reared under natural environment conditions of Mediterranean area. *Livest. Prod. Sci.* **2005**, *94*, 169–177. [CrossRef]
45. McNitt, J.I. Rabbit Management. In *Rabbit Production*; McNitt, J.I., Lukefahr, S.D., Cheeke, P.R., Patton, H.M., Eds.; CABI: Boston, MA, USA, 2013; pp. 42–61.
46. Pincus, G.; Enzemann, E. V Fertilisation in the Rabbit. *J. Exp. Biol.* **1932**, *9*, 403–408.
47. Duffy, D.M.; Ko, C.; Jo, M.; Brannstrom, M.; Curry, T.E. Ovulation: Parallels With Inflammatory Processes. *Endocr. Rev.* **2019**, *40*, 369–416. [CrossRef] [PubMed]

48. Braden, A.W.H. Distribution of sperms in the genital tract of the female rabbit after coitus. *Aust. J. Biol. Sci.* **1953**, *6*, 693–705. [CrossRef] [PubMed]
49. Morton, D.B.; Glover, T.D. Sperm transport in the female rabbit: The role of the cervix. *J. Reprod. Fertil.* **1974**, *38*, 131–138. [CrossRef] [PubMed]
50. Ashary, N.; Tiwari, A.; Modi, D. Embryo Implantation: War in Times of Love. *Endocrinology* **2018**, *159*, 1188–1198. [CrossRef] [PubMed]
51. Chen, X.; Man, G.C.W.; Liu, Y.; Wu, F.; Huang, J.; Li, T.C.; Wang, C.C. Physiological and pathological angiogenesis in endometrium at the time of embryo implantation. *Am. J. Reprod. Immunol.* **2017**, *78*, e12693. [CrossRef]
52. Waldhalm, S.J.; Dickson, W.M. Blood Flow to the Uterine Tube (Oviduct) of the Early Pregnant and Pseudopregnant Rabbit. *Am. J. Vet. Res.* **1976**, *37*, 817–821.

Publisher's Note: MDPI stays neutral with regard to jurisdictional claims in published maps and institutional affiliations.

 animals

Article

Antibacterial Activity of Some Molecules Added to Rabbit Semen Extender as Alternative to Antibiotics

María Pilar Viudes-de-Castro [1,*], Francisco Marco-Jimenez [2], José S. Vicente [2] and Clara Marin [3]

[1] Centro de Investigación y Tecnología Animal, Instituto Valenciano de Investigaciones Agrarias (CITA-IVIA), Polígono La Esperanza n° 100, 12400 Segorbe, Spain
[2] Instituto de Ciencia y Tecnología Animal, Universitat Politècnica de València, 46022 Valencia, Spain; fmarco@dca.upv.es (F.M.-J.); jvicent@dca.upv.es (J.S.V.)
[3] Departamento de Producción y Sanidad Animal, Salud Pública Veterinaria y Ciencia y Tecnología de los Alimentos, Instituto de Ciencias Biomédicas, Facultad de Veterinaria, Universidad Cardenal Herrera-CEU, CEU Universities, Avenida Seminario s/n, 46113 Moncada, Spain; clara.marin@uchceu.es
* Correspondence: viudes_mar@gva.es

Simple Summary: This study was conducted to evaluate the antibacterial activity of two aminopeptidase inhibitors and chitosan-based nanoparticles in liquid-stored rabbit semen. This study reports that the aminopeptidase inhibitors used to prevent bacterial growth could be used in semen extender as a suitable alternative to antibiotics.

Abstract: Although great attention is paid to hygiene during semen collection and processing, bacteria are commonly found in the semen of healthy fertile males of different species. As the storage of extended semen might facilitate bacterial growth, extenders are commonly supplemented with antibiotics. This study aimed to evaluate the antibacterial activity of ethylenediaminetetraacetic acid (EDTA), bestatin and chitosan-based nanoparticles added to rabbit semen extender and their effect on reproductive performance under field conditions. Four different extenders were tested, supplemented with antibiotics (TCG+AB), with EDTA and bestatin (EB), with EDTA, bestatin and chitosan-based nanoparticles (QEB) or without antibiotics (TCG-AB). Extended semen was cooled at 15 °C for three days. Cooled samples were examined for bacterial growth and semen quality every 24 h for 3 days. The enterobacteria count increased considerably during storage at 72 h in semen extended with TCG+AB and TCG-AB, while extenders EB and QEB showed a bacteriostatic effect over time. After 24, 48 and 72 h, quality characteristics were retained in all groups, with no significant motility differences, either in acrosome integrity, membrane functionality or the viability of spermatozoa. Additionally, bacterial concentration present in fresh semen did not affect reproductive performance. In conclusion, EDTA and bestatin exerted a potent bacteriostatic effect over time and could be used as an alternative to conventional antibiotics in rabbit semen extenders.

Keywords: artificial insemination; antibiotics; antibacterial activity; semen quality; reproductive performance

Citation: Viudes-de-Castro, M.P.; Marco-Jimenez, F.; Vicente, J.S.; Marin, C. Antibacterial Activity of Some Molecules Added to Rabbit Semen Extender as Alternative to Antibiotics. *Animals* **2021**, *11*, 1178. https://doi.org/10.3390/ani11041178

Academic Editors: Rosa María García-García and Maria Arias Alvarez

Received: 21 March 2021
Accepted: 17 April 2021
Published: 20 April 2021

1. Introduction

Artificial insemination (AI) is a highly efficient assisted reproductive technology used worldwide in animal breeding. Semen from healthy fertile males from different species contains bacteria stemming from natural colonisation in the male tract and the environment, despite the application of strict hygienic measures during collection and manipulation processes [1], so international directives stipulate the addition of antibiotics to semen extenders to prevent bacterial growth. However, as a consequence of excessive antibiotics use in different fields such as human medicine, veterinary medicine, livestock and fish production, agriculture and food technology, microbial resistance has emerged as one of the main concerns worldwide [2,3].

Semen contaminated with bacteria such as *Enterobacteriaceae* might adversely affect the quality of semen used for AI [4], in addition it is an important hygiene indicator [5–8]. Even though most of them are non-pathogenic bacteria, they can negatively influence sperm quality and longevity if present in high concentrations [6,7,9,10]. Additionally, several studies have shown that the seminal plasma of different species contains aminopeptidases [11–16]. Aminopeptidase activity promotes the proliferation of many bacteria that act as virulence factors, essential for the survival and maintenance of many microbial pathogens [17,18].

Nowadays, bacterial resistance to commonly used antibiotics and the global spread of resistance genes has become a serious health problem. Bacteria could counteract the actions of antimicrobials through different mechanisms, such as enzyme modification, alteration of the target binding sites, active efflux pumps or decreased permeability of bacterial membrane [19]. This resistance may occur through spontaneous mutations or by the horizontal transfer of mobile genetic elements from other bacteria, phages and/or the transmission of resistance genes from the environment [19,20]. In this sense, the transmission of resistance genes plays an important role in the spread of antimicrobial resistance among strains [19,20] and efforts need to be made to replace conventional antibiotics in the animal production industry. Some alternatives to conventional antibiotics in semen extender, such as colloid centrifugation [21–24] or removal of seminal plasma [25], have been investigated in different species. However, these techniques involve an increase in the processing time of semen. Others approaches for overcoming bacterial growth include the use of active molecules such as EDTA, [26], chitosan [27,28], nanoparticles [29], peptides [30,31] and aminopeptidase inhibitors [18,32], etc., with recognised antibacterial activity. EDTA is known to increase the outer cell wall permeability of Gram-negative bacteria, allowing other molecules easier access into the cell, facilitating an alteration or inhibition of its metabolism [33–35]. Chitosan, a biocompatible biodegradable and non-toxic polycationic copolymer extensively used as material for encapsulation and controlled release of chemicals [36], interacts with the bacterial cell membrane and causes cell lysis [37,38]. Furthermore, the encapsulation of GnRH in chitosan-based nanoparticles added to the extender supplemented with EDTA and bestatin can overcome the poor stability of the GnRH analogues in the presence of aminopeptidases [16,39] and allows to reduce the hormonal concentration used without affecting the reproductive performance of female rabbits [40]. This method of inducing ovulation in rabbit AI was developed to increase the welfare of rabbit insemination procedures and reduce the concentration of GnRH analogues added to semen extenders.

Against this background, this study aimed to evaluate the antibacterial activity of EDTA, bestatin and chitosan-based nanoparticles added to rabbit semen extender and their effect on the reproductive performance under field conditions, with a view to improving the sustainability of the rabbit production system.

2. Materials and Methods

The chemicals used in this study were purchased from Sigma-Aldrich (Merck Life Science S.L.U. Madrid, Spain). Animal housing and the protocols for semen collection and AI were approved by the Animal Care and Use Committee of Centro de Tecnología Animal, Instituto Valenciano de Investigaciones Agrarias. All animals were handled according to the European regulations for the care and use of animals for scientific purposes (European Commission Directive 2010/63/European Union).

2.1. Extenders Composition

Four different extenders were tested (Table 1). The solution used as a carrier for the molecules tested was Tris-citric acid-glucose (TCG extender; [41]). Chitosan and alginate were dissolved (0.05%) in the TCG supplemented with EDTA (20 mM) and bestatin (10 mM) according to Casares-Crespo et al. [40]. The nanoparticles were formed spontaneously in the coacervation process, directly after mixing the solutions of chitosan and alginate (4:1) through magnetic stirring (~600 rpm) for 30 min at room temperature.

Table 1. Semen extenders composition.

Group	Composition
TCG+AB	TCG extender supplemented with 100 IU/mL penicillin + 100 µg/mL streptomycin
EB	TCG supplemented with EDTA (20 mM) and bestatin (10 mM)
QEB	TCG supplemented with EDTA (20 mM), bestatin (10 mM) and nanoparticles of chitosan-alginate (0.05%)
TCG-AB	TCG extender without antibiotics

2.2. Experimental Design

2.2.1. Experiment 1: In Vitro Evaluation

Animals

Twelve males of New Zealand White origin were kept individually under similar conditions to those described by Viudes de Castro et al. [16].

Semen Collection

Semen from males was collected twice a week in three replicates. Strict attention was paid to the hygiene of collection equipment and semen samples were collected into sterile tubes. Semen collection and evaluation were conducted in accordance with Casares Crespo et al. [39]. Finally, all the ejaculates were pooled.

The pool was split into four equal fractions and diluted with the appropriate extender (dilution 1:10; *v:v*). The pools used in the experiment presented an average sperm concentration of 385 spermatozoa mL-1. Diluted samples were cooled at 15 °C for three days. Cooled samples were examined for bacterial growth, total motility, percentage of live sperm and membrane status every 24 h for 3 days.

Microbiological Analysis

Using Enterobacteriaceae as the sentinel, bacterial growth was evaluated by enumeration. Tenfold dilution series were performed in each extender to 10^{-6}, and 1000 µL of each dilution was then plated onto Violet Red Bile Dextrose Agar (VRBD agar, Scharlab®, Barcelona, Spain) per duplicate; after homogenisation of the plate, 10 mL of VRBD were added to seal the plate. Plates were incubated for 24–48 h at 37 ± 1 °C. Typical colonies were counted, and the least dilute pair of plates that contained an average of between 30 and 150 colonies was used to calculate the number of bacteria (CFU/mL).

Seminal Quality Evaluation

Percentage of total motile sperm was evaluated using a computer-assisted sperm analysis system (ISAS Proiser, Valencia, Spain) as described by Viudes-de-Castro et al. [13]. Briefly, ten microlitres of each sample was placed into a 10 mm deep Makler counting chamber (Sefi Medical Instruments, Haifa, Israel). Sperm motility was assessed at 37 °C by negative phase contrast objective at a magnification of X100 (NIKON E-400 microscope, Izasa Scientific, Barcelona, Spain). Six microscopic fields were captured for each sample. A minimum of 400 sperm were evaluated using the same criteria described by Casares Crespo et al. [39].

Flow cytometric analyses to assess viability (integrity of sperm membrane) and acrosome integrity were performed with a CytoFLEX Flow Cytometer (Beckman Coulter, S.L.U., Barcelona, Spain) equipped with red (638 nm), blue (488 nm) and violet (405 nm) lasers and operated by the CytExpert Software v.2.3 (Beckman Coulter, S.L.U., Barcelona, Spain). The cytometer was calibrated daily using specific calibration beads provided by the manufacturer. Data were collected from 10,000 events. Gating the spermatozoa population after Hoechst 33,342 staining eliminated non-sperm events. Doublets and clumps were further excluded by using a plot of side scatter area and side scatter height followed by a gate of simple events. A compensation overlap was performed before each experiment. A

FITC-PNA/PI/Hoechst triple staining method, validated for rabbit semen in our laboratory, was used to determine viability and acrosomal status. To this end, 100 μL of semen at 30×10^6 sperm/mL were stained with 0.5 μL Hoechst 33,342 (0.5 mg/mL) for 20 min at 37 °C without light. Subsequently, 1.5 μL FITC-PNA (1 mg/mL) and 0.5 μL PI (1 mg/mL) were added to the sample and incubated 10 min at 37 °C without light. Then 400 μL of TCG extender [36] were added to obtain a final concentration of 6×10^6 sperm/mL. PI-negative sperm were considered viable. The normal apical ridge (NAR) percentage was calculated as the proportion of acrosome intact sperm.

Membrane functionality analysis was assessed by hypo-osmotic swelling test (HOST). An aliquot of 100 μL of diluted semen was added to 1 mL of warmed 150 mOsm hypo-osmotic swelling solution containing sodium citrate (25 mmol/L) and fructose (75 mmol/L) and incubated for 30 min at 37 °C. Subsequently, 10 μL of each sample were placed on a clean glass slide with a coverslip, and sperm swelling was assessed under phase-contrast microscopy. For each sample, a total of 200 spermatozoa were examined.

2.2.2. Experiment 2: In Vivo Evaluation

Eight hundred and ninety-seven crossbreed females from a commercial farm (Altura, Castellón, Spain) were inseminated using fresh semen from 50 adult males belonging to a paternal rabbit line (Line R, [42]). Animal housing and seminal evaluation were similar to the previous experiment. All the ejaculates were pooled. The pool was split into four aliquots and diluted 1:10 with the four experimental extenders. After diluting the semen in the four experimental extenders, the insemination was initiated immediately. Each female was randomly assigned to one of the four experimental groups and was inseminated with 0.5 mL of semen using standard curved cannulas (24 cm). About 2 h elapsed between the first and the last inseminated female. At birth, pregnancy rate (number of kindlings/number of inseminated does) and prolificacy (total number of kits born) were evaluated.

2.3. *Statistical Analysis*

To analyse the effect of extender on Enterobacteriaceae growth, motility, viability, acrosome integrity and membrane functionality, a general linear model was used. The extender, refrigeration time and their interaction were taken as fixed effects and, in the case of seminal parameters, the corresponding parameter of the pool was introduced as a covariate in the analysis. A chi-square test was used to test differences in pregnancy rate at birth between groups. For the total number of kits born per litter, an ANOVA was performed, including as fixed effect the extender group and pool as covariate. All analyses were performed with the SPSS 26.0 software package (SPSS Inc., Chicago, IL, USA). Values were considered statistically different at $p < 0.05$.

3. Results

3.1. *Experiment 1: In Vitro Evaluation*

Results for bacterial growth are shown in Figure 1. The enterobacteria count increased considerably during storage at 72 h (Figure 1) in TCG+AB and TCG-AB groups. At 24 h, a significant increase of bacterial growth was observed in semen extended in TCG-AB (extender without antibiotics) compared to the rest of the groups. Additionally, at this time point, a significant decrease in bacterial growth was observed in the EB group versus the TCG+AB (extender with antibiotics) or QEB groups. At 48 h, it was found that there were no significant differences between EB and QEB, but there was significantly lower bacterial growth than TCG+AB and TCG-AB group. At 72 h, the trend was similar, with no significant differences between EB and QEB groups and showing a significantly lower bacterial growth than TCG+AB and TCG-AB groups. From 24 h, bacterial growth in the TCG+AB group increased over time, being three times higher at 72 h than that observed in groups EB and QEB, which maintained the same number of CFUs over time.

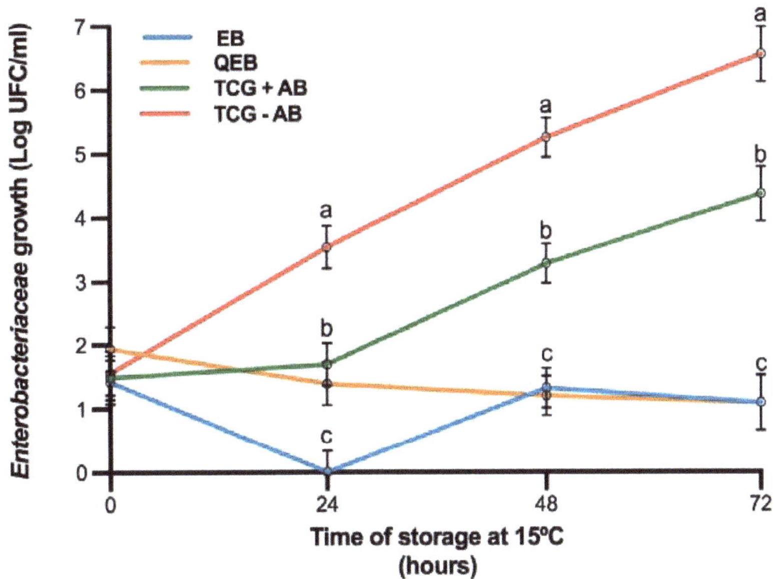

Figure 1. Bacterial growth (Log CFU/mL; mean ± SEM) in sperm samples from the rabbit where there was storage at 15 °C for 72 h in four extenders: EB: TCG supplemented with EDTA (20 mM) and bestatin (10 mM). QEB: TCG supplemented with EDTA (20 mM), bestatin (10 mM) and nanoparticles of chitosan-alginate (0.05%). TCG+AB: TCG extender supplemented with 100 IU/mL penicillin + 100 µg/mL streptomycin. TCG-AB: TCG extender without antibiotics. Different superscripts (a, b and c) indicate differences in values at the same time point ($p < 0.05$).

Seminal quality parameters of samples from the experimental extenders are shown in Table 2. There was no interaction between extender and refrigeration time. The presence of EDTA, bestatin and chitosan nanoparticles had no effect on total motility, acrosome integrity, membrane functionality or the viability of spermatozoa. Following 24, 48 and 72 h, quality characteristics were retained in all groups, with no significant differences in motility, acrosome integrity, membrane functionality or the viability of spermatozoa.

Table 2. Sperm quality in stored rabbit spermatozoa in four extenders (means ± standard deviation).

Extender Group	N	Total Mot (%)	HOST (%)	NAR (%)	Viability (%)
TCG+AB	9	73.9 ± 9.98	73.0 ± 9.44	94.2 ± 6.18	75.0 ± 8.38
EB	9	69.0 ± 12.42	73.4 ± 8.41	94.1 ± 5.48	72.0 ± 7.00
QEB	9	71.1 ± 9.48	72.6 ± 6.38	93.7 ± 5.89	72.7 ± 7.43
TCG-AB	9	68.8 ± 11.19	71.7 ± 8.47	94.6 ± 5.79	73.2 ± 6.09
Time					
24 h	12	73.8 ± 11.95	75.1 ± 7.41	96.1 ± 4.96	74.8 ± 6.82
48 h	12	72.0 ± 12.58	71.6 ± 8.78	94.5 ± 6.78	72.6 ± 8.61
72 h	12	66.3 ± 4.43	71.3 ± 7.59	91.8 ± 5.01	72.3 ± 5.76

UTCG+AB: TCG extender supplemented with 100 IU/mL penicillin + 100 µg/mL streptomycin. EB: TCG supplemented with EDTA (20 mM) and bestatin (10 mM). QEB: TCG supplemented with EDTA (20 mM), bestatin (10 mM) and nanoparticles of chitosan-alginate (0.05%). TCG-AB: TCG extender without antibiotics. N: number of seminal pools; Total Mot: total motility; HOST: hypo-osmotic swelling test; NAR: acrosome normality.

3.2. Experiment 2: In Vivo Evaluation

Pregnancy rate at birth and the total number of kits born are presented in Table 3. Neither pregnancy rate at birth nor prolificacy were affected by the experimental group, both parameters being similar between groups.

Table 3. Fertility (%) and prolificacy (means ± standard deviation) obtained from insemination of 897 females.

Extender Group	N	Pregnancy Rate (%)	Total Number of Kits Born
TCG+AB	228	90	10.1 ± 3.20
EB	225	88	10.0 ± 3.50
QEB	219	88	10.3 ± 3.17
TCG-AB	225	87	10.2 ± 3.25

TCG+AB: TCG extender supplemented with 100 IU/mL penicillin + 100 µg/mL streptomycin. EB: TCG supplemented with EDTA (20 mM) and bestatin (10 mM). QEB: TCG supplemented with EDTA (20 mM), bestatin (10 mM) and nanoparticles of chitosan-alginate (0.05%). TCG-AB: TCG extender without antibiotics. N: number of inseminated does.

4. Discussion

Bacterial contamination is of particular relevance in rabbit AI, where most inseminations are carried out with liquid semen storage at 15 °C [43]. Even though semen collection protocols in the livestock industry are very strict, semen collection is not a sterile process, and the addition of antibiotics to extenders to control contaminating bacterial populations is a routine fact at farm level. The efficacy of different antibiotics added to semen extenders in livestock has been widely demonstrated [44]. However, how quickly bacteria acquire tolerance and/or resistance to antibiotics is essential. Numerous studies show the critical resistance patterns found in semen samples of different species such as boars [45], humans [46] and bovine [47]. However, it can be suggested that the antibiotics currently used in routine practices in livestock, such as AI, may need to be modified to avoid future complications arising from bacterial resistance.

Previous study demonstrated that sperm microbiota diversity is influenced by host genetics [9]. The rabbit semen samples were contaminated with bacteria, especially those that belong to the *Enterobacteriaceae* family [44], ranging from 27.6% for Line V to 50.9% for Line R of semen samples analysed [9]. Bacteria contamination, such as *Enterobacteriaceae* family might adversely affect the quality of semen used for AI [4] and is an important hygiene indicator [5], although the majority of these bacterial strains are not currently considered pathogens [7]. The results of this study prove that the replacement of antibiotics in the current extenders by EDTA and bestatin prevent bacterial growth through 72 h in the rabbit doses. Therefore, the present study results validate the bacteriostatic effect of aminopeptidase inhibitors, such as bestatin and EDTA, and highlight the role of protease inhibitors in the control of seminal bacterial growth. This is in agreement with several studies in which an inhibitory activity of EDTA against Gram-negative bacteria, Gram-positive bacteria (staphylococci) and fungi (*Candida* spp.) was observed [33–35]. Likewise, some studies indicated that EDTA, alone or in combination, is an effective antibiofilm agent with a spectrum covering both Gram-positive and Gram-negative bacteria [35,48–52]. In addition, bacterial proteases participate in important metabolic pathways and have key roles in cell viability, stress response and pathogenicity [53]. On the other hand, despite the antimicrobial activity demonstrated by chitosan against several pathogens [54–56], in the present study, the presence of chitosan-alginate nanoparticles in the extender did not show a synergistic action with aminopeptidase inhibitors on the microbial growth, with both extenders showing similar results. A possible explanation for this is that the presence of alginate can interfere in the inhibition of bacterial growth by chitosan. As the presence of cationic charge situated in the amino group of chitosan is essential for exhibiting high antimicrobial properties [57], the ionic linkages between functional groups of the oppositely charged chitosan and alginate would result in low availability of unreacted positive amino groups of chitosan when nanoparticles were formed, which reduces the chances of interaction with negatively charged components of microbial cell membranes. Our results suggest that both aminopeptidase inhibitors (EDTA and bestatin), alone or in combination with nanoparticles of chitosan-alginate, maintained total motility, viability, acrosome status and functional integrity of the sperm plasma membrane for at least three days. Moreover, the use of both aminopeptidase inhibitors preserved the fertility and prolificacy under field conditions. Undoubtedly, this result could be used to improve the sustainability of the rabbit production system. In the present study, the classic combination of penicillin and streptomycin contributed to the diluent's efficacy in controlling the growth of *Enterobacteriaceae* only for up to 24 h. Nevertheless, from this moment on, this antibiotic cocktail cannot prevent the bacterial growth, which was probably due to an increase in tolerance of antibiotics by *Enterobacteriaceae* [58,59]. The common use of antibiotics in extenders is an important concern: apart from being prophylactic and non-therapeutic, and therefore going against the recommendations for prudent use of antimicrobials, they can cause increase antibiotic resistance in the bacteria commonly found in semen [24]. As expected, in the extender without antibiotics (TCG-AB), no bacteriostatic effect was observed throughout the entire cooling period, showing an increasing number of enterobacteria over time, highlighting the need to supplement AI extenders with substances that control bacterial growth. However, there were no effects of enterobacteria contamination on in vitro quality sperm in long-term stored samples. Although the presence of microorganisms in semen may reduce semen quality and fertilising capacity during preservation time, our results indicate that bacterial concentration present in fresh semen (0 to 4 h from semen collection) has no effect on reproductive performance, which is in agreement with Jäkel et al. [60] in pig, who reported similar reproductive performance at 24 h between groups inseminated with semen diluted in extender with or without antibiotics. Extenders and storage temperature are important factors to preserve the fertilising capacity of rabbit semen. As regards the sperm quality variables, all extenders used in the present study preserved the quality of rabbit semen throughout the cooling period. Several studies to evaluate stored rabbit semen have been carried out under different experimental conditions. On the one hand, several authors had observed that motility decreased when semen was stored at 5 °C during 48 h, irrespective of the extender used [61–63]. On the other hand, other authors have shown that rabbit semen stored at 15 °C up to 48 h retains fertility capacity [43,64]. However, further studies are needed to verify the reproductive performance of rabbit semen stored for 72 h in extenders supplemented with EDTA and bestatin.

5. Conclusions

In conclusion, we demonstrated in this study that the addition of EDTA and bestatin to semen extender exerted a potent bacteriostatic effect over time, effectively inhibiting the growth of Enterobacteriaceae, which suggests that EDTA and bestatin could be used as an alternative to conventional antibiotics in rabbit semen extenders.

Author Contributions: Conceptualisation M.P.V.-d.-C., C.M.; Methodology and data curation M.P.V.-d.-C., C.M., F.M.-J. and J.S.V.; writing—original draft preparation, M.P.V.-d.-C.; writing and analysis—review and editing, M.P.V.-d.-C., C.M., F.M.-J. and J.S.V. All authors have read and agreed to the published version of the manuscript.

Funding: Funding from the Ministry of Economy, Industry and Competitiveness (Research project: AGL2017-85162-C2-1-R) is acknowledged.

Institutional Review Board Statement: The study was conducted according to the guidelines of the Declaration of Helsinki, and approved by the Animal Care and Use Committee of Centro de Tecnología Animal, Instituto Valenciano de Investigaciones Agrarias (Protocol #2018/VSC/PEA/0116).

Acknowledgments: The authors thank Manuel Sierra and Felipe Lavara for offering their animal facility for conducting this study under field conditions.

Conflicts of Interest: The authors declare no conflict of interest.

Abbreviations

The following abbreviations are used in this manuscript:

AI	Artificial insemination
ANOVA	Analysis of variance
CASA	Computer assisted sperm analysis
CFU	Colony-forming unit
DE	Digestible energy
EB	Tris-citric acid-glucose supplemented with EDTA and bestatin
EDTA	Ethylenediaminetetraacetic acid
HOST	Hypo-osmotic swelling test
LSM	Least square means
NAR	Normal apical ridge
PI	Iodide propidium
QEB	Tris-citric acid-glucose supplemented with EDTA, bestatin and nanoparticles of chitosan-alginate
SEM	Standard error of the means
TCG	Tris-citric acid-glucose
TCG-AB	Tris-citric acid-glucose without antibiotics
TCG+AB	Tris-citric acid-glucose plus antibiotics
VRBD	Violet Red Bile Dextrose

References

1. Waberski, D.; Riesenbeck, A.; Schulze, M.; Weitze, K.F.; Johnson, L. Application of preserved boar semen for artificial insemination: Past, present and future challenges. *Theriogenology* **2019**, *137*, 2–7. [CrossRef]
2. Da Costa, P.M.; Loureiro, L.; Matos, A.J.F. Transfer of multidrug-resistant bacteria between intermingled ecological niches: The interface between humans, animals and the environment. *Int. J. Environ. Res. Public Health* **2013**, *10*, 278–294. [CrossRef]
3. Agrawal, Y.; Vanha-Perttula, T. Alanyl aminopeptidase of bovine seminal vesicle secretion. *Int. J. Biochem.* **1986**, *18*, 725–729. [CrossRef]
4. Althouse, G.C.; Lu, K.G. Bacteriospermia in extended porcine semen. *Theriogenology* **2005**, *63*, 573–584. [CrossRef] [PubMed]
5. Halkman, H.B.D.; Halkman, A.K. Indicator Organisms. In *Encyclopedia of Food Microbiology*, 2nd ed.; Elsevier Inc.: Amsterdam, The Netherlands, 2014; pp. 358–363. ISBN 9780123847331.
6. Asociación Española de Normalización y Certificación. Microbiology of the Food Chain. Horizontal Method for the Detection and Enumeration of Enterobacteriaceae. Part 2: Colony-Count Technique (ISO 21528-2:2017, Corrected Version 2018-06-01). 2018. Available online: https://www.en.une.org/_layouts/15/r.aspx?c=N0059874 (accessed on 20 March 2021).
7. European Food Safety Authority; European Centre for Disease Prevention and Control. The European Union Summary Report on Antimicrobial Resistance in Zoonotic and Indicator Bacteria from Humans, Animals and Food in 2017/2018. *EFSA J.* **2020**, *18*, e06007. [CrossRef]
8. Moretti, E.; Capitani, S.; Figura, N.; Pammolli, A.; Federico, M.G.; Giannerini, V.; Collodel, G. The presence of bacteria species in semen and sperm quality. *J. Assist. Reprod. Genet.* **2009**, *26*, 47–56. [CrossRef] [PubMed]
9. Úbeda, J.L.; Ausejo, R.; Dahmani, Y.; Falceto, M.V.; Usan, A.; Malo, C.; Perez-Martinez, F.C. Adverse effects of members of the Enterobacteriaceae family on boar sperm quality. *Theriogenology* **2013**, *80*, 565–570. [CrossRef] [PubMed]
10. Marco-Jiménez, F.; Borrás, S.; Garcia-Dominguez, X.; D'Auria, G.; Vicente, J.S.; Marin, C. Roles of host genetics and sperm microbiota in reproductive success in healthy rabbit. *Theriogenology* **2020**, *158*, 416–423. [CrossRef] [PubMed]
11. Althouse, G. Sanitary Procedures for the Production of Extended Semen. *Reprod. Domest. Anim.* **2008**, *43*, 374–378. [CrossRef] [PubMed]
12. Huang, K.; Takahara, S.; Kinouchi, T.; Takeyama, M.; Ishida, T.; Ueyama, H.; Nishi, K.; Ohkubo, I. Alanyl Aminopeptidase from Human Seminal Plasma: Purification, Characterization, and Immunohistochemical Localization in the Male Genital Tract. *J. Biochem.* **1997**, *122*, 779–787. [CrossRef] [PubMed]
13. Fernández, D.; Valdivia, A.; Irazusta, J.; Ochoa, C.; Casis, L. Peptidase activities in human semen. *Peptides* **2002**, *23*, 461–468. [CrossRef]
14. Irazusta, J.; Valdivia, A.; Fernández, D.; Agirregoitia, E.; Ochoa, C.; Casis, L. Enkephalin-degrading enzymes in normal and subfertile human semen. *J. Androl.* **2004**, *25*, 733–739. [CrossRef]
15. Osada, T.; Watanabe, G.; Kondo, S.; Toyoda, M.; Sakaki, Y.; Takeuchi, T. Male reproductive defects caused by puromycin-sensitive aminopeptidase deficiency in mice. *Mol. Endocrinol.* **2001**, *15*, 960–971. [CrossRef] [PubMed]
16. Viudes-de-Castro, M.P.; Mocé, E.; Lavara, R.; Marco-Jiménez, F.; Vicente, J.S. Aminopeptidase activity in seminal plasma and effect of dilution rate on rabbit reproductive performance after insemination with an extender supplemented with buserelin acetate. *Theriogenology* **2014**, *81*, 1223–1228. [CrossRef] [PubMed]

17. Bhosale, M.; Kadthur, J.C.; Nandi, D. Roles of Salmonella enterica serovar Typhimurium encoded Peptidase N during systemic infection of Ifnγ -/- mice. *Immunobiology* **2012**, *217*, 354–362. [CrossRef]
18. Correa, A.F.; Bastos, I.M.D.; Neves, D.; Kipnis, A.; Junqueira-Kipnis, A.P.; de Santana, J.M. The activity of a hexameric M17 metallo-aminopeptidase is associated with survival of Mycobacterium tuberculosis. *Front. Microbiol.* **2017**, *8*, 504. [CrossRef] [PubMed]
19. Agyare, C.; Etsiapa Boamah, V.; Ngofi Zumbi, C.; Boateng Osei, F. Antibiotic Use in Poultry Production and Its Effects on Bacterial Resistance. In *Antimicrobial Resistance—A Global Threat*; IntechOpen: London, UK, 2019.
20. Sharma, V.K.; Johnson, N.; Cizmas, L.; McDonald, T.J.; Kim, H. A review of the influence of treatment strategies on antibiotic resistant bacteria and antibiotic resistance genes. *Chemosphere* **2016**, *150*, 702–714. [CrossRef]
21. Nicholson, C.M.; Abramsson, L.; Holm, S.E.; Bjurulf, E. Bacterial contamination and sperm recovery after semen preparation by density gradient centrifugation using silane-coated silica particles at different g forces. *Hum. Reprod.* **2000**, *15*, 662–666. [CrossRef]
22. Morrell, J.M.; Klein, C.; Lundeheim, N.; Erol, E.; Troedsson, M.H.T. Removal of bacteria from stallion semen by colloid centrifugation. *Anim. Reprod. Sci.* **2014**, *145*, 47–53. [CrossRef]
23. Guimarães, T.; Lopes, G.; Pinto, M.; Silva, E.; Miranda, C.; Correia, M.J.; Damásio, L.; Thompson, G.; Rocha, A. Colloid centrifugation of fresh stallion semen before cryopreservation decreased microorganism load of frozen-thawed semen without affecting seminal kinetics. *Theriogenology* **2015**, *83*, 186–191. [CrossRef]
24. Martínez-Pastor, F.; Lacalle, E.; Martínez-Martínez, S.; Fernández-Alegre, E.; Álvarez-Fernández, L.; Martinez-Alborcia, M.-J.; Bolarin, A.; Morrell, J.M. Low density Porcicoll separates spermatozoa from bacteria and retains sperm quality. *Theriogenology* **2021**, *165*. [CrossRef]
25. Ramires Neto, C.; Sancler da Silva, Y.F.R.; Resende, H.L.; Guasti, P.N.; Monteiro, G.A.; Papa, P.M.; Dell'aqua Júnior, J.A.; Puoli Filho, J.N.P.; Alvarenga, M.A.; Papa, F.O. Control methods and evaluation of bacterial growth on fresh and cooled stallion semen. *J. Equine Vet. Sci.* **2015**, *35*, 277–282. [CrossRef]
26. Finnegan, S.; Percival, S.L. EDTA: An Antimicrobial and Antibiofilm Agent for Use in Wound Care. *Adv. Wound Care* **2015**, *4*, 415–421. [CrossRef]
27. Rabea, E.I.; Badawy, M.E.T.; Stevens, C.V.; Smagghe, G.; Steurbaut, W. Chitosan as antimicrobial agent: Applications and mode of action. *Biomacromolecules* **2003**, *4*, 1457–1465. [CrossRef]
28. Sahariah, P.; Másson, M. Antimicrobial Chitosan and Chitosan Derivatives: A Review of the Structure-Activity Relationship. *Biomacromolecules* **2017**, *18*, 3846–3868. [CrossRef] [PubMed]
29. Rudramurthy, G.R.; Swamy, M.K.; Sinniah, U.R.; Ghasemzadeh, A. Nanoparticles: Alternatives against drug-resistant pathogenic microbes. *Molecules* **2016**, *21*, 836. [CrossRef] [PubMed]
30. Bahar, A.; Ren, D. Antimicrobial Peptides. *Pharmaceuticals* **2013**, *6*, 1543–1575. [CrossRef] [PubMed]
31. Mahlapuu, M.; Håkansson, J.; Ringstad, L.; Björn, C. Antimicrobial peptides: An emerging category of therapeutic agents. *Front. Cell. Infect. Microbiol.* **2016**, *6*, 194. [CrossRef]
32. Dickneite, G.; Schorlemmer, H.U.; Hofstaetter, T.; Sedlacek, H.-H. Immunostimulation as a Therapeutic Principle in Bacterial Infections: The Effect of the Immunomodulator Bestatin on the Experimental Chronic E. coli Urinary Tract Infection. In *Experimentelle Urologie*; Springer: Berlin/Heidelberg, Germany, 1985; pp. 229–235.
33. Root, J.L.; McIntyre, O.R.; Jacobs, N.J.; Daghlian, C.P. Inhibitory effect of disodium EDTA upon the growth of Staphylococcus epidermidis in vitro: Relation to infection prophylaxis of Hickman catheters. *Antimicrob. Agents Chemother.* **1988**, *32*, 1627–1631. [CrossRef] [PubMed]
34. Gil, M.L.; Casanova, M.; Martínez, J.P. Changes in the cell wall glycoprotein composition of Candida albicans associated to the inhibition of germ tube formation by EDTA. *Arch. Microbiol.* **1994**, *161*, 489–494. [CrossRef] [PubMed]
35. Percival, S.L.; Kite, P.; Eastwood, K.; Murga, R.; Carr, J.; Arduino, M.J.; Donlan, R.M. Tetrasodium EDTA as a Novel Central Venous Catheter Lock Solution Against Biofilm. *Infect. Control Hosp. Epidemiol.* **2005**, *26*, 515–519. [CrossRef]
36. Li, J.; Cai, C.; Li, J.; Li, J.; Li, J.; Sun, T.; Wang, L.; Wu, H.; Yu, G. Chitosan-Based Nanomaterials for Drug Delivery. *Molecules* **2018**, *23*, 2661. [CrossRef] [PubMed]
37. Liu, X.F.; Guan, Y.L.; Yang, D.Z.; Li, Z.; Yao, K. Antibacterial action of chitosan and carboxymethylated chitosan. *J. Appl. Polym. Sci.* **2001**, *79*, 1324–1335. [CrossRef]
38. Hosseinnejad, M.; Jafari, S.M. Evaluation of different factors affecting antimicrobial properties of chitosan. *Int. J. Biol. Macromol.* **2016**, *85*, 467–475. [CrossRef]
39. Casares-Crespo, L.; Vicente, J.S.; Talaván, A.M.; Viudes-de-Castro, M.P. Does the inclusion of protease inhibitors in the insemination extender affect rabbit reproductive performance? *Theriogenology* **2016**, *85*, 928–932. [CrossRef] [PubMed]
40. Casares-Crespo, L.; Fernández-Serrano, P.; Viudes-de-Castro, M.P. Protection of GnRH analogue by chitosan-dextran sulfate nanoparticles for intravaginal application in rabbit artificial insemination. *Theriogenology* **2018**, *116*, 49–52. [CrossRef]
41. Viudes-De-Castro, M.P.; Vicente, J.S. Effect of sperm count on the fertility and prolificity rates of meat rabbits. *Anim. Reprod. Sci.* **1997**, *46*, 313–319. [CrossRef]
42. Estany, J.; Camacho, J.; Baselga, M.; Blasco, A. Selection response of growth rate in rabbits for meat production. *Genet. Sel. Evol.* **1992**, *24*, 527–537. [CrossRef]
43. Roca, J.; Martínez, S.; Vázquez, J.M.; Lucas, X.; Parrilla, I.; Martínez, E.A. Viability and fertility of rabbit spermatozoa diluted in Tris-buffer extenders and stored at 15 °C. *Anim. Reprod. Sci.* **2000**, *64*, 103–112. [CrossRef]

44. Duracka, M.; Lukac, N.; Kacaniova, M.; Kantor, A.; Hleba, L.; Ondruska, L.; Tvrda, E. Antibiotics versus natural biomolecules: The case of in vitro induced bacteriospermia by enterococcus faecalis in rabbit semen. *Molecules* **2019**, *24*, 4329. [CrossRef] [PubMed]
45. Bennemann, P.E.; Machado, S.A.; Girardini, L.K.; Sonálio, K.; Tonin, A.A. Bacterial contaminants and antimicrobial susceptibility profile of boar semen in Southern Brazil Studs. *Rev. MVZ Córdoba* **2018**, *23*, 6637–6648. [CrossRef]
46. Machen, G.L.; Bird, E.T.; Brown, M.L.; Ingalsbe, D.A.; East, M.M.; Reyes, M.; Kuehl, T.J. Time trends for bacterial species and resistance patterns in semen in patients undergoing evaluation for male infertility. *Baylor Univ. Med. Cent. Proc.* **2018**, *31*, 165–167. [CrossRef] [PubMed]
47. Jayarao, B.M.; Oliver, S.P. Aminoglycoside-Resistant Streptococcus and Enterococcus Species Isolated from Bovine Mammary Secretions. *J. Dairy Sci.* **1992**, *75*, 991–997. [CrossRef]
48. Raad, I.; Hanna, H.; Dvorak, T.; Chaiban, G.; Hachem, R. Optimal antimicrobial catheter lock solution, using different combinations of minocycline, EDTA, and 25-percent ethanol, rapidly eradicates organisms embedded in biofilm. *Antimicrob. Agents Chemother.* **2007**, *51*, 78–83. [CrossRef] [PubMed]
49. Kite, P.; Eastwood, K.; Sugden, S.; Percival, S.L. Use of in vivo-generated biofilms from hemodialysis catheters to test the efficacy of a novel antimicrobial catheter lock for biofilm eradication in vitro. *J. Clin. Microbiol.* **2004**, *42*, 3073–3076. [CrossRef] [PubMed]
50. Banin, E.; Brady, K.M.; Greenberg, E.P. Chelator-induced dispersal and killing of Pseudomonas aeruginosa cells in a biofilm. *Appl. Environ. Microbiol.* **2006**, *72*, 2064–2069. [CrossRef]
51. Sherertz, R.J.; Boger, M.S.; Collins, C.A.; Mason, L.; Raad, I.I. Comparative in vitro efficacies of various catheter lock solutions. *Antimicrob. Agents Chemother.* **2006**, *50*, 1865–1868. [CrossRef]
52. Al-Bakri, A.G.; Othman, G.; Bustanji, Y. The assessment of the antibacterial and antifungal activities of aspirin, EDTA and aspirin-EDTA combination and their effectiveness as antibiofilm agents. *J. Appl. Microbiol.* **2009**, *107*, 280–286. [CrossRef]
53. Culp, E.; Wright, G.D. Bacterial proteases, untapped antimicrobial drug targets. *J. Antibiot.* **2017**, *70*, 366–377. [CrossRef]
54. Goy, R.C.; Morais, S.T.B.; Assis, O.B.G. Evaluation of the antimicrobial activity of chitosan and its quaternized derivative on E. Coli and S. aureus growth. *Rev. Bras. Farmacogn.* **2016**, *26*, 122–127. [CrossRef]
55. Erdem, B.; Kariptaş, E.; Kaya, T.; Tulumoğlu, Ş.; Görgülü, Ö. Factors ınfluencing antibacterial activity of chitosan against Aeromonas hydrophila and Staphylococcus aureus. *Int. Curr. Pharm. J.* **2016**, *5*, 45–48. [CrossRef]
56. Li, J.; Zhuang, S. Antibacterial activity of chitosan and its derivatives and their interaction mechanism with bacteria: Current state and perspectives. *Eur. Polym. J.* **2020**, *138*, 109984. [CrossRef]
57. Holappa, J.; Hjálmarsdóttir, M.; Másson, M.; Rúnarsson, Ö.; Asplund, T.; Soininen, P.; Nevalainen, T.; Järvinen, T. Antimicrobial activity of chitosan N-betainates. *Carbohydr. Polym.* **2006**, *65*, 114–118. [CrossRef]
58. Levin-Reisman, I.; Brauner, A.; Ronin, I.; Balaban, N.Q. Epistasis between antibiotic tolerance, persistence, and resistance mutations. *Proc. Natl. Acad. Sci. USA* **2019**, *116*, 14734–14739. [CrossRef] [PubMed]
59. Windels, E.M.; Michiels, J.E.; van den Bergh, B.; Fauvart, M.; Michiels, J. Antibiotics: Combatting tolerance to stop resistance. *MBio* **2019**, *10*, e02095-19. [CrossRef] [PubMed]
60. Jäkel, H.; Scheinpflug, K.; Mühldorfer, K.; Gianluppi, R.; Lucca, M.S.; Mellagi, A.P.G.; Bortolozzo, F.P.; Waberski, D. In vitro performance and in vivo fertility of antibiotic-free preserved boar semen stored at 5 °C. *J. Anim. Sci. Biotechnol.* **2021**, *12*, 1–12. [CrossRef]
61. El-Gaafary, M.N. Quality and fertility of cooled rabbit semen supplemented with cyclic-AMP stimulators. *Anim. Reprod. Sci.* **1994**, *34*, 307–313. [CrossRef]
62. El-Kelawy, H.M.; Tawfeek, M.I.; El-Gaafary, M.N.; Ibrahim, H. Viability and Fertilizing Ability of Extended Rabbit Semen Stored at 5 °C. In Proceedings of the 10th World Rabbit Congress, Sharm EL-Sheikh, Egypt, 3–6 September 2012; pp. 285–289.
63. Rosato, M.P.; Iaffaldano, N. Effect of Chilling Temperature on the Long-Term Survival of Rabbit Spermatozoa held Either in a Tris-Based or a Jellified Extender. *Reprod. Domest. Anim.* **2011**, *46*, 301–308. [CrossRef]
64. López-Gatius, F.; Sances, G.; Sancho, M.; Yániz, J.; Santolaria, P.; Gutiérrez, R.; Núñez, M.; Núñez, J.; Soler, C. Effect of solid storage at 15 °C on the subsequent motility and fertility of rabbit semen. *Theriogenology* **2005**, *64*, 252–260. [CrossRef]

 *animals* 

Article

Influence of Different Regimes of Moderate Maternal Feed Restriction during Pregnancy of Primiparous Rabbit Does on Long-Term Metabolic Energy Homeostasis, Productive Performance and Welfare

Carlota Fernández-Pacheco [1], Pilar Millán [1], María Rodríguez [2,†], Nora Formoso-Rafferty [2], Ana Sánchez-Rodríguez [1,‡], Pedro L. Lorenzo [1], María Arias-Álvarez [3], Rosa M. García-García [1] and Pilar G. Rebollar [2,*]

[1] Department of Physiology, Faculty of Veterinary, Complutense University of Madrid, Avenida Puerta de Hierro s/n, 28040 Madrid, Spain; cafpmartorell@ucm.es (C.F.-P.); pmillanp@vet.ucm.es (P.M.); anasanro.vet@gmail.com (A.S.-R.); plorenzo@vet.ucm.es (P.L.L.); rosa.garcia@vet.ucm.es (R.M.G.-G.)
[2] Department of Agrarian Production, ETSIAAB, Technical University of Madrid, Ciudad Universitaria s/n, 28040 Madrid, Spain; maria.rodriguez@pigchamp-pro.com (M.R.); nora.formosorafferty@upm.es (N.F.-R.)
[3] Department of Animal Production, Faculty of Veterinary, Complutense University of Madrid, Avenida Puerta de Hierro s/n, 28040 Madrid, Spain; m.arias@vet.ucm.es
* Correspondence: pilar.grebollar@upm.es
† Current address: Department of R+D & Projects, PigCHAMP Pro Europa S.L., 40006 Segovia, Spain.
‡ Current address: Department of Biodiversity and Evolutionary Biology, Museo Nacional de Ciencias, Naturales (CSIC), 28006 Madrid, Spain.

Citation: Fernández-Pacheco, C.; Millán, P.; Rodríguez, M.; Formoso-Rafferty, N.; Sánchez-Rodríguez, A.; Lorenzo, P.L.; Arias-Álvarez, M.; García-García, R.M.; Rebollar, P.G. Influence of Different Regimes of Moderate Maternal Feed Restriction during Pregnancy of Primiparous Rabbit Does on Long-Term Metabolic Energy Homeostasis, Productive Performance and Welfare. *Animals* **2021**, *11*, 2736. https://doi.org/10.3390/ani11092736

Academic Editor: Víctor Hugo Parraguez Gamboa

Received: 1 September 2021
Accepted: 16 September 2021
Published: 19 September 2021

Simple Summary: In rabbit farms, the main production costs come directly from food supplies. Although the reproductive outcomes in this species are acceptable, the results are worse when it comes to primiparous rabbits, so it is recommended that insemination be carried out post-weaning. By avoiding the overlap of the second gestation and the first lactation, better fertility results are expected. Still, despite this, the rabbits garner more adipose tissue than desired, and the productive efficiency deteriorates in the long term. The purpose of this study was to evaluate the influence of different periods of moderate feed restriction (one, two or three weeks) applied during the second pregnancy of primiparous does. We studied fetoplacental development, productive parameters, metabolism and possible stress indicators. Results showed that the voluntary feed intake of dams increased right after feed restriction. No permanent alterations were found in reproductive outcome, metabolism or welfare of does, meaning that this feeding strategy could be successfully applied in rabbit farms.

Abstract: In this study, a maternal feed restriction (MFR; 105 g/d) in primiparous rabbit does was applied from day 0 to 7 post artificial insemination (AI) (R07, n = 96), from day 7 to 21 post AI (R721, n = 92), from day 0 to 21 post AI (R021, n = 94) or fed ad libitum during whole pregnancy (Control, n= 92). Feed intake (FI) was measured after MFR was over. On day 28 of gestation, fetoplacental development was evaluated (n = 11/group) and the productive parameters of the remaining dams were analyzed. Plasma free tri-iodothyronine (T3) and thyroxine, glucose, insulin, non-esterified fatty acids (NEFA), and corticosterone were analyzed during gestation and lactation (n = 5/group). After MFR, all groups significantly increased their voluntary FI. The longer MFR was, the lower the weight and length of the fetuses, but no long-term effects over litter performance were observed. R021 groups had the lowest T3 and the highest NEFA concentrations during pregnancy and showed insulin resistance at the end of gestation, but during lactation, energy homeostasis was balanced in all groups. MFR did not affect corticosterone concentrations. In conclusion, the ration setting applied slightly involved the energy homeostasis and metabolism of the animals, but their overall metabolic condition, productive performance and welfare were not compromised.

Keywords: rabbit; feed intake; free tri-iodothyronine; thyroxine; insulin; glucose; corticosterone; NEFA; fetus; placenta

1. Introduction

In rabbit farms, dam feeding represents roughly a third of the total feed costs (3.7% and 31.7% for replacement and reproductive does, respectively) [1]. Moreover, the number of kits born alive and the feed conversion rate during fattening are of high economic importance. These traits depend directly on maternal nutrition during pregnancy and lactation. Artificial insemination (AI) of does is usually applied on Day 4 or 11 post-partum, while females are lactating [2]. Alternative breeding strategies can be applied to alleviate the negative energy balance that befalls after first parturition with concurrent gestation and lactation of primiparous does, which are still completing their body growth [3]. One of them is delaying the interval parturition-AI until weaning (Day 30 post-partum), applying a reproductive extensive rhythm [4]. This improves reproductive outcome [5], but considering the first weeks of pregnancy do not involve a very high energy expenditure and it is considered an anabolic period [6,7], the important risk of fattening exists due to ad libitum feeding of post-weaning inseminated does. Nonetheless, it is still a critical phase that can influence the rate of embryonic implantation and the development of the placenta, which are limiting factors for good fetal growth. Moreover, the mentioned fattening risk increases if these dams do not become pregnant (pregnancy diagnosis is usually performed on Day 10–14 post AI). Thus, feeding strategies as a subject of study in these animals include applying different levels of maternal feed restriction (MFR) during diverse periods of pregnancy.

MFR has had diverse targets to avoid fattening and high mortality around parturition [8], to determine the competition for materno-fetal resource partitioning [9], to increase voluntary feed intake at the beginning of lactation, or to allow a longer productive life of rabbit does [10]. López-Tello et al. [9,11] verified that undernourishment in dams (50% MFR) during the entire pregnancy or during preimplantation period (1st week) has adverse effects on the fetoplacental unit (reduced fetal crown-rump lengths, asymmetrical growth and high apoptotic rates at the decidua and labyrinth zone). In this sense, it has also been described that if the daily amount of food provided increases to 60% of the dam's total voluntary intake, negative effects are observed only in the second half of gestation. In contrast, if MFR is applied in the first 15 days of pregnancy, no adverse effects are observed on the viability or weight of the kits at birth [12]. Moreover, ad libitum refeeding in the last third of gestation has been shown to improve the energetic status of the mothers before parturition [13], precisely when fetal needs are most significant. It also reduces the negative effects of subsequent lactation on their reproductive function (poor ovulatory response and increased delivery-fertile insemination interval) and improves the weight and viability of the kits.

What we can introduce about the application of MFR is largely based upon our previous experimental studies [14,15], in which we have described the metabolic consequences observed in mothers and offspring subjected to a 60% MFR during three weeks of pregnancy (from day 0 to 21). This MFR led to a compensatory feed intake during the last week of pregnancy that offset the effects in the live body weight (LBW) and body reserves of the dams at parturition and the fetal body weight and phenotype at day 28 of pregnancy. These studies on MFR applied during three weeks of pregnancy have been limited to specific comparisons of the metabolic profiles of mothers and fetuses at a particular moment of gestation (day 28).

Thus, in order to gain further understanding of the metabolic changes and energetic modulations of primiparous does subjected to MFR, the novelty of this study has been to apply shorter and more specific periods of restriction during gestation studying the long-term consequences on additional hormones and metabolites. Among others, thyroid hormones are key to the regulation of metabolism and adaptation to fasting. They contribute to both mandatory and adaptive thermogenesis, regulating appetite and energy expenditure [16]. On the other hand, glucose is central to energy consumption. Carbohydrates, lipids, and proteins all ultimately break down into glucose, which serves as the main metabolic fuel of mammals' cells and the prevalent fuel of the fetus [17]. Likewise, insulin is a primary anabolic hormone secreted in response to increased blood glucose and amino acids following feeding. The major action of insulin is to stimulate and control glucose consumption and exertion. Like other hormones, insulin performs its activities by binding to specific receptors present on many cells distributed all through the body [18]. Additionally, non-esterified fatty acids (NEFA) concentrations reflect the mobilization of body reserves since they are generated from lipolysis in fatty tissue and are an indicators of negative energy balance [19]. Briefly, the main interest of this study was to measure these metabolic pathways and hormones since they are known to have different behaviors when adapting to the specific needs of gestation, fetal development and milk production [20]. Furthermore, and considering animal welfare standards, food or water deprivation is a procedure that can cause pain or distress [21], making the analysis of stress-related hormones mandatory when a feed restriction is applied. It should be noted that in rabbits and rodents, unlike other mammals, the main glucocorticoid analyzed is usually corticosterone rather than cortisol [22].

Therefore, the main goal of this study was to determine in primiparous lactating does the comparative effect of MFR applied during one, two or three weeks of pregnancy followed by ad libitum refeeding on the feed intake of the dams, feto-placental development on day 28 of pregnancy, productive performance, the weekly plasma concentrations of thyroid hormones and corticosterone, and the glucose and lipid metabolism during pregnancy and consequent lactation.

2. Materials and Methods

2.1. Experimental Design

The animals were housed at the animal facilities of the Technical University of Madrid (Spain), which meet the local, national and European requirements for Scientific Procedure Establishments (RD 53/2013; PROEX 302/15). The experimental design is displayed in Figure 1. New Zealand × California rabbits (18 weeks old), were fed ad libitum during their first pregnancy (length 31 days) and lactation (length 30 days) with a diet containing 16% crude protein, 37% crude fiber, 3.7% fat and 2400 kcal/kg of digestible energy (NANTA, Madrid, Spain). During the first pregnancy, the feed intake of each doe was recorded daily and established in 175 g per animal and day.

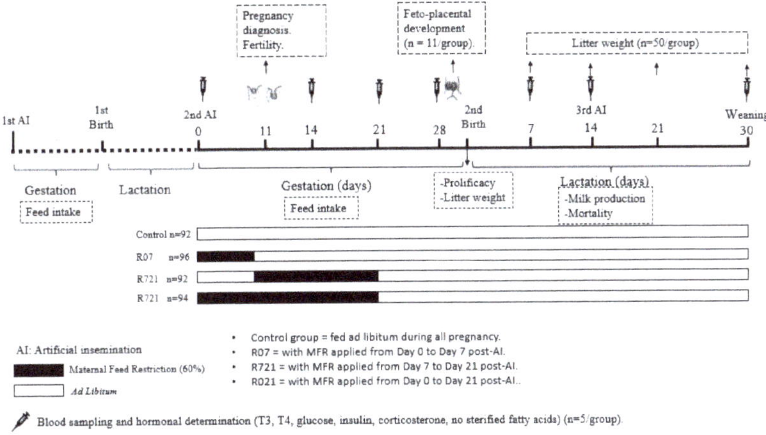

Figure 1. Experimental design.

After parturition and weaning (day 30 post-partum) these does started a second cycle and were artificially inseminated (AI) with fresh diluted semen (commercial extender, MA 24, Ovejero, León, Spain). Each dose contained at least 20 million spermatozoa in 0.5 mL of semen diluent. Ovulation was induced with gonadoreline at the time of AI (20 µg/doe, i.m.; Inducel-GnRH, Ovejero, León, Spain). At this time (day 0), a total of 374 does were randomly allocated into four groups:

- Control group (n = 92) always fed ad libitum.
- R07 (n = 96) with MFR applied from day 0 to day 7 post-AI.
- R721 (n = 92) with MFR applied from day 7 to day 21 post-AI.
- R021 (n = 94) with MFR applied from day 0 to day 21 post-AI.

MFR was set at 60% of the measured voluntary feed intake of the first pregnancy (105 g/day), in order to maintain their basal metabolic needs. In the fourth week of pregnancy, all animals were fed ad libitum until the end of the study. Pregnancy was diagnosed at day 11 post-AI by means of abdominal palpation and fertility ([no. of does pregnant at day 11 after AI/no. of does inseminated] × 100) was calculated. Feed intake was weekly calculated exclusively in the pregnant does from day 0 until the day of parturition.

2.2. Fetoplacental Study

On day 28 of gestation, 44 pregnant does (n = 11 from each group) were weighed and euthanized with an overdose barbiturate (Dolethal, Lab. Vetoquinol, Madrid, Spain) to study fetoplacental development. A ventral laparotomy was performed to count the corpora lutea present in the ovaries and calculate the ovulation rate per animal. Both uterine horns were opened to extract the fetuses and placentas, calculating the rate of viable structures [(number of viable fetuses/number total of fetuses) × 100]. In morphologically viable fetuses, crown–rump length (CRL), biparietal (BPD), occipito-nasal (OND), and thoracic (TD) diameters were measured. Whole fetuses were weighed, and after decapitation, heads and trunks were weighed separately. Fetal organs' (brain, liver, heart, lungs, digestive tract and kidneys) weights and their ratios (organ weight/fetus weight) were calculated to assess fetal growth patterns. Placentas were weighed both as a whole and separately for the maternal (decidua) from the fetal (labyrinth) part. The length, width and thickness of each of the pieces were recorded. Placental efficiency was calculated by dividing the weight of the fetus between the weight of the whole placenta.

2.3. Productive Outcome

The gestation of the remaining rabbit does was carried out, and prolificacy (no. of kits born alive and stillborn per doe) and litter weights at birth were recorded. Litter size was subsequently standardized to 8–12 kits by removing or adding kits within each experimental group.

During lactation, a total number of 50 litters from each experimental group were weighed at day 7, 14 and 21 post-partum. Milk production was estimated using the regression equation developed by Helfenstein et al. [23] as follows: milk production (kg) = 0.75 ± 1.75 LW21 (kg) where LW21 corresponds to litter weight at 21 days of lactation. The day of weaning (day 30 post-partum) the number of weaned kits per doe, litter weight and mortality during lactation period were assessed. At the end of the study, the percentage of culled does (abortions or sudden death) in each experimental group was calculated.

2.4. Long-Term Maternal Metabolic and Hormonal Study

In addition, the evaluation of metabolic and hormonal parameters was carried out on blood samples obtained from the marginal ear vein of randomly chosen females. An initial blood sample was taken at the moment of AI (day 0 of pregnancy) from 10 dams before being allocated into the experimental groups. After pregnancy diagnosis, five pregnant animals per group were sampled at three different moments of gestation: day 14 (n = 20), day 21 (n = 20) and day 28 (n = 20) of gestation. Following parturition, samples were taken from lactating females (five per group) at three moments of lactation: day 7 (n = 20) and day 14 (n = 20). On day 14 post-partum, all does were re-inseminated as previously described. Then, on day 30 post-partum, weaning was performed and blood samples were taken from pregnant (n = 20) and non-pregnant (n = 20) does from the third AI. Blood samples were placed in tubes with EDTA as an anticoagulant and centrifuged for 15 min at 1200 g to obtain plasma.

Moreover, the evaluation of energetic homeostasis was performed by measuring free triiodothyronine (T3) and thyroxine (T4) in plasma samples using commercial immunoassays (Demeditec Diagnostics GmbH, Kiel, Germany). The assay sensibility was 0.05 pg/mL for both T3 and T4, and the inter-assay and intra-assay variation coefficients were 9.8% and 3.6% for T3 and 4.9% and 3.3% for T4. Glycemic metabolism was assessed by the determination of insulin and glucose levels. Insulin was determined using a commercial immunoassay (Mercodia Ultrasensitive Insulin ELISA, Mercodia AB, Uppsala, Sweden) with an assay sensitivity of 0.071 mU/L and inter-assay and intra-assay variation coefficients of 6.2% and 4.2%, respectively. Insulin sensitivity was calculated by glucose-to-insulin ratio and the homeostasis model assessment for insulin resistance (HOMA-IR) using the following equation: [insulin (mU/L) × (glucose (mg/dL)/18)]/22.5 [23]. Low HOMA-IR values (<1.96) indicate high insulin sensitivity, whereas high HOMA-IR values (>3) indicate low insulin sensitivity (insulin resistance). Non-esterified fatty acids (NEFA) were selected as the main indicator of lipid metabolism and were measured using a clinical biochemistry assay (Biolabo SAS, Maizy, France), with detection in the range of 0.01–3.0 mmol/L and assay sensitivity 0.050 mmol/L. Corticosterone was used as a stress indicator [22] and was determined using a commercial immunoassay (Demeditec Diagnostics GmbH, Kiel, Germany). The assay sensitivity was 1.63 nmol/L and the inter-assay intra-assay variation coefficients were 6.5% and 2.8%.

2.5. Statistical Analysis

SAS software (Statistical Analysis System Institute Inc.; Cary, NC, USA, 2001) was used for the statistical analysis of the data. In all analyses, the doe was considered the experimental unit. Differences between groups in the total feed intake were analyzed by a one-way ANOVA (Proc GLM) with the feed regime as the main effect. The effect of MFR on weekly feed intake during gestation was analyzed by repeated measure analysis (MIXED), with feed regime (Control, R07, R721 and R021), time (weeks 1, 2, 3 and 4), and their interaction as main effects. Ovulation rate, live body weight of does, number, weights and measurements of fetuses and placentas, and placental efficiency on day 28 of pregnancy were analyzed (Proc GLM) with feed regime as main effect and considering the litter size as a covariate. With the same procedure, prolificacy and milk production were analyzed with feed regime as the main effect. A chi squared test (Proc CATMOD) was used to evaluate the effect of MFR on fertility and percentage of culled does. Litter weights during the lactation period were analyzed by repeated measure analysis (Proc MIXED), considering the MFR, the time (birth, 7, 14, 21, and 30 days post-partum) and their interaction in the model. For hormonal and metabolic parameters, due to the blood sampling being carried out in independent randomly chosen rabbit does, a GLM procedure was performed, considering as main effects the treatment (MFR), the time (pregnancy: day 14, 21 and 28 of gestation; and lactation day 7, 14 and 30 post-partum), and the interaction between both. If significant main effects were detected, a Tukey test (for parametric variables) or Kruskal–Wallis (non-parametric variables) were used to compare means among groups, considering the existence of significant differences for a p value of less than 0.05. All data are presented as least squares means.

3. Results

3.1. Feed Intake during Pregnancy

As expected, whole feed intake during pregnancy was different between groups ($p < 0.0001$). The Control and R07 groups consumed 5854 ± 124.3 g and 5529 ± 109 g, respectively, followed by R721 does (4518 ± 112 g) and the R021 group (4007 ± 106.3 g). The weekly evolution of daily feed intake is shown in Figure 2. Interestingly, the week after MFR, we observed that all groups increased their voluntary feed intake, surpassing the control group, like group R07 during the second week, and groups R721 and R021 during the fourth week of pregnancy.

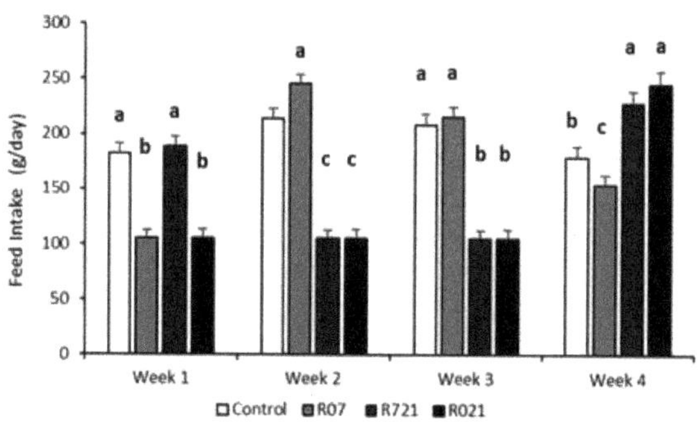

Figure 2. Weekly evolution of the daily feed intake (g/day) in pregnant rabbit does fed ad libitum (Control), restricted to 60% one week from day 0 to 7 (R07), two weeks from day 7 to 21 (R721) and three weeks from day 0 to 21 (R021) of gestation. (a–c): Different letters represent significant differences between experimental groups ($p < 0.0001$).

3.2. Fetoplacental Study

Results obtained on day 28 of pregnancy are shown in Table 1. This table is quite revealing in several ways. First, regarding body reserves of does at the end of pregnancy, does restricted for two weeks (R721) presented a lower LBW than Control and R07 groups ($p = 0.0254$), whereas does restricted for three weeks (R021) had an intermediate body weight.

From a reproductive point of view, the number of corpora lutea, total fetuses per doe, implantation rate and fetal viability were high and similar in all groups. In R07 does, there is a tendency (p = 0.0824) to have more fetuses but with lower viability.

Feed restriction significantly affected ($p < 0.01$) the total, head and trunk weights and CRL of fetuses, being lower in those from R721 and R021 does than in the control group, and intermediate in R07 females. Regarding fetal cephalic diameters, BPD was similar in all groups ($p > 0.05$), however OND was higher in Control and R07 does than in R021, being intermediate in R721 ($p = 0.0032$). Fetal thoracic diameter was greater in R07 compared to the other three groups ($p = 0.0034$). Regarding fetal organs, only liver, lung and kidney weights were different between groups ($p = 0.0097$, $p = 0.0166$, $p = 0.0001$, respectively), with those of the Control group being heavier except for the liver.

Weights and dimensions of labyrinth and decidua from euthanized does at day 28 of pregnancy are shown in Table 2. Regarding these data, it is interesting to indicate that the whole placenta of R07 does weighed more than those from the R721 and R021, and the Control group had an intermediate value ($p < 0.0001$). Furthermore, the labyrinth of R07 does tended to be the heaviest ($p = 0.0809$), whilst that of R021 does was the shortest ($p < 0.0001$) and that from R721 the thinnest ($p < 0.0053$). In does subjected to MFR for two weeks (R721), decidua had lower weight ($p < 0.0001$) than the other three groups. The longest decidua was observed in Control does ($p < 0.0001$) and the thinnest was observed in R721 ($p = 0.0069$). Finally, the lowest placental efficiency was observed in R07 does, the highest in R721, and intermediate in Control and R021 does ($p = 0.0162$).

Table 1. Number and biometric parameters of fetuses of 28 days of gestational age in rabbits fed ad libitum (Control), restricted to 60% one week from day 0 to 7 (R07), two weeks from day 7 to 21 (R721) and three weeks from day 0 to 21 (R021) of gestation.

	Control n = 11	R07 n = 11	R721 n = 11	R021 n = 11	SEM	*p* Value
LBW of does (g)	4638 [a]	4640 [a]	4342 [b]	4413 [ab]	80.90	0.0254
Corpora lutea	12.2	12.3	13.1	13.4	0.58	0.4350
Total fetuses/doe	11.7	13.2	12.5	12.5	0.38	0.0814
Implantation rate [1] (%)	92.5	97	95.2	91	2.57	0.3684
Fetal Viability [2] (%)	94.6	84.9	90.3	90.1	9.13	0.0824
Fetuses weights (g)						
Total	39.7 [a]	38.4 [ab]	37.4 [b]	37.0 [b]	1.78	0.0028
Head	9.32 [a]	9.06 [ab]	8.75 [b]	8.71 [b]	0.35	0.0001
Trunk	29.0 [a]	28.0 [ab]	27.0 [b]	26.8 [b]	1.43	0.0014
Fetuses measures (mm)						
CRL	101.4 [a]	99.5 [ab]	98.7 [b]	98.6 [b]	1.83	0.0018
OND	29.0 [a]	29.2 [a]	28.7 [ab]	28.3 [b]	0.61	0.0032
BPD	19.3	19.2	19.3	19.0	0.52	0.5801
TD	20.6 [b]	21.3 [a]	20.8 [b]	20.2 [b]	0.71	0.0034
Organs weights (g)						
Brain	0.913	0.901	0.892	0.901	0.04	0.6395
Liver	2.64 [a]	2.39 [b]	2.36 [b]	2.39 [b]	0.18	0.0097
Heart	0.221	0.207	0.219	0.209	0.02	0.0734
Lungs	0.134 [a]	0.126 [a]	0.122 [b]	0.126 [a]	0.08	0.0166
Kidneys	0.359 [a]	0.337 [ab]	0.323 [b]	0.307 [b]	0.02	0.0001
Digestive tract	0.208	0.215	0.207	0.197	0.17	0.0952
Ratios						
Brain	0.024	0.024	0.024	0.025	0.00	0.4033
Liver	0.066 [a]	0.061 [b]	0.064 [ab]	0.066 [a]	0.00	0.0033
Heart	0.006 [a]	0.005 [b]	0.006 [a]	0.006 [a]	0.00	0.0028
Lungs	0.034	0.033	0.033	0.034	0.00	0.0942
Kidneys	0.009 [a]	0.009 [a]	0.009 [a]	0.008 [b]	0.00	0.0001
Digestive tract	0.053	0.056	0.056	0.053	0.00	0.1759
Brain: Liver	0.367	0.395	0.400	0.395	0.03	0.2273

LBW: live body weight. [1] (number of fetuses/number of corpora lutea) × 100; [2] (number of viable fetuses/total number of fetuses) × 100; CRL: crown–rump length; TD: thoracic diameter; BPD: biparietal diameter; OND: occipito-nasal diameter. Data are shown as least squares means. Ratio: organ weight/fetus weight. Values in the same row with different letters are significantly different. SEM: standard error of mean.

Table 2. Weight and dimensions of placentas from pregnant rabbits on day 28, fed ad libitum (Control), restricted to 60% one week from day 0 to 7 (R07), two weeks from day 7 to 21 (R721) and three weeks from day 0 to 21 (R021) of gestation.

	Control n = 11	R07 n = 11	R721 n = 11	R021 n = 11	SEM	*p* Value
Whole placenta (g)	5.16 [ab]	5.21 [a]	4.60 [c]	4.81 [bc]	0.30	0.0001
Labyrinth						
Weight (g)	3.27	3.51	3.30	3.33	0.80	0.0809
Length (mm)	37.40 [a]	35.80 [a]	35.50 [a]	33.90 [b]	4.28	0.0001
Width (mm)	28.50	28.40	27.70	27.40	3.72	0.1286
Thickness (mm)	5.16 [a]	5.17 [a]	4.70 [b]	5.05 [ab]	0.99	0.0053
Decidua						
Weight (g)	1.48 [a]	1.55 [a]	1.22 [b]	1.40 [a]	0.11	0.0001
Length (mm)	39.61 [a]	37.32 [b]	37.30 [b]	36.09 [b]	1.41	0.0001
Width (mm)	17.84 [a]	18.76 [a]	16.42 [b]	18.39 [a]	0.95	0.0001
Thickness (mm)	3.45 [a]	3.52 [a]	3.11 [b]	3.27 [ab]	0.26	0.0069
Placental efficiency [1]	7.52 [ab]	7.34 [b]	7.86 [a]	7.45 [ab]	0.34	0.0162

[1] Fetus weight/whole placenta weight. Data are shown as least squares means. Values in the same row with different letters are significantly different. SEM: standard error of mean.

3.3. Productive Outcome

Productive outcome is shown in Table 3.

Table 3. Productive parameters of rabbits fed ad libitum (Control), or restricted to 60% of their needs, one week from day 0 to 7 (R07), two weeks from day 7 to 21 (R721) and three weeks from day 0 to 21 (R021) in its second gestation.

	Control n = 92	R07 n = 96	R721 n = 92	R021 n = 94	RMSE	*p* Value
Fertility [1] (%)	71.11	76.04	78.26	78.72		0.6129
Parturitions	59	69	67	68		
Litter size						
Born alive	11.01	10.98	10.05	11.02	3.17	0.2904
Stillborn	0.50	0.43	0.32	0.24	1.00	0.5514
Weaned (30 days post-partum)	10.20 [a]	10.54 [a]	9.59 [b]	10.04 [ab]	1.14	0.0004
Litter weight (g) [2]						
At birth	605.70	619.34	580.93	643.98	142.6	0.1276
At 7 days post-partum	1306	1361	1272	1288	216.2	0.1646
At 14 days post-partum	2290	2398	2260	2276	318.1	0.1040
At 21 days p post-partum p	3314	3394	3227	3304	381.2	0.1613
At 30 days post-partum (weaning)	5782	5708	5447	5484	1129	0.3592
Milk production (kg) [3]	8.29	8.49	8.07	8.26	9.53	0.1613
Mortality in lactation (%) [4]	5.13 [ab]	2.83 [b]	3.43 [ab]	6.32 [a]	6.77	0.0325

[1] Number of does pregnant after pregnancy diagnosis at 11 days post AI/total does inseminated × 100. [2] This parameter was obtained from 50 litters per experimental group. [3] Milk production = 0.75 ± 1.75 × LW21 (kg) where LW21 corresponds to litter weight at 21 days of lactation. [4] Mortality: [100−(kits weaned/no. of kits after litter adjustment to 8–12 kits) × 100]. Data are shown as least squares means. RMSE: root mean square error. Values in the same row with different letters are significantly different. pp: post-partum.

Fertility, kits born alive and stillborn, and litter weight at birth and during lactation were similar between groups ($p > 0.05$). However, after litter adjustment, the effect of MFR on the number of weaned kits and mortality of kits during lactation was significantly different between groups ($p = 0.0004$ and $p = 0.0325$, respectively). Control and R07 does weaned more kits than R721 does, whereas R021 does had an intermediate result. The highest and lowest mortalities during lactation were observed in R021 and R07 litters, respectively.

At the end of the study, the percentage of culled does was 7.6, 4.2, 5.4 and 6.4% in Control, R07, R721 and R021 groups, respectively ($p > 0.05$).

3.4. Long-Term Maternal Metabolic and Hormonal Study

Endocrine and metabolic parameters obtained in the experimental groups during pregnancy are shown in Table 4.

Regarding thyroid hormones, the animals subjected to the most extended restriction treatment (R021) had lower plasma T3 concentrations than Control and R07 does ($p = 0.0196$) but similar to R721 ones. Plasma T3 concentrations decreased significantly on days 14 and 21 of pregnancy in relation to day 0 and returned to intermediate values on day 28 ($p = 0.0026$). Plasma T4 concentrations on Days 14 and 21 of pregnancy were also significantly lower than on day 28 ($p = 0.0001$) but similar to that on day 0. MFR did not affect T4 concentrations, T3 to T4 ratio, plasma glucose and HOMA-IR ($p > 0.05$).

Table 4. Effect of maternal feed restriction (MFR) and time on endocrine and metabolic parameters during pregnancy of rabbits fed ad libitum (Control), restricted to 60% one week from day 0 to 7 (R07), two weeks from day 7 to 21 (R721) and three weeks from day 0 to 21 (R021) of gestation.

	Maternal Feed Restriction				Days of Pregnancy					p Value		
Pregnancy	Control n = 5	R07 n = 5	R721 n = 5	R021 n = 5	0 n = 10	14 n = 20	21 n = 20	28 n = 20	RMSE	p_{MFR}	p_{Time}	$p_{MFR \times Time}$
T3 (pg/mL)	3.76 [a]	3.77 [a]	3.19 [ab]	3.06 [b]	4.16 [a]	3.07 [b]	3.02 [b]	3.54 [ab]	0.79	0.0196	0.0026	0.0353
T4 (pg/mL)	4.74	4.24	5.00	5.16	5.89 [ab]	3.66 [b]	2.58 [b]	7.00 [a]	3.11	0.8525	0.0001	0.5731
T3 to T4 ratio	1.45	1.05	1.01	1.12	0.91 [ab]	1.00 [a]	2.06 [a]	0.66 [b]	1.24	0.8079	0.0022	0.5237
Insulin (mU/L)	15.73 [a]	15.71 [a]	9.47 [b]	12.45 [ab]	16.26	10.10	12.80	14.20	6.57	0.0225	0.0871	0.0055
Glucose (mg/dL)	79.94	78.00	74.49	84.13	63.82 [b]	61.72 [b]	76.64 [b]	114.38 [a]	22.34	0.6207	0.0001	0.7501
Glucose-to-insulin ratio	6.79 [b]	7.12 [b]	12.93 [a]	9.17 [ab]	4.59	9.56	10.26	11.70	6.79	0.0482	0.0802	0.2518
HOMA-IR	2.90	3.11	1.75	2.87	2.63 [ab]	1.44 [b]	2.40 [b]	4.15 [a]	1.79	0.1140	0.0001	0.0049
NEFAS (mmol/L)	0.42 [ab]	0.33 [b]	0.62 [a]	0.52 [ab]	0.54	0.54	0.46	0.35	0.27	0.0158	0.0781	0.0100
Corticosterone (ng/mL)	257.45	274.56	256.02	251.15	293.30 [a]	363.82 [b]	192.20 [c]	195.85 [c]	12.13	0.7344	0.0001	0.2519

NEFAS: no esterified fatty acids. RMSE: root mean square error. HOMA-IR: homeostasis model assessment for insulin resistance [Insulin × (glucose/18)/22.5]. Values in the same row with different letters (a, b, c) are significantly different.

However, insulinemia was lower, and glucose-to-insulin ratio was higher in R721 does than in Control and R07 does, but similar to R021 ones ($p = 0.0225$ and $p = 0.0482$, respectively). In addition, plasmatic glucose increased dramatically at the end of pregnancy in all groups ($p < 0.0001$).

The highest and lowest NEFA concentrations were obtained in R721 and R07 does, respectively, and Control and R021 does showed intermediate values ($p = 0.0158$). No significant differences in NEFA levels throughout the pregnancy period were observed ($p > 0.05$).

Finally, plasma corticosterone concentration was only affected by time. The lowest concentrations of this stress indicator were observed at 21 and 28 days of pregnancy ($p < 0.0001$).

MFR and time had significant effects on plasma concentrations of T3 ($p = 0.0353$), insulin ($p = 0.0055$), NEFA ($p = 0.0100$) and HOMA index ($p = 0.0049$) (Figure 3A–D, respectively). On day 21 of pregnancy, R021 does showed a dramatic reduction in plasma concentrations of T3 compared to R07 groups, but it was similar to Control and R721 does. The lowest plasma insulin concentrations were observed on day 21 of pregnancy in the rabbit females restricted at this time (R721 and R021). Control and R07 does had the lowest NEFA concentrations on day 14 of pregnancy. During the first half of pregnancy, HOMA-IR remained lower than 3 and by day 21 of gestation its values rose above the limit established to consider insulin resistance in Control and R07 groups. On day 28 of pregnancy, all groups showed HOMA-IR values over 3 except for group R721.

As shown in Table 5, MFR did not affect any maternal endocrine or metabolic parameters in the experimental groups during lactation ($p > 0.05$). As the lactation period progressed, T3 and glucose concentrations remained at similar values during the first two weeks of lactation, but insulin and HOMA-IR decreased on day 14 ($p = 0.0006$) compared to day 7. On weaning day, non-pregnant does had lower T3 concentrations than pregnant ones ($p < 0.0056$), but HOMA-IR, glucose and insulin concentrations were similar in both ($p > 0.05$).

Energy mobilization measured by means of plasma NEFA concentrations was similar in all groups and during all lactation. Plasma corticosterone concentrations as welfare indicators were not affected by MFR. However, significant increases in this hormone were observed as lactation progressed. No significant interactions between MFR and time in any of the studied variables were detected in this period.

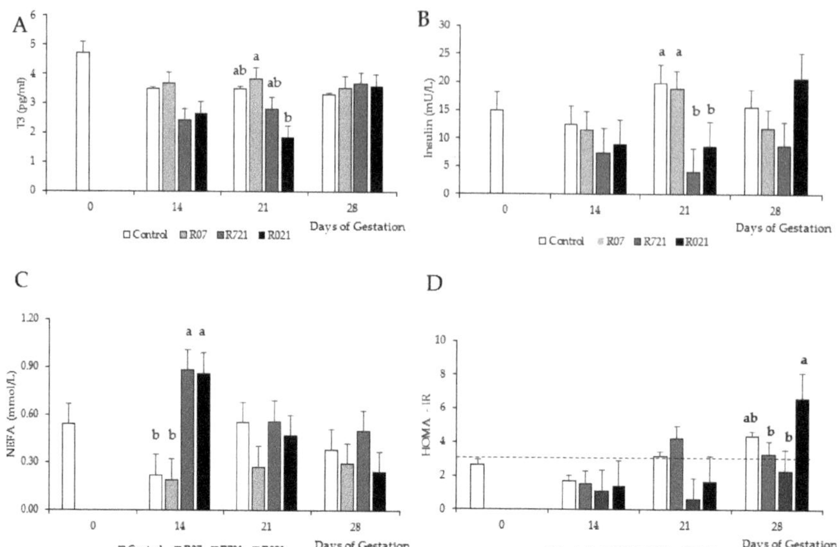

Figure 3. Effect of Maternal Feed Restriction and Time on (**A**) plasma triiodothyronine (T3), (**B**) insulin, (**C**) NEFA concentrations and (**D**) HOMA-IR during pregnancy of rabbits fed ad libitum (Control), restricted to 60% one week from day 0 to 7 (R07), two weeks from day 7 to 21 (R721) and three weeks from day 0 to 21 (R021) of gestation. (a–c): different letters indicate differences between experimental groups on each day of pregnancy ($p < 0.05$). Each bar on day 14, 21 and 28 shows the mean of five does, and on day 0 is the mean of 10 does. The dashed line in (D) represents the value above which HOMA-IR indicates the presence of insulin resistance.

Table 5. Effect of Maternal Feed Restriction (MFR) and Time on endocrine and metabolic parameters during lactation of rabbits fed ad libitum (Control), restricted to 60% one week from day 0 to 7 (R07), two weeks from day 7 to 21 (R721) and three weeks from day 0 to 21 (R021) of gestation.

	Maternal Feed Restriction				Days of Lactation					*p* Value		
Lactation	Control	R07	R721	R021	7	14 [1]	30 [2]					
	n = 5	n = 5	n = 5	n = 5	n = 20	n = 20	(P) n = 20	(NP) n = 20	RMSE	p_{MFR}	p_{Time}	$p_{MFR \times Time}$
T3 (pg/mL)	3.47	3.41	3.48	3.40	3.70 [a]	3.56 [a]	3.54 [a]	2.95 [b]	0.66	0.9830	0.0056	0.0878
T4 (pg/mL)	10.10	12.83	11.26	11.38	7.84	12.45	13.18	12.10	5.76	0.6012	0.0753	0.5611
T3 to T4 ratio	0.57	0.36	0.40	0.72	0.74	0.67	0.33	0.31	0.58	0.2853	0.0847	0.2714
Insulin (mU/L)	16.36	12.09	13.36	14.23	21.48 [a]	12.83 [b]	10.05 [b]	11.68 [b]	7.41	0.4223	0.0006	0.2539
Glucose (mg/dL)	114.33	109.93	109.31	109.51	116.51 [a]	111.28 [ab]	112.09 [ab]	103.20 [b]	12.40	0.6352	0.0199	0.2822
Glucose to Insulin ratio	10.31	21.19	11.05	17.95	6.73 [b]	15.13 [b]	23.34 [a]	15.30 [b]	15.83	0.1555	0.0469	0.0802
HOMA IR	4.73	3.32	3.70	3.95	6.26 [a]	3.58 [b]	2.82 [b]	3.05 [b]	2.17	0.3178	0.0003	0.2801
NEFAS (mmol/L)	0.26	0.28	0.25	0.29	0.24	0.26	0.27	0.30	0.12	0.8113	0.4605	0.2305
Corticosterone (ng/mL)	239.11	234.36	215.89	246.54	151.11 [a]	217.52 [b]	272.84 [c]	294.44 [c]	11.51	0.4158	0.0001	0.2334

HOMA-IR: homeostasis model assessment for insulin resistance. NEFA: no esterified fatty acids. [1] 3rd artificial insemination was carried out on day 14 of lactation. [2] Weaning day. P: pregnant. NP: non-pregnant. RMSE: root mean square error. Values in the same row with different letters are significantly different.

4. Discussion

This work deals with the potential consequences of moderate MFR in different periods (1, 2 or 3 weeks) throughout gestation on reproductive outcome, including fetoplacental development and post-natal growth and survival of litters from primiparous rabbit does, and introduces further insight involving endocrine and metabolic parameters during pregnancy and the following lactation.

After different MFR periods, when fed ad libitum, all pregnant does significantly increased their feed intake. These results match with those observed in our previous studies [9,14,15]. Dams from R721 and R021 groups altogether consumed 1.3 kg and 1.8 kg less feed than the Control group during gestation, respectively. Those in the R07 group reduced their feed intake by about 300 g compared to the Control group. Considering the feed intake of Control females during the second and third weeks as the average regular consumption (227 ± 6.0 g/d and 228.3 ± 6.1 g/d), R721 and R021 does ingested 46% of the food that the Controls did. Consequently, LBW of these does was impaired, showing around a 5.2% decrease in LBW compared to Control or R07.

As expected, the number of corpora lutea assessed in the rabbits on day 28 of gestation was similar in all groups, as MFR was applied right after AI, with no impact on the ovulation rate. Similar results have been described in our previous studies where only Control and R021 treatments were compared [14,15]. Nonetheless, R07 does tended to have a higher number of fetuses, although with worse viability. In rabbits, the implantation and placentation events occur after day 6–7 post-AI [24]. In the current study, at that time R07 females experienced a significant compensatory increase in feed intake, surpassing the Control group. The slight increase in implanted fetuses observed in this group could lead to greater competition between them for uterine space and nutrients, ultimately leading to lower viability [25].

The dimensions of the fetuses were affected by the restriction, so that the longer it was, the more weight and length of the fetuses decreased, although the TD and the OND did not seem to have the same response. MFR usually causes vital organs, such as the brain or the liver, to increase their relative weight to the detriment of other organs [9,26]. This trend was not observed in this study, and other organs such as the kidneys and the liver reduced their weight in the fetuses of more restricted females. Several studies [12,27,28] assert that the longer and greater the feed deprivation is, the greater the probability of finding impaired fetal growth and development. Cappon et al. [29] performed an experiment evaluating MFR levels from slight to severe (110 g/d down to 15 g/d) and from days 7 to 19 of gestation (a period that registers from implantation to closure of the hard palate), recording the effect on fetal development. These authors showed that in severe feed deprivation, the size and weight of the fetuses were reduced, and the incidence of abortions and inadequate ossification increased. These results are in accordance with other studies [27,30] that evaluated the incidence of abortions and impairment of fetal development/growth due to feed restriction, where authors agreed that the main factors to be considered are the degree of intake restriction, its length, and more critically the period of gestation when it is applied [20], highlighting the second half of gestation as that where the energy intake of the mothers is most critical. The findings regarding fetal development comparing organ sizes of the fetuses between experimental groups at day 28 of gestation, suggest that the ration setting and periods of feed restriction applied have no long-term effects over the fetuses and later litter performance. In fact, in these experimental conditions, the brain/liver ratio was the same in all groups, and no critical variations in the relative weights of the fetal organs were observed due to the feed restriction, which supports the idea that the feed restriction applied is to be considered moderate compared to the existing records and does not induce severe impairment in fetal growth nor organogenesis.

In addition, MFR for 1, 2 or 3 weeks affected the growth of the placental structures. The largest placentas were those of R07 does, but this group was one of the least efficient, meaning that even though fetal growth directly depends on the placenta [31], in this group placentas were bigger than their fetuses. However, R721 does showed the smallest placenta because the mothers were restricted in the period when this tissue is developed. These results suggest that although placental efficiency was low in R07 and high in R721, the amount of energy provided during the period when does were fed ad libitum was enough to maintain gestation adequately, since does of these two groups were able to achieve a similar litter size at parturition, a high number of weaned kits (after adjustment), and the lowest lactation mortality rates. In this sense, Rommers et al. [28] described that a moderate decrease in energy intake during the first week of gestation, provided it is above maintenance needs, does not affect implantation rate, while in the fourth week of pregnancy, energy deprivation can generate more problems in the long-term.

Furthermore, during lactation and in the juvenile stage, our previous studies [14,15] that focused on the likely consequences on progeny of a three-week MFR (equal to the current R021 group) evidenced that the surviving offspring did not manifest any outstanding alterations in growth, serum or metabolic parameters nor feed intake during juvenile phase until puberty.

Regarding the effects of MFR on fertility and prolificacy of does, there were no adverse consequences. Although litter weight and milk production were similar between groups during lactation, a significant increase in kit mortality was observed in both the Control and R021 groups. It may be that litter adjustment performed after birth affected these results, but it is difficult to explain. Interestingly, does in group R07, despite previous lower fetal viability detected during pregnancy, had a better weaning outcome and lower kit mortality, outpacing the results of the other groups.

Concerning the mortality rate of the dams during the study (culled due to experimental needs or for other reasons such as low prolificacy, abortions, sudden deaths or mastitis), the data collected did not show significant differences between groups. They were consistent with the experimental farm's historical data, so these casualties can be considered incidental rather than related to the experimental conditions.

As to the effects of MFR on energy metabolism, thyroid hormones were determined to evaluate metabolic adaptation to energy intake deprivation. In the current study, the mothers that underwent a longer MFR presented lower T3 plasma concentrations. During the first three weeks of gestation, a decrease in T3 and T4 was also noted, along with an increase at the end when the females were fed ad libitum. The significant treatment $\times$ time interaction observed in T3 concentrations was due to the R021 group having the lowest T3 values on day 21 of pregnancy (after three continuous weeks of moderate restriction), but by the end of gestation all groups had similar concentrations. The decrease in free T3 during periods of feed restriction usually reduces the basal metabolic rate, resulting in energy savings for the animal [32], but of all metabolism-involved hormones, T3 is the fastest to recover [33]. Thus, in the 4th week of pregnancy and during lactation, when all groups were already fed ad libitum, energy homeostasis was balanced even in the case of T3. T4 plasma concentrations during lactation rose significantly more than those observed during gestation, indicating a total recovery of the energy metabolism after MFR.

The studied relation between insulin and glucose profiles suggests that females in the first half of pregnancy, being subjected to MFR, manage the diet's energy and body reserves. Insulin is an intermediate metabolic hormone, involved in many metabolic pathways to maintain energy homeostasis by coordinating the different axes and systems [34,35]. Furthermore, it should not be forgotten that insulin undergoes postprandial alterations and at day 14 of pregnancy the Control and R07 groups are both being fed ad libitum since day 0 and 7 of pregnancy, respectively. Afterwards, on day 21 of pregnancy, insulin increases and insulin resistance manifests with the rise in HOMA-IR in these groups, which Menchetti et al. [32] described as an adaptation to the energy demand produced by the fetuses. Following the same dynamic, decreased insulin plasma concentrations on day 21 of pregnancy in groups R721 and R021 reinforces the role of insulin in the gear of metabolism, since Buczkowska and Jarosz-Chobot [36] reported, in the same way as other authors [35], that reduced insulin plasma concentrations promotes lipolysis, which makes sense since by day 21 of pregnancy these dams had been restricted for two and three weeks, respectively, and needed to obtain energy from their body reserves. On the other hand, Fortun-Lamothe [37] described that the fetuses request high amounts of glucose at the end of gestation. Therefore, as MFR was over in R721 and R021 groups by the 4th week of gestation, these dams were obtaining glucose directly from the diet and their voluntary feed intake was considerably increased to compensate for the time of restriction, which would explain the glucose rise detected at this point, particularly in group R021, which underwent the longer MFR.

Moreover, HOMA-IR results showed significant differences between groups by the end of gestation (day 28). Particularly, R021 does displayed a critical rise in HOMA-IR values, while other groups still did not exceed the limit to be considered insulin resistant (R07 and R721). This could be due to the continuous consumption of glucose by the fetuses that forces an increase in the mother's hepatic production of glucose by the liver during fasting to maintain plasmatic glucose sufficient to meet the needs of glucose-dependent tissues [38], which would also explain the slight but not significant rise of glucose in group R021 on day 28. Since circulating NEFA concentrations are not high enough in group R021 at this point to consider body reserve mobilization, it can be assumed that the main source of glucose was energy obtained from the diet, which at this time was being provided ad libitum for all groups.

In humans and other species, the second half of pregnancy is defined by a progressive increase in resistance to the action of insulin, so that in the third-trimester insulin sensitivity is about a third of normal and insulin levels are increased by about four times [39,40], which in our study would be equivalent to the 3rd and 4th weeks of gestation. Other authors have also obtained these same results and appreciations in rabbits [41], which concur with the data obtained in this study.

As shown by the data, MFR during gestation affected the overall energetic balance of the does and high NEFA plasma concentrations are a manifestation of body reserves mobilization. In this study, MFR affected plasma NEFA concentrations, and the highest values were observed on day 14 in the mothers with a more prolonged period of restriction at this time. This is in accordance with other studies where authors determined higher NEFA plasma concentrations in does undergoing slight to moderate MFR during similar gestation periods [32,42]. Additionally, lower circulating NEFA concentrations were observed on day 14 of gestation in Control and R07 groups, which supports the previous statement and suggests a redirection of metabolism towards a reduction of lipolysis due to higher energy intake in these groups.

Conductive to the lactation period, the main difference between groups occurred with T3 concentrations in pregnant and non-pregnant females, with lower levels in the latter. Since the third AI was performed on day 14 after parturition, increased T3 in lactating-pregnant dams may be explained by its role in tissue synthesis, particularly in the first half of pregnancy [43]. Moreover, Bober et al. [44] reported that thyroid hormones are considered necessary for cellular metabolism of the mammary gland and energy utilization, which may be an important factor in fetal development during pregnancy and milk biosynthesis, which is relevant for our study since pregnant does are overlapping lactation and 3rd pregnancy. Highlighting the slight insulin and HOMA-IR decrease that occurred on day 14 compared to day 7 post-partum, this could be explained by the stressful event of AI, which implies manipulation of the animals (removing the females from their cages, holding them by the tail, inserting the insemination catheter and injecting GnRH). When stressful situations occur, insulin levels fall, glucagon and epinephrine (adrenaline) levels rise and more glucose is released from the liver [45].

Surprisingly, no differences due to MFR between experimental groups were observed during pregnancy and lactation, something unexpected since corticosterone was assessed as a hormonal measure of stress in the animals. Some authors contend that plasma corticosterone concentrations correspond more with metabolic stress than with behavioral stress [33], which would support the assertion that in this study, MFR is only moderate and implies no metabolic depletion of the dams, since corticosterone was lower by the end of gestation. On the other hand, Menchetti et al. [32] measured cortisol levels during the pregnancies of does subjected to MFR and determined that this parameter was not altered during gestation, which can be attributed to changes in the body's regulation during pregnancy that favor the maintenance of glucocorticoid levels, which is essential for the correct maturation of many fetal organs [46–49].

However, it is interesting to note that at the end of lactation, corticosterone levels are higher than those on day 7 of lactation. This could be explained due to the size of the kits towards the end of lactation. Kits begin to leave the nest at 14–16 days of age [50], and other authors have described that higher housing density could increase corticosterone circulating levels [51]. In these experimental conditions, by the end of lactation there are approximately 10 kits per doe weighing around 500 g, which leave the nest and occupy a lot of space in the cage, try to continue suckling and can increase the mother's level of discomfort and explain the rise in corticosterone. Nonetheless, further studies are needed to corroborate this hypothesis.

5. Conclusions

The present study states that a moderate MFR (60% of the estimated voluntary feed intake) applied during 1, 2 or 3 weeks of gestation followed by ad libitum refeeding confirms previous results regarding compensatory food intake by mothers, which helps to preserve maternal energy homeostasis (thyroid hormones, glucose and insulin) and lipid metabolism (NEFA). This allowed the necessary energy supply to be delivered to the fetuses to conduct fetal growth and development satisfactorily, confirming previously obtained results as no major long-term effects in the early life development of the offspring were observed. Moreover, the compensatory feed intake granted the maintenance of an adequate productive performance in the current pregnancy and showed no effect whatsoever in the amount of milk produced, the viability of litters or the mortality rate of the dams. In addition, the corticosterone concentrations observed throughout pregnancy and lactation seem to indicate that, in our experimental conditions, the moderate MFR applied seems not to induce additional stress in dams along the gestation and lactation periods studied. Moreover, it has proven that this strategy can be profitable for the farmer as overall feed intake was lower in all experimental groups compared to the Control group. In summary, these results reveal that as long as the feed supply covers maintenance needs, moderate MFR is a suitable strategy to reduce production costs, since it implies no adverse effects on the metabolic, energetic and welfare status of the dams, and on pre- and post-natal growth of fetuses and litters, respectively.

Author Contributions: Conceptualization, P.L.L., M.A.-Á., R.M.G.-G. and P.G.R.; data curation, M.R. and P.G.R.; formal analysis, C.F.-P., P.M., M.R. and P.G.R.; funding acquisition, P.L.L., M.A.-Á., R.M.G.-G. and P.G.R.; investigation, C.F.-P., P.M., M.R., N.F.-R., A.S.-R., M.A.-Á., R.M.G.-G. and P.G.R.; writing—original draft, C.F.-P., P.M. and P.G.R.; writing—review and editing, C.F.-P., P.M., M.R., N.F.-R., A.S.-R., P.L.L., M.A.-Á., R.M.G.-G. and P.G.R. All authors have read and agreed to the published version of the manuscript.

Funding: This research was funded by the SPANISH MINISTRY OF ECONOMY AND COMPETITIVENESS (Grant Number: AGL2015-65572 and RTI2018-094404-B).

Institutional Review Board Statement: The study was conducted according to the guidelines of The Declaration of Helsinki and approved by the Ethics Committee of the TECHNICAL UNIVERSITY OF MADRID (protocol code PROEX 302/15 approved in 2015).

Data Availability Statement: The data presented in this study are available on request from the corresponding author.

Acknowledgments: We thank B. Velasco for her assistance with the animal husbandry and maintenance.

Conflicts of Interest: The authors declare no conflict of interest.

References

1. Cartuche, L.; Pascual, M.; Gómez, E.A.; Blasco, A. Economic weights in rabbit meat production. *World Rabbit. Sci.* **2014**, *22*, 165–177. [CrossRef]
2. Theau-Clément, M. Preparation of the rabbit doe to insemination: A review. *World Rabbit. Sci.* **2010**, *15*, 61–80. [CrossRef]
3. Fortun-Lamothe, L.; Rochambeau, H.; Lebas, F.; Tudela, F. Influence of the number of suckling young on reproductive performance in intensively rabbit does. In Proceedings of the 7th World Rabbit Congress, Valencia, Spain, 4–7 July 2000; pp. 125–132.
4. Arias-Álvarez, M.; García-García, M.R.; Rebollar, P.G.; Revuelta, L.; Millán, P.; Lorenzo, P.L. Influence of metabolic status on oocyte quality and follicular characteristics at different post-partum periods in primiparous rabbit does. *Theriogenology* **2009**, *72*, 612–623. [CrossRef] [PubMed]
5. Sakr, O.G.; García-García, R.M.; Arias-Álvarez, M.; Lorenzo, P.L.; Millán, P.; Velasco, B.; Rebollar, P.G. Métodos de sincronización de celo en conejas primiparas lactantes a 25 días post-parto. *Rev. Complut. Cienc. Vet.* **2012**, *6*, 6–13.

6. Partridge, G.G.; Lobley, G.E.; Fordyce, R.A. Energy and nitrogen metabolism of rabbits during pregnancy, lactation, and concurrent pregnancy and lactation. *Br. J. Nutr.* **1986**, *56*, 199–207. [CrossRef]
7. Bauman, D.E.; Currie, W.B. Partitioning of Nutrients During Pregnancy and Lactation: A Review of Mechanisms Involving Homeostasis and Homeorhesis. *J. Dairy Sci.* **1980**, *63*, 1514–1529. [CrossRef]
8. Rommers, J.M.; Kemp, B.; Meijerhof, R.; Noordhuizen, J.P.T.M. The effect of litter size before weaning on subsequent body development, feed intake, and reproductive performance of young rabbit does. *J. Anim. Sci.* **2001**, *79*, 1973–1982. [CrossRef] [PubMed]
9. Lopez-Tello, J.; Arias-Álvarez, M.; Jimenez-Martinez, M.A.; Garcia-Garcia, R.M.; Rodriguez, M.; Gonzalez, P.L.L.; Bermejo-Poza, R.; Gonzalez-Bulnes, A.; Rebollar, P.G. Competition for Materno-Fetal Resource Partitioning in a Rabbit Model of Undernourished Pregnancy. *PLoS ONE* **2017**, *12*, e0169194. [CrossRef]
10. Partridge, G.G.; Daniels, Y.; Fordyce, R.A. The effects of energy intake during pregnancy in doe rabbits on pup birth weight, milk output and maternal body composition change in the ensuing lactation. *J. Agric. Sci.* **1986**, *107*, 697–708. [CrossRef]
11. López-Tello, J.; Arias-Álvarez, M.; Jiménez-Martínez, M.A.; Barbero-fernábdez, A.; García, R.M.; Rodríguez, M.; Lorenzo, P.L.; Torres-Rovira, L.; Astiz, S.; González-Bulnes, A.; et al. The effects of sildenafil citrate and haemodynamics in a rabbit model of intaruterine growth restriction. *Reprod. Fertil. Dev.* **2016**, *29*, 1239–1248.
12. Nafeaa, A.; Ahmed, S.A.E.; Fat Hallah, S. Effect of feed restriction during pregnancy on performance and productivity of New Zealand white rabbit does. *Vet. Med. Int.* **2011**, *2011*, 839737. [CrossRef]
13. Manal, A.F.; Tony, M.A.; Ezzo, O.H. Feed restriction of pregnant nulliparous rabbit does: Consequences on reproductive per-formance and maternal behaviour. *Anim. Reprod. Sci.* **2010**, *120*, 179–186. [CrossRef]
14. Garcia-Garcia, R.; Arias-Alvarez, M.; Millan, P.; Francisco, M.R.; Rodriguez, A.S.; Lorenzo, P.; Rebollar, P. Gestation Food Restriction and Refeeding Compensate Maternal Energy Status and Alleviate Metabolic Consequences in Juvenile Offspring in a Rabbit Model. *Nutrition* **2021**, *13*, 310. [CrossRef]
15. García-García, R.M.; Arias-Álvarez, M.; Rodríguez, M.; Sánchez-Rodríguez, A.; Formoso-Rafferty, N.; Lorenzo, P.L.; Rebollar, P.G. Effects of feed restriction during pregnancy on maternal reproductive outcome, foetal hepatic IGF gene expression and offspring performance in the rabbit. *Animal* **2021**, in press.
16. Mullur, R.; Liu, Y.-Y.; Brent, G.A. Thyroid Hormone Regulation of Metabolism. *Physiol. Rev.* **2014**, *94*, 355–382. [CrossRef] [PubMed]
17. Nakrani, M.N.; Wineland, R.H.; Anjum, F. Physiology, Glucose Metabolism. In *StatPearls*; StatPearls [Internet]; StatPearls Publishing: Treasure Island, FL, USA, 2021.
18. Aronoff, S.L.; Berkowitz, K.; Shreiner, B.; Want, L. Glucose Metabolism and Regulation: Beyond Insulin and Glucagon. *Diabetes Spectr.* **2004**, *17*, 183–190. [CrossRef]
19. Emery, R.S.; Liesman, J.S.; Herdt, T.H. Metabolism of Long Chain Fatty Acids by Ruminant Liver. *J. Nutr.* **1992**, *122*, 832–837. [CrossRef]
20. Brecchia, G.; Menchetti, L.; Cardinali, R.; Polisca, A.; Troisi, A.; Maranesi, M.; Boiti, C. Effects of fasting during pregnancy in rabbit does. In Proceedings of the 10th World Rabbit Congress, Sharm El-Sheikh, Egypt, 3–6 September 2012; pp. 341–345.
21. Webster, J. Animal Welfare: Freedoms, Dominions and "A Life Worth Living". *Animals* **2016**, *6*, 35. [CrossRef]
22. Woodman, D. *Reproductive Hormones In: Laboratory Animal Endocrinology*; Wiley: Hoboken, NJ, USA, 1997; pp. 453–499.
23. Helfenstein, T.; Fonseca, F.A.; Ihara, S.S.; Bottós, J.M.; Moreira, F.T.; Pott, H., Jr.; Farah, M.E.; Martins, M.C.; Izar, M.C. Impaired glucose tolerance plus hyperlip-idaemia induced by diet promotes retina microaneurysms in New Zealand rabbits. *Int. J. Exp. Pathol.* **2011**, *92*, 40–49. [CrossRef] [PubMed]
24. Enders, A.C.; Schlafke, S. Penetration of the uterine epithelium during implantation in the rabbit. *Am. J. Anat.* **1971**, *132*, 219–239. [CrossRef] [PubMed]
25. Argente, M.J.; Santacreu, M.; Climent, A.; Blasco, A. Effects of intrauterine crowding on available uterine space per fetus in rabbits. *Livest. Sci.* **2008**, *114*, 211–219. [CrossRef]
26. Godfrey, K.M.; Barker, D.J.; Aranceta, J.; Serra-Majem, L.; Ribas, L.; Pérez-Rodrigo, C. Fetal programming and adult health. *Public Health Nutr.* **2001**, *4*, 611–624. [CrossRef] [PubMed]
27. Matsuoka, T.; Mizoguchi, Y.; Serizawa, K.; Ishikura, T.; Mizuguchi, H.; Asano, Y. Effects of Stage and Degree of Restricted Feeding on Pregnancy Outcome in Rabbits. *J. Toxicol. Sci.* **2006**, *31*, 169–175. [CrossRef]
28. Rommers, J.M.; Meijerhof, R.; Noordhuizen, J.P.T.M.; Kemp, B. The effect of level of feeding in early gestation on reproductive success in young rabbit does. *Anim. Reprod. Sci.* **2004**, *81*, 151–158. [CrossRef] [PubMed]
29. Cappon, G.; Fleeman, T.; Chapin, R.; Hurtt, M. Effects of feed restriction during organogenesis on embryo-fetal development in rabbit. *Birth Defects Res. Part B Dev. Reprod. Toxicol.* **2005**, *74*, 424–430. [CrossRef]
30. Clark, R.L.; Robertson, R.T.; Peter, C.P.; Bland, J.A.; Nolan, T.E.; Oppenheimer, L.; Bokelman, D.L. Association between adverse maternal and embryo-fetal effects in norfloxacin-treated and food-deprived rabbits. *Fundam. Appl. Toxicol.* **1986**, *7*, 272–286. [CrossRef]
31. Robinson, J.S.; Owens, J.A.; Owens, P.C. *Fetal Growth and Growth Retardation. Textbook of Fetal Physiology*; Thorburn, G.D., Harding, R., Eds.; Oxford University Press: New York, NY, USA, 1994; pp. 83–94.

32. Menchetti, L.; Brecchia, G.; Canali, C.; Cardinali, R.; Polisca, A.; Zerani, M.; Boiti, C. Food restriction during pregnancy in rabbits: Effects on hormones and metabolites involved in energy homeostasis and metabolic programming. *Res. Vet.-Sci.* **2015**, *98*, 7–12. [CrossRef] [PubMed]
33. Brecchia, G.; Bonanno, A.; Galeati, G.; Federici, C.; Maranesi, M.; Gobbetti, A.; Zerani, M.; Boiti, C. Hormonal and metabolic adaptation to fasting: Effects on the hypothalamic–pituitary–ovarian axis and reproductive performance of rabbit does. *Domest. Anim. Endocrinol.* **2006**, *31*, 105–122. [CrossRef]
34. Hornick, J.; Van Eenaeme, C.; Gérard, O.; Dufrasne, I.; Istasse, L. Mechanisms of reduced and compensatory growth. *Domest. Anim. Endocrinol.* **2000**, *19*, 121–132. [CrossRef]
35. Rommers, J.M.; Boiti, C.; Brecchia, G.; Meijerhof, R.; Noordhuizen, J.P.T.M.; Decuypere, E.; Kemp, B. Metabolic adaptation and hormonal regulation in young rabbit does during long-term caloric restriction and subsequent compensatory growth. *Anim. Sci.* **2004**, *79*, 255–264. [CrossRef]
36. Buczkowska, E.O.; Jarosz-Chobot, P. Insulin effect on metabolism in skeletal muscles and the role of muscles in regulation of glucose homeostasis. *Przegl. Lek.* **2001**, *58*, 782–787.
37. Fortun-Lamothe, L. Energy balance and reproductive performance in rabbit does. *Anim. Reprod. Sci.* **2006**, *93*, 1–15. [CrossRef]
38. Garcia-Benasach, F. Diabetes Gestacional Análisis de la Influencia de Parámetros Clínicos y Ecográficos en los Resultados pe-rinatales. Ph.D. Thesis, Autonomic University of Madrid, Madrid, Spain, 2012.
39. Buchanan, T.A.; Metzger, B.E.; Freinkel, N.; Bergman, R.N. Insulin sensitivity and B-cell responsiveness to glucose during late pregnancy in lean and moderately obese women with normal glucose tolerance or mild gestational diabetes. *Am. J. Obstet. Gynecol.* **1990**, *162*, 1008–1014. [CrossRef]
40. Père, M.-C.; Etienne, M. Insulin sensitivity during pregnancy, lactation, and post-weaning in primiparous gilts1. *J. Anim. Sci.* **2007**, *85*, 101–110. [CrossRef] [PubMed]
41. Menchetti, L.; Andoni, E.; Barbato, O.; Canali, C.; Quattrone, A.; Vigo, D.; Codini, M.; Curone, G.; Brecchia, G. Energy homeostasis in rabbit does during pregnancy and pseudopregnancy. *Anim. Reprod. Sci.* **2020**, *218*, 106505. [CrossRef]
42. Martínez-Paredes, E.; Ródenas, L.; Martínez-Vallespín, B.; Cervera, C.; Blas, E.; Brecchia, G.; Boiti, C.; Pascual, J.J. Effects of feeding programme on the performance and energy balance of nulliparous rabbit does. *Animals* **2012**, *6*, 1086–1095. [CrossRef] [PubMed]
43. Habeeb, A.A.; El-Masry, K.A. Hormonal pattern in pregnant rabbits and some productive aspects as affected by litter size at birth. 1991. *Egypt J. Poult. Sci.* **1991**, *1*, 429.
44. Bober, M.A.; Dollah, A.; Becker, B.A.; Johnson, H.D. The influence of exogenous T3 on plasma prolactin, T3, TSH and milk production during heat stress in Holstein cows. *J. Anim. Sci.* **1980**, *55*, 339.
45. Ganong, W.F. The Stress Response—A Dynamic Overview. *Hosp. Pract.* **1988**, *23*, 155–171. [CrossRef]
46. Brunton, P.J.; Russell, J.A.; Douglas, A.J. Adaptive responses of the maternal hypothalamic-pituitary-adrenal axis during pregnancy and lactation. *J. Neuroendocrinol.* **2008**, *20*, 764–776. [CrossRef]
47. Entringer, S.; Buss, C.; Shirtcliff, E.A.; Cammack, A.; Yim, I.S.; Chicz-DeMet, A.; Sandman, C.A.; Wadhwa, P.D. Attenuation of maternal psychophysiological stress responses and the maternal cortisol awakening response over the course of human pregnancy. *Stress* **2009**, *13*, 258–268. [CrossRef] [PubMed]
48. Tamashiro, K.L.; Moran, T.H. Perinatal environment and its influences on metabolic programming of offspring. *Physiol. Behav.* **2010**, *100*, 560–566. [CrossRef]
49. Mastorakos, G.; Ilias, I. Maternal and Fetal Hypothalamic-Pituitary-Adrenal Axes During Pregnancy and Postpartum. *Ann. N. Y. Acad. Sci.* **2003**, *997*, 136–149. [CrossRef] [PubMed]
50. Szendrő, Z.; Trocino, A.; Hoy, S.; Xiccato, G.; Villagrá, A.; Maertens, L. A review of recent research outcomes on the housing of farmed domestic rabbits: Reproducing does. *World Rabbit. Sci.* **2019**, *27*, 1–14. [CrossRef]
51. Kalaba, Z.M. Physiological Response and Stress Indicators of California Rabbits under Intensive Conditions in Egypt. *Asian J. Poult. Sci.* **2012**, *6*, 65–78. [CrossRef]

Article

Impact of Goji Berries (*Lycium barbarum*) Supplementation on the Energy Homeostasis of Rabbit Does: Uni- and Multivariate Approach

Laura Menchetti [1,2], Giulio Curone [3], Egon Andoni [4], Olimpia Barbato [2,*], Alessandro Troisi [5], Bernard Fioretti [6], Angela Polisca [2], Michela Codini [7], Claudio Canali [2], Daniele Vigo [3] and Gabriele Brecchia [3,*]

1 Department of Agricultural and Agri-food Sciences and Technologies, University of Bologna, Viale Fanin 46, 40138 Bologna, Italy; laura.menchetti7@gmail.com
2 Department of Veterinary Medicine, University of Perugia, Via San Costanzo 4, 06126 Perugia, Italy; angela.polisca@unipg.it (A.P.); claudio.canali@unipg.it (C.C.)
3 Department of Veterinary Medicine, University of Milano, Via dell'Università 6, 26900 Lodi, Italy; giulio.curone@unimi.it (G.C.); daniele.vigo@unimi.it (D.V.)
4 Faculty of Veterinary Medicine, Agricultural University of Albania, Rr Paisi Vodica, Koder, 1029 Kamez, Albania; eandoni@ubt.edu.al
5 School of Biosciences and Veterinary Medicine, University of Camerino, Via 9 Circonvallazione 93/95-62024 Matelica, Italy; alessandro.troisi75@gmail.com
6 Department of Chemistry, Biology and Biotechnologies, University of Perugia, Via Elce di Sotto 8, 06123 Perugia, Italy; bernard.fioretti@unipg.it
7 Department of Pharmaceutical Sciences, University of Perugia, Via A. Fabretti 48, 06123 Perugia, Italy; michela.codini@unipg.it
* Correspondence: olimpia.barbato@unipg.it (O.B.); gabriele.brecchia@unimi.it (G.B.)

Received: 12 October 2020; Accepted: 28 October 2020; Published: 30 October 2020

Simple Summary: The energy balance during the reproduction cycle is a problematic issue for livestock species because it has consequences not only on animal welfare but also on the profitability of the farm. The adoption of new nutritional strategies could improve both of these aspects. In the present study, the supplementation with goji berries was proposed and evaluated on the rabbit, which is both a livestock animal and a useful animal model. Goji berry is the fruit of *Lycium barbarum* that is a natural resource made up of several compounds with biological activities and their consumption could be beneficial for the health and the general well-being of humans and animals. Its effect on several hormones and metabolites involved on energy balance of rabbit doe were evaluated by using both uni- and multivariate approach. Our finding, in addition to describing the intricate relationships between body conditions, hormones and metabolites during pregnancy and lactation, suggested that the supplementation with goji berry in the rabbit diet at low percentage could improve some aspects of energy metabolism and, in particular, doe's insulin sensitivity. Conversely, the intake of high doses of goji raises concerns due to the risk of excessive fattening and worsening of insulin resistance.

Abstract: This study examined the effects of goji berries dietary supplementation on the energetic metabolism of doe. Thirty days before artificial insemination, 75 New Zealand White does were assigned to three different diets: commercial standard diet (C) and supplemented with 1% (LG) and 3% (HG) of goji berries, respectively. Body conditions, hormones and metabolites were monitored until weaning. Body weight and BCS were higher in HG than C ($p < 0.05$). LG showed lower T3/T4 ratio and cortisol concentrations ($p < 0.05$) and tended to have lower indices of insulin resistances ($p < 0.1$) than HG. Compared to control, leptin was higher in HG at AI ($p < 0.01$) and in LG during lactation ($p < 0.05$). Two principal components were extracted by multivariate analysis describing the relationships between (1) non-esterified fatty acids, insulin and glucose levels, and (2) body conditions and leptin metabolism. The first component highlighted the energy deficit and the insulin resistance of the does during pregnancy and lactation. The second one showed that leptin, body weight and Body Condition Score (BCS) enhance as levels of goji berries in the diet increase. Thus, the effects of goji supplementation are dose-dependent: an improvement on energy metabolism was achieved with a low-dose while the highest dose could determine excessive fattening and insulin resistance in does.

Keywords: goji berries; rabbit; insulin resistance; leptin; non-esterified fatty acids; pregnancy; lactation; body condition score; principal component analysis

---

1. Introduction

Recently, research has turned towards natural products, rich in compounds with high biological activity that have favourable effects on health with low production expenses and reduced side effects for the prevention and treatment of various pathologies. In this contest, goji berry, the fruit of *Lycium barbarum* L., consumed in several Asian countries for a long time as a traditional tonic food and natural herbal remedy, has attracted a lot of attention also in Western countries in the last decades [1,2]. In fact, there are a lot of evidences showing that the berry has numerous potential beneficial effects for the general well-being of the individual and for the prevention and treatment of numerous pathologies [1–4]. Beside several biological active compounds such as carotenoids, vitamins (riboflavin, thiamin and ascorbic acid) and flavonoids, goji berry is primarily rich in polysaccharides [1] which are responsible for the main beneficial pharmacological effects of the fruit both in vitro [5–8] and in vivo in various laboratory animal species [9,10] and in clinical trials in humans [11,12]. The effects of the goji berries are mainly studied in laboratory animals such as mice and rats [9,10,13] and only a few trials were conducted using the rabbit [14–16] although it is considered a useful experimental animal model [17–22]. Moreover, only a limited number of researches evaluated the effects of goji berry on the reproductive and productive performance, other than on the quality of meat, in livestock animals, rabbits included [15,16,23–25].

During pregnancy, the energy demands increase to favor the growth of fetuses and prepare the mother to the subsequent lactation. For these reasons, some changes in the energy homeostasis are necessary. In fact, during this physiological status, the pregnant animal enhances the feed intake and the mobilization of the body reserves as an adaption mechanisms to the modification in energetic request [26–28]. Different hormones (insulin, leptin, T3, T4 and cortisol) and metabolites such as glucose and non-esterified fatty acids (NEFA) are implicated in maintaining energy homeostasis during pregnancy in different species [29–32]. To our knowledge, no studies have evaluated the impact of the goji berries enriched diet on the hormonal control of the energetic homeostasis during pregnancy of the rabbit does.

Therefore, the aims of this research were to evaluate the effects of the integration of the rabbit diet with goji berries at two different concentrations, 1% and 3%, on the body conditions as well as on the levels of several metabolic hormones and metabolites involved in energy homeostasis in pregnant and lactating rabbits does. First, the patterns of these parameters in the three groups was assessed individually using a univariate approach. Subsequently, a multivariate approach was used to identify the main hormonal and metabolic profiles and the effects of goji berries inclusion in the feed on these frameworks.

2. Materials and Methods

The study was performed at the facilities of the Faculty of Veterinary Medicine, of the Agricultural University of Tirana, Albania. The experimental protocol was in agreement with the national rules on the use of livestock animal for experimental and other scientific purposes, their handling, protection, and welfare. Every effort has been made to reduce animal discomfit and to use only the number of animals sufficient to produce valid results.

2.1. Animals and Experimental Design

Seventy-five New Zealand White nulliparous does were maintained in single cage in controlled environmental conditions: temperature ranged from +18 to +23 °C, relative humidity being from 60% to 75%, and the lighting schedule of 16 L:8 D. Rabbit does, on the base of a random chose, were assigned to three groups depending by the diet administered (n = 25/group): Control (C), 1% goji (LG) and 3% goji (HG). Does were fed with commercial feed (control) or the same feed integrated with 1% and 3% of goji berries in LG and HG, respectively (Table 1 [16]). The animals were undergoing artificial insemination (AI) after 30 days of nutritional adaptation to the experimental dietary regime. During this period, they received 150 g/d of feed, while after AI, the rabbits had ad libitum access to feed. Water was always administered ad libitum. An injection of 0.8 µg of synthetic GnRH (Receptal, Hoechst-Roussel Vet, Milan, Italy) just before AI was used to induce the ovulation [33]. AI was executed with 0.5 mL of diluted fresh semen. Pregnancy was determined by abdominal palpation 10 days after AI. Eleven does for each group were monitored during pregnancy and lactation until weaning. Lactation was controlled, the does had access to the nest once a day for ten minutes until day 18 post-partum. The nest was open on the 19th day of lactation. Weaning occurred at 35 days of age separating the little rabbits from their mothers.

Table 1. Formulation and chemical composition (as fed) of control (C) and experimental diets supplemented with goji berries. LG = diet supplemented with 1% of goji berries; HG = diet supplemented with 3% of goji berries.

Parameter	Unit	Diet		
		C	LG	HG
Ingredients				
Wheat bran	%	30.0	29.5	29.0
Dehydrated alfalfa meal	%	42.0	41.5	41.0
Barley	%	9.5	9.5	9.0
Sunflower meal	%	4.5	4.5	4.2
Rice bran	%	4.0	4.0	3.9
Soybean meal	%	4.0	4.0	3.9
Calcium carbonate	%	2.2	2.2	2.2
Cane molasses	%	2.0	2.0	2.0
Vitamin-mineral premix *	%	0.4	0.4	0.4
Soybean oil	%	0.4	0.4	0.4
Salt	%	0.3	0.3	0.3
Goji berries	%	-	1.0	3.0
Analytical data				
Crude Protein	%	15.74	15.64	15.66
Ether extract	%	2.25	2.23	2.47
Ash	%	9.28	9.36	9.25
Starch	%	16.86	17.07	16.99
NDF	%	38.05	38.55	37.49
ADF	%	19.54	19.60	19.01
ADL	%	4.01	4.31	3.98
Digestible Energy **	Kcal/kg	2464	2463	2459

* Per kg diet: vitamin A 11,000 IU; vitamin D3 2000 IU; vitamin B1 2.5 mg; vitamin B2 4 mg; vitamin B6 1.25 mg; vitamin B12 0.01 mg; alpha-tocopherol acetate 50 mg; biotine 0.06 mg; vitamin K 2.5 mg; niacin 15 mg; folic acid 0.30 mg; D-pantothenic acid 10 mg; choline 600 mg; Mn 60 mg; Fe 50 mg; Zn 15 mg; I 0.5 mg; Co 0.5 mg. ** Estimated by Maertens et al. [34].

2.2. Body Conditions

During the experimental period, the feed intake was registered daily while body weight (BW) and body condition score (BCS) were recorded weekly, between 7:30 and 9:00 AM, from time 0 (before the administration of the three experimental diets) until day 35 post-partum (weaning). Weights and BCS were also measured on the day of the IA. The aggregated BCS (from 0 to 4) was obtained by summing the score (0–2) estimated for the rump and the loin [28].

2.3. Hormone and Metabolite Assays

Blood samples were collected to evaluate the metabolic hormones (insulin, leptin, T3, T4 and cortisol) and metabolites (glucose and NEFA) concentrations. These were withdrawn at the time 0 (basal), at AI, post-partum, 20th day of lactation (top lactation), and at the post-weaning. The samples were extracted from the central ear vein into tubes containing EDTA, and instantly centrifuged at 3000× *g* for 20 min; subsequently, plasma was frozen and stored until evaluated for hormones and metabolites. Radioimmunological procedures (RIA) were used to evaluate plasma cortisol, insulin, leptin, triiodothyronine (T3) and thyroxine (T4) concentrations as previously reported [26,27,33]. Leptin concentrations were determined by using the multispecies leptin kit based on a double antibody RIA method (Linco Research Inc., St. Charles, MO, USA). The intra- and inter-assay coefficients of variations were 3.4% and 8.7%, respectively while the limit of sensitivity was 1.0 ng/mL.

A porcine insulin RIA kit was used to quantify the plasma levels of insulin making use of double antibody/PEG technique (Linco Research Inc., Saint Charles, MO, USA). A purified recombinant human insulin was used to produce the standard curve as well as labelled antigen while the antibody against the porcine insulin (antiserum) was produced in the guinea pig. The minimum level of insulin detected by the kit was 2 μU/mL. The coefficients of variations were 6.8% for the intra-assay and 9.2% for the interassay.

Thyroid hormones (T3 and T4) were assayed using the procedure reported by the manufacturer on the brochure of the kit (Immunotech, Prague, Czech Republic). The sensitiveness of the analysis was 0.26 nmol/L for T3 and 10.63 nmol/L for T4. The coefficients of variations of T3 were 6.3% for the intraassay and 7.7%, inter-assay; whereas, for T4 were 3.29% for the intraassay and 7.53% for the inter-assay.

CORT kit was used to determine the cortisol plasma concentrations (Immunotech, Prague, Czech Republic). The lower level of cortisol detected by the kit was 2.5 nM and intra-assay and inter-assay coefficients of variations were 5.8% and 9.2%, respectively.

The metabolites glucose and NEFA were assayed by colorimetric methods according to Menchetti et al. [31], and García-García et al. [35] respectively. The NEFA plasma concentrations were evaluated using a colorimetric assay from Wako that use two enzymatic reactions (NEFA-C, Wako Chemicals GmbH, Neuss, Germany), and based on the ability of NEFA to acylate coenzyme A in the presence of CoA synthetase.

A glucose Infinity kit from Sigma (Sigma Diagnostic Inc., St. Louis, MO, USA), that take advantage of the glucose oxidase method, was used to assess the glucose plasma levels.

Finally, the homeostasis model assessment for insulin resistance (HOMA) was used as index of insulin-resistance and it was calculated using a formula previously reported [27,31]: [insulin concentration × (glucose concentration/18)]/22.5. Insulin resistance (low insulin sensibility) is evident when the HOMA-IR is high while low HOMA-IR values are expression of high insulin sensitivity.

2.4. Statistical Considerations and Data Analysis

Statistical analyses were performed with SPSS Statistics version 25 (IBM, SPSS Inc., Chicago, IL, USA). We defined $p \leq 0.05$ as significant and a p value between 0.1 and 0.05 as a trend.

First, data were analysed by a univariate approach using the Linear Mixed Model. The models evaluated the effects of the Group (three levels: C, LG or HG), Time (as days, depending on the dependent variable) and Group × Time interaction. The Time was included as a repeated factor. The Sidak method was used for multiple comparisons. We verified assumptions and outliers by using diagnostic graphics. Results were expressed as estimated marginal means ± standard error (SE) but raw data were presented in figures.

Then, the principal component analysis (PCA) was used to describe relationships between body condition, hormones, and metabolites and convert the numerous variables into a few profiles describing the hormonal and metabolic framework. The very low or very high correlations between variables were checked using correlation matrices in order to identify suitable variables for the PCA [36–38]. Moreover, the Determinant of the correlation matrix and the Bartlett's test of Sphericity were also used. The sampling adequacy was tested using a Kaiser–Meyer–Olkin (KMO) value >0.6 as index of adequate factorability. The number of components (PCs) to retain was chosen according to the Kaiser's criterion (eigenvalues > 1) and Varimax rotation was applied. The interpretation of PCs was mainly based on items having loadings greater than |0.4|. Then, the scores were calculated for each PC using regression techniques to create two new variables (PC1 and PC2) as linear combinations of the indices of body conditions, hormones, and metabolites.

Finally, the effect of percentage of goji berries inclusion on these PCs was evaluated using Generalized Linear Models (GLMs). Normal and Identity were set as the probability distribution and the link function, respectively, while the Time was included as Within-Subject Effects. In the GLMs, the two PCs were evaluated separately as dependent variables and the percentage of goji berries inclusion was included as predictor [16]. In addition, the number of days of supplementation was included as covariate. Results were expressed as b coefficient with SE and p value from Wald Chi-square statistics [16,39].

The G*Power® software was used to calculate the achieved power [40]. An F test was chosen for this power analysis by setting $\alpha = 0.05$ and the effect size = 0.25 (medium effect size) [41]. For the 3 experimental groups (C, LG or HG) and the five repeated measures (basal, AI, post-partum, 20th day of lactation, and post-weaning), a power of 90.5% was reached using a total sample size of 33 animals.

3. Results

3.1. Univariable Approach

3.1.1. Body Conditions

BW was influenced by time (F = 9.59; $p < 0.001$) and group (F = 5.43; $p = 0.005$). In supplemented rabbits, the BW increase was early, starting from the first week of pregnancy in LG (4044 ± 128 g; $p < 0.05$) and from AI in HG (4054 ± 88 g; $p < 0.01$) groups. After pregnancy, it returned at baseline values after one week of lactation in C (3969 ± 151 g) and LG (4000 ± 128 g) while, in the HG group, it remained elevated up to 20 d of lactation (4220 ± 139 g) (Figure 1).

The aggregate BCS showed significant effects for time (F = 9.51; $p < 0.001$), group (F = 4.19; $p = 0.016$) and their interaction (F = 1.82; $p = 0.007$). Marginal means were 1.1 ± 0.1, 1.3 ± 0.1, and 1.5 ± 0.1 in C, LG, and HG, respectively. HG group showed higher BCS values than C starting from week 3 of supplementation, ($p < 0.001$; Figure 2). From mid-pregnancy onwards, the differences were no longer significant, perhaps also due to the high variability of the data.

Food intake did not differ between groups during pregnancy (144 ± 2 g/d, 146 ± 2 g/d, and 148 ± 3 g/d for C, LG, and HG, respectively; F = 0.78; $p = 0.459$) while it was greater in the control group than supplemented rabbits during lactation (320 ± 2 g/d, 301 ± 2 g/d, and 287 ± 3 g/d for C, LG, and HG, respectively; F = 42.00; $p < 0.001$).

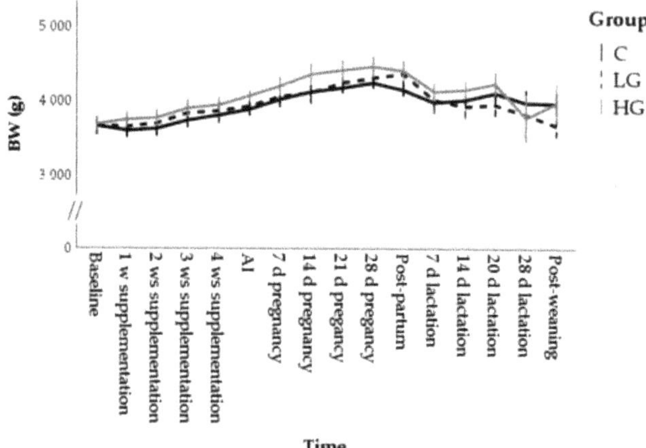

Figure 1. Changes in the body weight (BW) during nutrition adaptation and productive cycle of rabbit does. C = control diet; LG = diet supplemented with 1% of goji berries; HG = diet supplemented with 3% of goji berries. AI = artificial insemination.

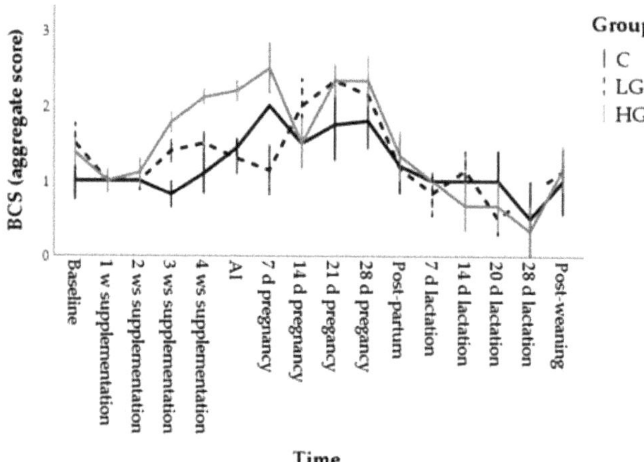

Figure 2. Changes in the body condition score (BCS) during nutrition adaptation and productive cycle of rabbit does. C = control diet; LG = diet supplemented with 1% of goji berries; HG = diet supplemented with 3% of goji berries. AI = artificial insemination.

3.1.2. Hormone and Metabolite

Overall, insulin concentrations reduced from baseline (6.4 ± 0.4 µU/mL) to post-partum (3.7 ± 0.6 µU/mL; F = 6.71; $p < 0.001$). They increased during lactation (6.5 ± 0.6) but, at weaning, lower concentrations than baseline were found (3.7 ± 0.8 µU/mL; $p = 0.003$). With regards the group effect, only a trend was found at post-partum, when values of LG tended to be lower than HG ($p = 0.081$; Figure 3).

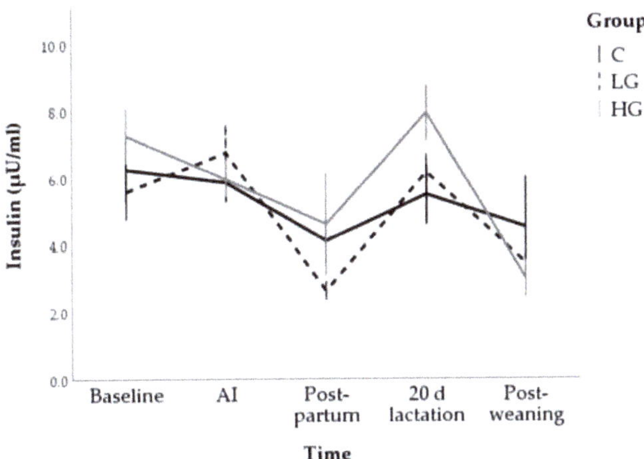

Figure 3. Changes in insulin concentrations during nutrition adaptation and productive cycle of rabbit does. C = control diet; LG = diet supplemented with 1% of goji berries; HG = diet supplemented with 3% of goji berries. AI = artificial insemination.

The HG group showed the highest leptin concentrations at AI (2.9 ± 0.1 ng/mL; $p < 0.01$). Subsequently the concentrations decreased ($p < 0.05$) and returned to baseline values after 20 d of lactation (2.5 ± 0.3 ng/mL). Constant values over time were found in C group (marginal mean: 2.3 ± 0.1 ng/mL) while LG showed greater leptin concentrations during lactation (2.7 ± 0.2 ng/mL) compared to baseline (2.2 ± 0.1 ng/mL; $p = 0.013$; Figure 4).

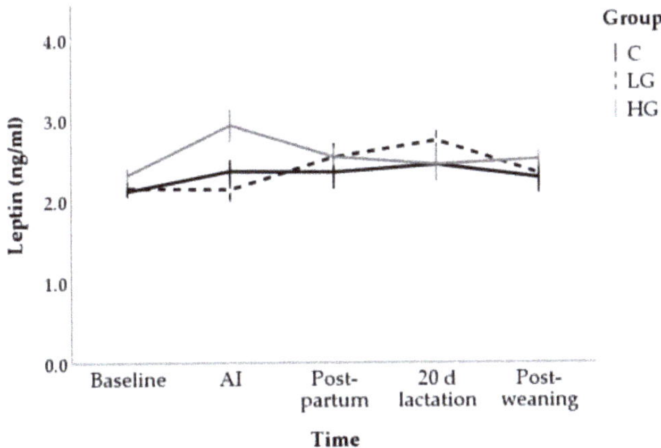

Figure 4. Changes in leptin concentrations during nutrition adaptation and productive cycle of rabbit does. C = control diet; LG = diet supplemented with 1% of goji berries; HG = diet supplemented with 3% of goji berries. AI = artificial insemination.

A progressive reduction until postpartum was observed in T3 concentrations (F = 9.17; $p < 0.001$). This reduction was more marked in the LG group which showed lower marginal means (1.9 ± 0.1 nmol/L and 1.6 ± 0.1 nmol/L in C and LG groups; respectively; $p < 0.05$) and lower postpartum values (1.7 ± 0.2 nmol/L and 1.0 ± 0.2 in C and LG groups, respectively; $p < 0.05$) than the control. After weaning, baseline values were found for both C and LG while HG showed lower T3 concentration (2.4 ± 0.2 nmol/L and 1.3 ± 0.3 nmol/L at time 0 and post weaning, respectively; $p < 0.01$; Figure 5).

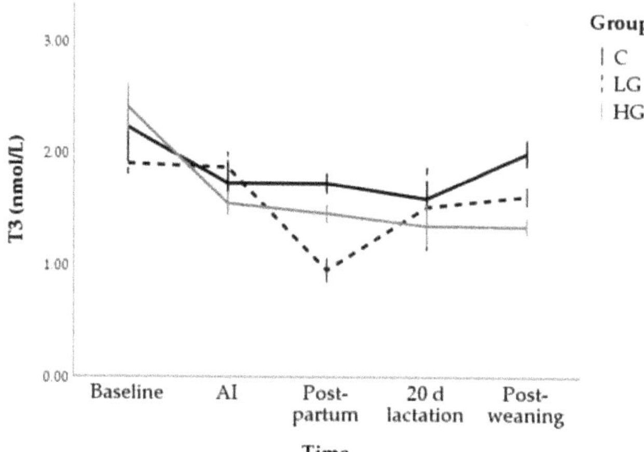

Figure 5. Changes in T3 concentrations during nutrition adaptation and productive cycle of rabbit does. C = control diet; LG = diet supplemented with 1% of goji berries; HG = diet supplemented with 3% of goji berries. AI = artificial insemination.

Only changes over time ($F = 5.97$; $p < 0.001$) were found for T4 which decreased until post-partum (from 51.8 ± 1.9 nmol/L at T0 to 38.8 ± 2.6 nmol/L at post-partum; $p < 0.001$); subsequently, it increased (46.5 ± 0.2 nmol/L at day 20 of lactation) but its values on the last time, (42.9 ± 3.8 nmol/L) were lower than in the first sampling time ($p < 0.05$). No group-related differences were significant ($F = 0.75$; $p = 0.475$; Figure 6).

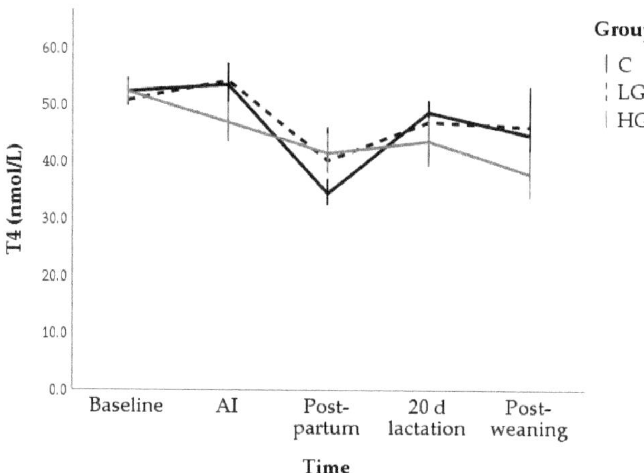

Figure 6. Changes in T4 concentrations during nutrition adaptation and productive cycle of rabbit does. C = control diet; LG = diet supplemented with 1% of goji berries; HG = diet supplemented with 3% of goji berries. AI = artificial insemination.

Changes over time in T3/T4 ratio were only found for the Control group ($p = 0.043$) which showed significantly higher values than the LG group at the postpartum (0.034 ± 0.003 and 0.029 ± 0.005 for C and LG, respectively; $p = 0.010$; Figure 7).

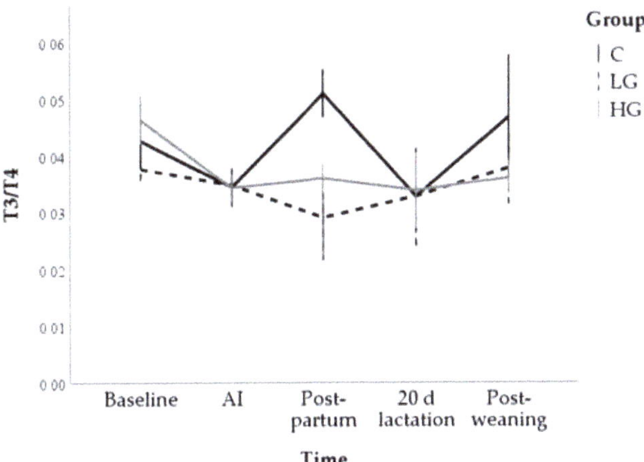

Figure 7. Changes in T3/T4 ratio during nutrition adaptation and productive cycle of rabbit does. C = control diet; LG = diet supplemented with 1% of goji berries; HG = diet supplemented with 3% of goji berries. AI = artificial insemination.

Cortisol concentrations progressively decreased in all groups, from 334.3 ± 7.4 nmol/L at baseline to 258 ± 14.4 nmol/L at post-weaning ($F = 12.63$; $p < 0.001$). The LG group (272.5 ± 7.0 nmol/L; $p < 0.01$) had, on average, the lowest concentrations of cortisol but no multiple comparisons reached statistical significance (Figure 8).

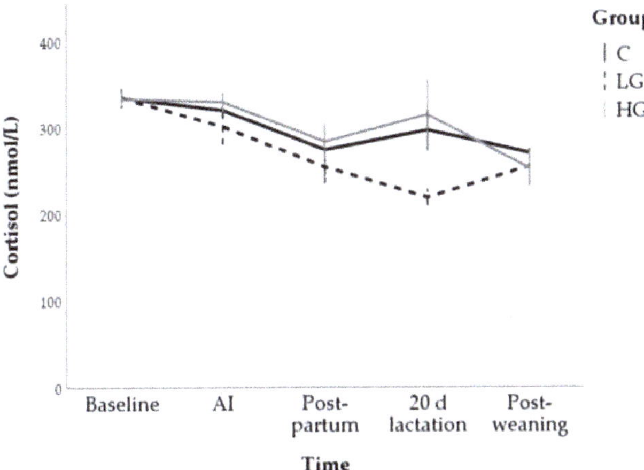

Figure 8. Changes in cortisol concentrations during nutrition adaptation and productive cycle of rabbit does. C = control diet; LG = diet supplemented with 1% of goji berries; HG = diet supplemented with 3% of goji berries. AI = artificial insemination.

No change over time was observed in the control group while a reduction in glucose concentrations compared to baseline (5.7 ± 0.4 mmol/L) was observed post-partum (4.2 ± 0.4 mmol/L; $p < 0.001$) and during lactation (4.3 ± 0.4 mmol/L; $p < 0.001$) in the LG group. Moreover, marginal means of LG group (5.5 ± 0.2 mmol/L) was lower than C (5.8 ± 0.2 mmol/L; $p < 0.05$; Figure 9).

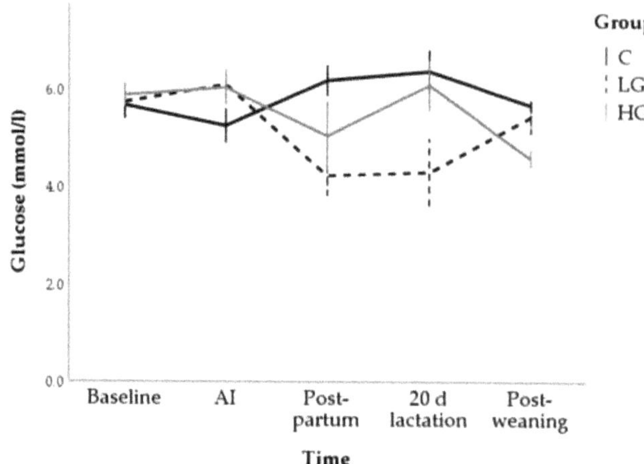

Figure 9. Changes in glucose concentrations during nutrition adaptation and productive cycle of rabbit does. C = control diet; LG = diet supplemented with 1% of goji berries; HG = diet supplemented with 3% of goji berries. AI = artificial insemination.

Conversely, NEFA showed significant changes over time (F = 30.67; $p < 0.001$) with no difference between groups (F = 1.81; $p = 0.167$). In particular, greater values on their concentration were found at post-partum (0.33 ± 0.02 mmol/L; $p < 0.001$) and post-weaning (0.37 ± 0.03; $p < 0.001$) compared to baseline (0.14 ± 0.02 mmol/L; Figure 10).

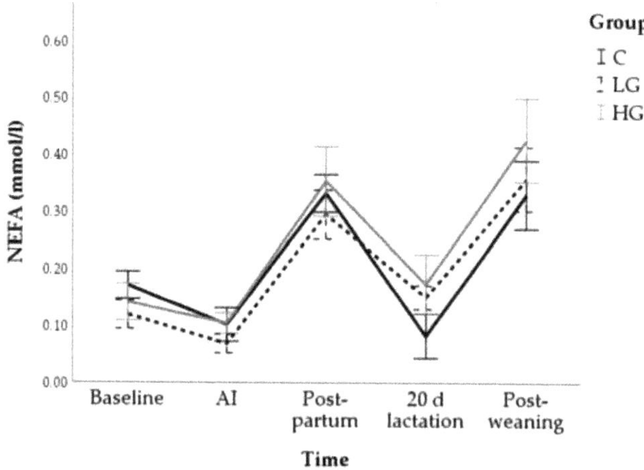

Figure 10. Changes in non-esterified fatty acids (NEFA) concentrations during nutrition adaptation and productive cycle of rabbit does. C = control diet; LG = diet supplemented with 1% of goji berries; HG = diet supplemented with 3% of goji berries. AI = artificial insemination.

Overall, a reduction in HOMA was found at post-partum (0.05 ± 0.01; $p < 0.01$) and post-weaning (0.05 ± 0.01; $p < 0.01$) compared to baseline values (0.09 ± 0.01). This index tended to be lower in LG (0.06 ± 0.01) than HG (0.08 ± 0.01; $p = 0.063$) and, in particular, lower values were found in the post-partum ($p = 0.072$) and at day 20 of lactation ($p = 0.030$; Figure 11).

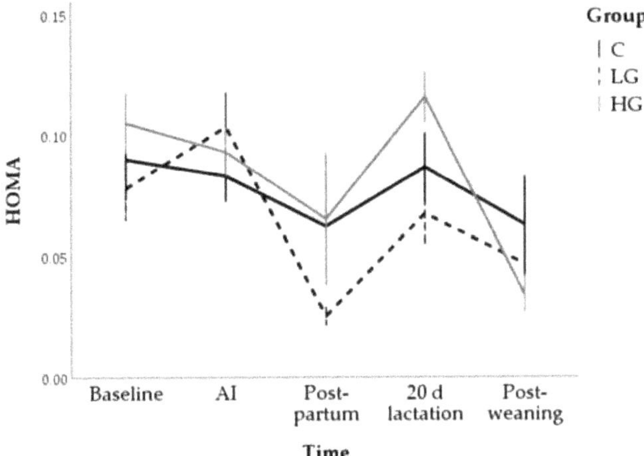

Figure 11. Changes in homeostasis model assessment for insulin resistance index (HOMA) during nutrition adaptation and productive cycle of rabbit does. C = control diet; LG = diet supplemented with 1% of goji berries; HG = diet supplemented with 3% of goji berries. AI = artificial insemination.

3.2. Multivariable Approach

According to the correlation matrix, all the variables indicating body condition and hormonal and metabolic profiles were included in the PCA except for HOMA (Table 2). The Bartlett's Test of Sphericity ($p < 0.001$) indicated that this dataset was suitable for a data reduction technique and the KMO (=0.70) confirmed the sampling adequacy of the PCA. The PCA extracted two PCs, which together account for 46.1% of the variance (Table 2). The PC1 was a bipolar component including NEFA with negative loading and insulin, cortisol, glucose and thyroid hormones with positive loadings. In the PC2, the highest loadings were found for leptin and body condition indices, BW and BCS (Table 2 and Figure 12).

Table 2. Loadings of factors extracted with the principal component analysis.

Item	Component	
	PC1	PC2
NEFA	**−0.744**	−0.051
Insulin	**0.729**	0.054
Cortisol	**0.658**	−0.075
Glucose	**0.616**	0.257
T3	**0.575**	−0.316
T4	**0.414**	−0.399
BW	−0.046	**0.741**
Leptin	−0.052	**0.725**
BCS	0.291	**0.499**
% Variance explained	27.7%	18.4%
Cumulative % variance explained	46.1%	

Factor loadings with an absolute value greater than 0.4 are in bold. NEFA = non-esterified fatty acids; T3 = triiodothyronine; T4 = thyroxine; BW = body weight; BCS = Body Condition Score.

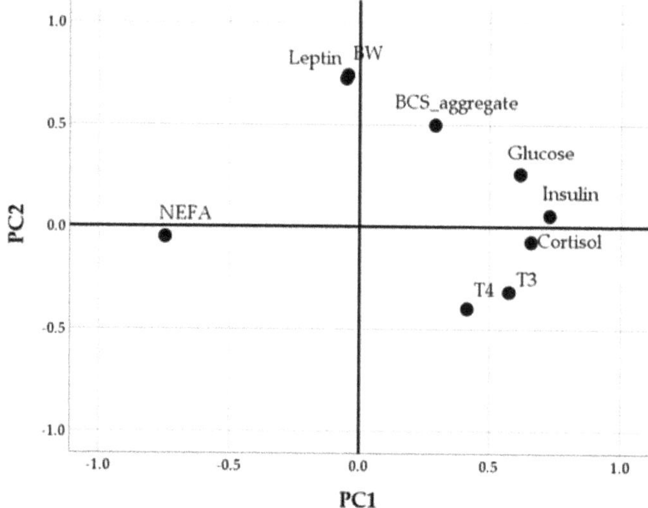

Figure 12. Component plot in rotate space (Varimax rotation). The position of the variables in the Cartesian graph indicates that PC1 (x-axis) was positively associated with insulin, cortisol, glucose, and thyroid hormones while negatively with NEFA levels. The PC2 (y-axis) was positively associated with body weight (BW), leptin, and BCS.

The two PCs were then analysed by regression techniques to evaluate how the two hormonal-metabolic profiles change as the concentration of supplementation increases (Table 3). Independently from the duration of supplementation, for every 1% increase in the goji concentration, the PC2 score increased by 0.275 ($p < 0.001$). This result indicated that the effects of goji on body conditions and leptin levels are dose-dependent: as the percentage of goji supplementation increases, the BW, the leptin, and the BCS of does tend to increase (Figure 13). In contrast, there was no linear relationship between goji berries inclusion in the diet and PC1 ($b = 0.019$, $p > 0.1$), which included NEFA, insulin, cortisol, glucose, and thyroid hormones.

Table 3. Results of regression analyses including percentage of goji berries inclusion in feed as independent variables and principal component as outcome. Days of supplementation were included as covariate.

Dependent Variable/Outcome	Independent Variable/Predictor	B Coefficient	Standard Error	*p* Value
PC1	**Constant**	0.611	0.1511	<0.001
	Goji berries concentration in feed	0.019	0.0663	0.776
	Days of supplementation	−0.017	0.0024	<0.001
PC2	**Constant**	−0.619	0.1373	<0.001
	Goji berries concentration in feed	0.275	0.0765	<0.001
	Days of supplementation	0.007	0.0027	0.006

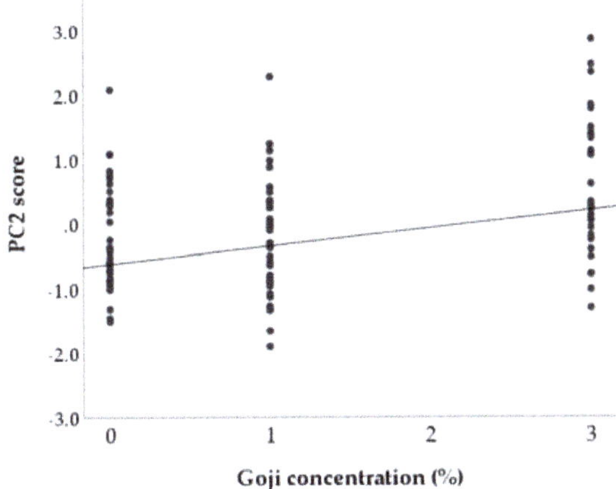

Figure 13. Scatter plot showing the relationship between the percentage of goji berries inclusion in feed (independent variable, x-axis) and PC2 (dependent variable, y-axis). The PC2 was positively associated with body weight, leptin, and BCS. The parameters of regression line were estimated by a generalized linear model including the days of supplementation as covariate.

Both components changed linearly over time ($p < 0.01$). The signs of b indicated a progressive reduction of PC1 (b = −0.017, $p < 0.001$) while, with lower slope and significance, an increase over time for PC2 (b = 0.007, $p < 0.01$).

4. Discussion

Energy homeostasis is finely regulated by many factors which, in a synergistic way, have the task of maintaining a constant body weight. Maintaining this balance is never easy and, in fact, there are frequent alterations that lead to pathological conditions such as obesity and diabetes. During pregnancy, energy homeostasis is even more complicated because the body must also ensure the growth of the foetus and prepare for the next lactation [27,31,32]. The energy balance during a reproduction cycle is one of the most problematic issues for rabbit breeding because it has consequences not only on the animal welfare but also on the profitability of the farm [28,29]. The adoption of new nutritional strategies for the rabbit doe could, therefore, improve both of these aspects. In the present study, a product which is receiving growing interest as a nutraceutical [1,2] and that has shown promising results for the rabbit's growth performance [15,25] was proposed: the goji berry. Two percentages of supplementation were evaluated in order to find the optimal concentration for goji inclusion in the food of rabbit doe.

The effects of goji supplementation were evaluated on different hormones and metabolites responsible for the maintenance of the energy homeostasis: leptin, insulin, cortisol, thyroid hormones, NEFA, and glucose. All of them are involved in the changes of energy intake, efficiency, and expenditure but the success is only assured if their actions are integrated and strictly coordinated. For this reason, in addition to the description of the individual patterns of these hormones and metabolites, an assessment of the overall picture during the doe's reproduction cycle was proposed by using multivariate analysis techniques.

In the group fed with standard diet, lower insulin, T3, and cortisol values were found at the end of the production cycle compared to the pre-pregnancy. The T4 remarkably decreased at the post-partum meanwhile the T3/T4 ratio increased. Later, an increase in T4 concentrations was observed but, after weaning, its values were lower than in the first sampling time. Conversely, significant increases were observed for NEFA concentrations with peaks at the post-partum and post-weaning. The concentrations of leptin and glucose did not undergo great changes in the timing of collection.

Finally, HOMA insulin sensitivity index showed a fluctuating trend but, at the end of the observation period, its values had decreased.

These hormones and metabolites have recently been evaluated by Menchetti et al. [31] in pregnant and pseudopregnant rabbits. However, this previous study focused on changes during the 32 days post insemination, the sampling was weekly, and no observations were done after the birth. By using weekly sampling, these authors highlighted short-term changes during pregnancy, especially in insulin and leptin. These changes suggested a condition of insulin and leptin resistance that characterizes the pregnancy of many mammals including women [30,32,42–45]. In the present study, the changes were assessed at longer intervals but the concentrations at early and late pregnancy were comparable to those of Menchetti et al. [31]. Moreover, the longer-term evaluations of the present study allowed to highlight the physiological conditions of the lactating rabbits. In particular, the post-partum increases in NEFA and insulin resistance confirmed the state of negative energy balance during early lactation. In this phase, in fact, there is the big gap between the energy the energy required for maintenance and milk production and that taken with the diet. Several factors can contribute to this severe energy deficit of the does, such as its prolificacy, the quantity and characteristics of the milk, and the intense reproductive rhythm [29,46].

This state is characterized by an intense mobilization of the body reserves which increase the NEFA concentrations. At the same time, the increase in insulin was associated with moderate concentrations of glucose triggering a peripheral insulin resistance status [47] which could contribute to raise NEFA levels. Insulin is, indeed, an anti-lipolytic hormone and the insulin resistance can favor lipolysis and fatty acid availability.

The condition of insulin resistance was exacerbated when the highest goji supplementation was included in the feed (3%, HG group). A similar condition has been found in lactating dairy cows consuming excessive concentrate [47,48]. The insulin resistance develops as a homeorhetic adaptation to ensure a sufficient glucose supply for the lactating mammary gland [49] but it has been negatively associated with milk production, reproduction performance, and animal health [50,51]. The insulin resistance observed in the present study could therefore explain the low milk yields reported by Menchetti et al. [15] in rabbits receiving 3% of goji supplementation. Goji berries contain high amounts of polysaccharides and polyphenols [1]. Although most of the studies reported that these compounds exert beneficial effects on metabolic diseases [4,8,52], their excessive consumption during pregnancy and lactation is debated. For example, Caimari et al. [53] demonstrated that supplementations with a grape seed extract (rich in polyphenols) in pregnant and lactating rats induced insulin and adiponectin resistance. On the other hand, other authors report several positive effects of the administration of polyphenols in lactation: they seem to increase the immunoglobulins content in colostrum of sows and improve the preweaning survivability [54], to protect bovine mammary epithelial cells from oxidative stress [55] and also to exert positive effects on the adult offspring's metabolism [56,57].

Thus, our findings suggest that the quantity and source of energy during lactation affect milk yield in rabbit doe by modulating body mobilization and insulin sensitivity. Moreover, they raise concerns about the excessive consumption of polyphenol-enriched foods during lactation although dose, source, species, and duration of polyphenols administration should be evaluated from time to time.

Menchetti et al. [15] also reported greater BWs in rabbits fed with the 3% of goji. In the present study, a greater BCS was found in the same group of does. This could suggest that high-doses of Goji supplementation also leads to excessive fattening which, in rabbit does, is negatively associated with reproductive performance [58].

The effects of a lower supplementation diverged: the does supplemented with 1% goji showed lower T3 at post-partum as well as lower cortisol levels during lactation; their leptin concentrations were lower at AI and higher during lactation while insulin, glucose and HOMA values suggested a higher insulin sensitivity during lactation. This hormonal framework would seem to favour the milk yield as high productions were recorded by Menchetti et al. [15] with the low-dose goji supplementation. For example, the low levels of T3 at the beginning of lactation could decrease thermogenesis diverting nutrients to the mammary glands. As mentioned above, the positive association between insulin sensitivity and lactation has been already showed in ruminants and humans [59]. These findings suggest that the effects of goji berries on insulin resistance are dose-dependent and a moderate consumption could improve insulin sensitivity during the reproductive cycle.

The increase in leptin concentrations during lactation could explain the differences in food intake between groups as this is the main function of leptin, as well as the most studied [60]. Some authors reported hypoleptinemia during lactation and have hypothesized that leptin resistance may also continue after giving birth [61,62]. However, our results do not seem to support these theses. On the other hand, maternal leptin circulating during lactation may be important for the neurological development of the new-borns because it is secreted into milk and it is taken up during suckling [63]. As observed in several species including humans, the leptin ingested trough suckling could have a role not only in the changes during the perinatal period but also in the metabolic programming of adulthood phenotype [64,65]. Finally, Koch et al. [66] demonstrated that rabbit mammary tissue expresses leptin mRNA and protein during lactation suggesting that it could be also involved in the mammary gland development.

In the present study, all the variables described so far were then included in the multivariate analysis to identify the hormonal and metabolic profiles characterizing rabbit pregnancy and lactation. Two dimensions were extracted.

The conditions of insulin resistance and energy deficit were well described by the first component extracted by the principal component analysis. In this component, insulin and glucose had the same sign (both positive loadings) and this could be explained by the reduction in insulin sensitivity: insulin rises as the glucose increases but, under conditions of insulin resistance, cells fail to respond normally and high blood glucose levels are still maintained. In the first dimension, instead, the insulin was opposed to NEFA levels. The anti-lipolytic effect of insulin could explain the opposite signs of these items. Indeed, the known effect of systemic insulinization is a decline in circulating NEFA concentrations [67]. However, rabbits' pregnancy and lactation are characterized by elevated NEFA levels. To explain the negative association between NEFA and insulin levels in this context, it can also be hypothesized that chronically elevated NEFA levels reduce the insulin-secretory capacity in rabbit does, as already proposed for high-yielding dairy cows [51].

Overall, the first dimension progressively reduced, suggesting that insulin, cortisol, glucose, and thyroid hormones decreased during the productive cycle of the doe while, in the meantime, the NEFAs increased. Similar profiles are described in lactating cows [47].

The second dimension extracted by the principal component analysis included leptin, a hormone secreted mainly by adipose tissue [60]. Therefore, its positive associations with body weight and BCS found in this component are not surprising. The positive loadings of glucose and insulin in this dimension also confirm that they could be factors implicated in leptin regulation [60]. Of major interest is the positive relationship between this second component and goji supplementation indicating that leptin, BW and BCS enhance as the percentage of goji berries inclusion in the diet increases. In agreement with the univariate results, this finding suggests that there is a risk of excessive fattening in rabbit does supplemented with goji and that, probably, a high percentage of inclusion ultimately determines detrimental reproductive effects.

It is worthwhile noting that 3% of goji supplementation had a different impact in growing rabbits: it guaranteed good growth performance and an improvement in the meat quality [15,25]. The use of goji in rabbit diet was also positively perceived by consumers who probably appreciate the use of nutraceutical products also in animal nutrition [25]. Overall, the goji inclusion in the feed could be a promising nutritional strategy from both a marketing and animal welfare point of view. However, the dosage of goji berries should be differentiated according to the category of animals and physiological status.

5. Conclusions

The rabbit doe is subjected to a severe energy deficit during reproductive cycle so that several physiological phenomena, such as the increase in NEFA and insulin resistance, are comparable to those of lactating cows. The quantity and source of energy during pregnancy and lactation modulate body mobilization, insulin sensitivity and leptin levels that, in turn, could affect milk yield. This balance is so delicate that the effects of goji supplementation diverged according to the dose: a moderate consumption seems improve insulin resistance while a high-dose could make animals excessively fat and reduce insulin sensitivity. This work has proved that goji berry could play a role in rabbit nutrition, but the formulation should be carefully modulated according to the physiological state.

Author Contributions: Conceptualization, L.M. and G.B.; methodology, L.M., G.C., D.V. and G.B.; software, L.M.; formal analysis, L.M.; investigation, M.C., O.B., C.C. and E.A.; resources, E.A. and O.B.; data curation, L.M.; writing—original draft preparation, L.M. and G.B.; writing—review and editing, L.M., G.C., D.V. and G.B.; visualization, L.M.; supervision, G.B.; project administration, D.V. and G.B.; funding acquisition, O.B., B.F., A.P. and A.T. All authors have read and agreed to the published version of the manuscript.

Funding: This study has been financially supported by Regione Umbria (Italy) (Grant PSR 2007/2013; no. 44750050831).

Acknowledgments: The authors wish to thank Giovanni Migni for his excellent technical assistance and Margherita Angelucci for the English revision.

Conflicts of Interest: The authors declare no conflict of interest.

References

1. Cheng, J.; Zhou, Z.-W.W.; Sheng, H.-P.P.; He, L.-J.J.; Fan, X.-W.W.; He, Z.-X.X.; Sun, T.; Zhang, X.; Zhao, R.J.; Gu, L.; et al. An evidence-based update on the pharmacological activities and possible molecular targets of Lycium barbarum polysaccharides. *Drug Des. Devel. Ther.* **2015**, *9*, 33–78. [PubMed]
2. Potterat, O. Goji (Lycium barbarum and L. chinense): Phytochemistry, pharmacology and safety in the perspective of traditional uses and recent popularity. *Planta Med.* **2010**, *76*, 7–19. [CrossRef] [PubMed]
3. Ma, Z.F.; Zhang, H.; Teh, S.S.; Wang, C.W.; Zhang, Y.; Hayford, F.; Wang, L.; Ma, T.; Dong, Z.; Zhang, Y.; et al. Goji Berries as a potential natural antioxidant medicine: An insight into their molecular mechanisms of action. *Oxid. Med. Cell. Longev.* **2019**, *2019*, 1–9. [CrossRef] [PubMed]
4. Cai, H.; Liu, F.; Zuo, P.; Huang, G.; Song, Z.; Wang, T.; Lu, H.; Guo, F.; Han, C.; Sun, G. Practical application of antidiabetic efficacy of Lycium barbarum polysaccharide in patients with type 2 diabetes. *Med. Chem.* **2015**, *11*, 383–390. [CrossRef] [PubMed]
5. Georgiev, K.D.; Slavov, I.J.; Iliev, I.A. Antioxidant activity and antiproliferative effects of Lycium barbarum's (Goji berry) fractions on breast cancer cell lines. *Folia Med. (Plovdiv)* **2019**, *61*, 104–112. [CrossRef]
6. Ceccarini, M.R.; Vannini, S.; Cataldi, S.; Moretti, M.; Villarini, M.; Fioretti, B.; Albi, E.; Beccari, T.; Codini, M. Effect of Lycium barbarum berries cultivated in Umbria (Italy) on human hepatocellular carcinoma cells. *J. Biotechnol.* **2016**, *231*, S26–S27. [CrossRef]
7. You, J.; Chang, Y.; Zhao, D.; Zhuang, J.; Zhuang, W. A mixture of functional complex extracts from Lycium barbarum and grape seed enhances immunity synergistically in vitro and in vivo. *J. Food Sci.* **2019**, *84*, 1577–1585. [CrossRef]
8. Zhao, R.; Li, Q.; Xiao, B. Effect of Lycium barbarum polysaccharide on the improvement of insulin resistance in NIDDM rats. *Yakugaku Zasshi* **2005**, *125*, 981–988. [CrossRef]
9. Chien, K.J.; Horng, C.T.; Huang, Y.S.; Hsieh, Y.H.; Wang, C.J.; Yang, J.S.; Lu, C.C.; Chen, F.A. Effects of Lycium barbarum (goji berry) on dry eye disease in rats. *Mol. Med. Rep.* **2018**, *17*, 809–818. [CrossRef]
10. Kang, Y.; Yang, G.; Zhang, S.; Ross, C.F.; Zhu, M.J. Goji berry modulates gut microbiota and alleviates colitis in IL-10-deficient mice. *Mol. Nutr. Food Res.* **2018**, *62*, e1800535. [CrossRef]
11. Seeram, N. Berry fruits: Compositional elements, biochemical activities, and the impact of their intake on human health, performance, and disease. *J. Agric. Food Chem.* **2008**, *56*, 627–629. [CrossRef]
12. Bucheli, P.; Vidal, K.; Shen, L.; Gu, Z.; Zhang, C.; Miller, L.E.; Wang, J. Goji berry effects on macular characteristics and plasma antioxidant levels. *Optom. Vis. Sci.* **2011**, *88*, 257–262. [CrossRef]

13. Wang, K.; Xiao, J.; Peng, B.; Xing, F.; So, K.-F.; Tipoe, G.L.; Lin, B. Retinal structure and function preservation by polysaccharides of wolfberry in a mouse model of retinal degeneration. *Sci. Rep.* **2014**, *4*, 7601. [CrossRef] [PubMed]
14. Zhao, Q.; Li, J.; Yan, J.; Liu, S.; Guo, Y.; Chen, D.; Luo, Q. Lycium barbarum polysaccharides ameliorates renal injury and inflammatory reaction in alloxan-induced diabetic nephropathy rabbits. *Life Sci.* **2016**, *157*, 82–90. [CrossRef]
15. Menchetti, L.; Vecchione, L.; Filipescu, I.; Petrescu, V.F.; Fioretti, B.; Beccari, T.; Ceccarini, M.R.; Codini, M.; Quattrone, A.; Trabalza-Marinucci, M.; et al. Effects of Goji berries supplementation on the productive performance of rabbit. *Livest. Sci.* **2019**, *220*, 123–128. [CrossRef]
16. Menchetti, L.; Brecchia, G.; Branciari, R.; Barbato, O.; Fioretti, B.; Codini, M.; Bellezza, E.; Trabalza-Marinucci, M.; Miraglia, D. The effect of Goji berries (Lycium barbarum) dietary supplementation on rabbit meat quality. *Meat Sci.* **2020**, *161*, 108018. [CrossRef] [PubMed]
17. Brecchia, G.; Menchetti, L.; Cardinali, R.; Castellini, C.; Polisca, A.; Zerani, M.; Maranesi, M.; Boiti, C. Effects of a bacterial lipopolysaccharide on the reproductive functions of rabbit does. *Anim. Reprod. Sci.* **2014**, *147*, 128–134. [CrossRef] [PubMed]
18. Martínez-Paredes, E.; Ródenas, L.; Martínez-Vallespín, B.; Cervera, C.; Blas, E.; Brecchia, G.; Boiti, C.; Pascual, J.J. Effects of feeding programme on the performance and energy balance of nulliparous rabbit does. *Animal* **2012**, *6*, 1086–1095. [CrossRef] [PubMed]
19. Boiti, C.; Guelfi, G.; Zerani, M.; Zampini, D.; Brecchia, G.; Gobbetti, A. Expression patterns of cytokines, p53 and nitric oxide synthase isoenzymes in corpora lutea of pseudopregnant rabbits during spontaneous luteolysis. *Reproduction* **2004**, *127*, 229–238. [CrossRef]
20. Foote, R.H.; Carney, E.W. The rabbit as a model for reproductive and developmental toxicity studies. *Reprod. Toxicol.* **2000**, *14*, 477–493. [CrossRef]
21. Menchetti, L.; Barbato, O.; Sforna, M.; Vigo, D.; Mattioli, S.; Curone, G.; Tecilla, M.; Riva, F. Brecchia Effects of diets enriched in linseed and fish oil on the expression pattern of Toll-Like receptors 4 and proinflammatory cytokines on gonadal axis and reproductive organs in rabbit buck. *Oxid. Med. Cell. Longev.* **2020**, *2020*. [CrossRef] [PubMed]
22. Menchetti, L.; Barbato, O.; Filipescu, I.E.; Traina, G.; Leonardi, L.; Polisca, A.; Troisi, A.; Guelfi, G.; Piro, F.; Brecchia, G. Effects of local lipopolysaccharide administration on the expression of Toll-like receptor 4 and pro-inflammatory cytokines in uterus and oviduct of rabbit does. *Theriogenology* **2018**, *107*, 162–174. [CrossRef] [PubMed]
23. Yang, Q.; Xing, Y.; Qiao, C.; Liu, W.; Jiang, H.; Fu, Q.; Zhou, Y.; Yang, B.; Zhang, Z.; Chen, R. Semen quality improvement in boars fed with supplemental wolfberry (Lycium barbarum). *Anim. Sci. J.* **2019**, *90*, 1517–1522. [CrossRef]
24. Bai, X.; Yan, X.; Xie, L.; Hu, X.; Lin, X.; Wu, C.; Zhou, N.; Wang, A.; See, M.T. Effects of pre-slaughter stressor and feeding preventative Chinese medicinal herbs on glycolysis and oxidative stability in pigs. *Anim. Sci. J.* **2016**, *87*, 1028–1033. [CrossRef] [PubMed]
25. Castrica, M.; Menchetti, L.; Balzaretti, C.; Branciari, R.; Ranucci, D.; Cotozzolo, E.; Vigo, D.; Curone, G.; Brecchia, G.; Miraglia, D. Impact of dietary supplementation with goji berries (Lycium barbarum) on microbiological quality, physico-chemical, and sensory characteristics of rabbit meat. *Foods* **2020**, *9*, 1480. [CrossRef] [PubMed]
26. Menchetti, L.; Brecchia, G.; Canali, C.; Cardinali, R.; Polisca, A.; Zerani, M.; Boiti, C. Food restriction during pregnancy in rabbits: Effects on hormones and metabolites involved in energy homeostasis and metabolic programming. *Res. Vet. Sci.* **2015**, *98*, 7–12. [CrossRef]
27. Menchetti, L.; Canali, C.; Castellini, C.; Boiti, C.; Brecchia, G. The different effects of linseed and fish oil supplemented diets on insulin sensitivity of rabbit does during pregnancy. *Res. Vet. Sci.* **2018**, *118*, 126–133. [CrossRef]
28. Menchetti, L.; Brecchia, G.; Cardinali, R.; Polisca, A.; Boiti, C. Feed restriction during pregnancy: Effects on body condition and productive performance of primiparous rabbit does. *World Rabbit Sci.* **2015**, *23*, 1–8. [CrossRef]
29. Fortun-Lamothe, L. Energy balance and reproductive performance in rabbit does. *Anim. Reprod. Sci.* **2006**, *93*, 1–15. [CrossRef]

30. Troisi, A.; Cardinali, L.; Menchetti, L.; Speranza, R.; Verstegen, J.P.; Polisca, A. Serum concentrations of leptin in pregnant and non-pregnant bitches. *Reprod. Domest. Anim.* **2020**, *55*, 454–459. [CrossRef]
31. Menchetti, L.; Andoni, E.; Barbato, O.; Canali, C.; Quattrone, A.; Vigo, D.; Codini, M.; Curone, G.; Brecchia, G. Energy homeostasis in rabbit does during pregnancy and pseudopregnancy. *Anim. Reprod. Sci.* **2020**, *218*, 106505. [CrossRef] [PubMed]
32. Ladyman, S.R.; Augustine, R.A.; Grattan, D.R. Hormone interactions regulating energy balance during pregnancy. *J. Neuroendocrinol.* **2010**, *22*, 805–817. [CrossRef] [PubMed]
33. Brecchia, G.; Bonanno, A.; Galeati, G.; Federici, C.; Maranesi, M.; Gobbetti, A.; Zerani, M.; Boiti, C. Hormonal and metabolic adaptation to fasting: Effects on the hypothalamic-pituitary-ovarian axis and reproductive performance of rabbit does. *Domest. Anim. Endocrinol.* **2006**, *31*, 105–122. [CrossRef] [PubMed]
34. Maertens, L.; Moermans, R.; De Groote, G. Prediction of the apparent digestible energy content of commercial pelleted feeds for rabbits. *J. Appl. Rabbit Res.* **1988**, *11*, 60–67.
35. García-García, R.M.; Rebollar, P.G.; Arias-Álvarez, M.; Sakr, O.G.; Bermejo-Álvarez, P.; Brecchia, G.; Gutierrez-Adan, A.; Zerani, M.; Boit, C.; Lorenzo, P.L. Acute fasting before conception affects metabolic and endocrine status without impacting follicle and oocyte development and embryo gene expression in the rabbit. *Reprod. Fertil. Dev.* **2011**, *23*, 759–768. [CrossRef] [PubMed]
36. Garson, G.D. *Factor Analysis*; Statistical Associates Publishers: Asheboro, NC, USA, 2013.
37. Righi, C.; Menchetti, L.; Orlandi, R.; Moscati, L.; Mancini, S.; Diverio, S. Welfare Assessment in Shelter Dogs by Using Physiological and Immunological Parameters. *Animals* **2019**, *9*, 340. [CrossRef] [PubMed]
38. Agradi, S.; Curone, G.; Negroni, D.; Vigo, D.; Brecchia, G.; Bronzo, V.; Chiesa, L.; Peric, T.; Danes, D.; Menchetti, L. Determination of fatty acids profile in Original Brown cows dairy products and relationship with alpine pasture farming system. *Animals* **2020**, *10*, 1231. [CrossRef]
39. Menchetti, L.; Calipari, S.; Guelfi, G.; Catanzaro, A.; Diverio, S. My dog is not my cat: Owner perception of the personalities of dogs and cats living in the same household. *Animals* **2018**, *8*, 80. [CrossRef]
40. Faul, F.; Erdfelder, E.; Lang, A.G.; Buchner, A. G*Power 3: A flexible statistical power analysis program for the social, behavioral, and biomedical sciences. *Behav. Res. Methods* **2007**, *39*, 175–191. [CrossRef]
41. Cohen, J. *Statistical Power Analysis for the Behavioural Sciences*, 2nd ed.; Lawrence Erlbaum Associates Inc.: Mahwah, NJ, USA, 1988.
42. Kawai, M.; Yamaguchi, M.; Murakami, T.; Shima, K.; Murata, Y.; Kishi, K. The placenta is not the main source of leptin production in pregnant rat: Gestational profile of leptin in plasma and adipose tissues. *Biochem. Biophys. Res. Commun.* **1997**, *240*, 798–802. [CrossRef]
43. Block, S.S.; Butler, W.R.; Ehrhardt, R.A.; Bell, A.W.; Van Amburgh, M.E.; Boisclair, Y.R. Decreased concentration of plasma leptin in periparturient dairy cows is caused by negative energy balance. *J. Endocrinol.* **2001**, *171*, 339–348. [CrossRef] [PubMed]
44. Cardinali, L.; Troisi, A.; Verstegen, J.P.; Menchetti, L.; Elad Ngonput, A.; Boiti, C.; Canello, S.; Zelli, R.; Polisca, A. Serum concentration dynamic of energy homeostasis hormones, leptin, insulin, thyroid hormones, and cortisol throughout canine pregnancy and lactation. *Theriogenology* **2017**, *97*, 154–158. [CrossRef] [PubMed]
45. Ciampelli, M.; Lanzone, A.; Caruso, A. Insulin in obstetrics: A main parameter in the management of pregnancy. *Hum. Reprod. Update* **1998**, *4*, 904–914. [CrossRef]
46. Maertens, L.; Lebas, F.; Szendrö, Z. Rabbit milk: A review of quantity, quality and non-dietary affecting factors. *World Rabbit Sci.* **2006**, *14*, 205–230. [CrossRef]
47. Baruselli, P.S.; Vieira, L.M.; Sá Filho, M.F.; Mingoti, R.D.; Ferreira, R.M.; Chiaratti, M.R.; Oliveira, L.H.; Sales, J.N.; Sartori, R. Associations of insulin resistance later in lactation on fertility of dairy cows. *Theriogenology* **2016**, *86*, 263–269. [CrossRef]
48. Leiva, T.; Cooke, R.F.; Brandão, A.P.; Aboin, A.C.; Ranches, J.; Vasconcelos, J.L.M. Effects of excessive energy intake and supplementation with chromium propionate on insulin resistance parameters, milk production, and reproductive outcomes of lactating dairy cows. *Livest. Sci.* **2015**, *180*, 121–128. [CrossRef]
49. De Koster, J.D.; Opsomer, G. Insulin resistance in dairy cows. *Vet. Clin. N. Am. Food Anim. Pract.* **2013**, *29*, 299–322. [CrossRef] [PubMed]
50. Leblanc, S. Monitoring metabolic health of dairy cattle in the transition period. *J. Reprod. Dev.* **2010**, *56*, S29–S35. [CrossRef]

51. Bossaert, P.; Leroy, J.L.M.R.; De Vliegher, S.; Opsomer, G. Interrelations between glucose-induced insulin response, metabolic indicators, and time of first ovulation in high-yielding dairy cows. *J. Dairy Sci.* **2008**, *91*, 3363–3371. [CrossRef]
52. Munir, K.M.; Chandrasekaran, S.; Gao, F.; Quon, M.J. Mechanisms for food polyphenols to ameliorate insulin resistance and endothelial dysfunction: Therapeutic implications for diabetes and its cardiovascular complications. *Am. J. Physiol. Endocrinol. Metab.* **2013**, *305*, E679–E686. [CrossRef]
53. Caimari, A.; Mariné-Casadó, R.; Boqué, N.; Crescenti, A.; Arola, L.; Del Bas, J.M. Maternal intake of grape seed procyanidins during lactation induces insulin resistance and an adiponectin resistance-like phenotype in rat offspring. *Sci. Rep.* **2017**, *7*, 12573. [CrossRef] [PubMed]
54. Wang, X.; Jiang, G.; Kebreab, E.; Yu, Q.; Li, J.; Zhang, X.; He, H.; Fang, R.; Dai, Q. Effects of dietary grape seed polyphenols supplementation during late gestation and lactation on antioxidant status in serum and immunoglobulin content in colostrum of multiparous sows. *J. Anim. Sci.* **2019**, *97*, 2515–2523. [CrossRef] [PubMed]
55. Ma, Y.; Zhao, L.; Gao, M.; Loor, J.J. Tea polyphenols protect bovine mammary epithelial cells from hydrogen peroxide-induced oxidative damage in vitro. *J. Anim. Sci.* **2018**, *96*, 4159–4172. [CrossRef] [PubMed]
56. Kataoka, S.; Norikura, T.; Sato, S. Maternal green tea polyphenol intake during lactation attenuates kidney injury in high-fat-diet-fed male offspring programmed by maternal protein restriction in rats. *J. Nutr. Biochem.* **2018**, *56*, 99–108. [CrossRef] [PubMed]
57. Tanaka, M.; Kita, T.; Yamasaki, S.; Kawahara, T.; Ueno, Y.; Yamada, M.; Mukai, Y.; Sato, S.; Kurasaki, M.; Saito, T. Maternal resveratrol intake during lactation attenuates hepatic triglyceride and fatty acid synthesis in adult male rat offspring. *Biochem. Biophys. Reports* **2017**, *9*, 173–179. [CrossRef]
58. Castellini, C.; Dal Bosco, A.; Cardinali, R. Long term effect of post-weaning rhythm on the body fat and performance of rabbit doe. *Reprod. Nutr. Dev.* **2006**, *46*, 195–204. [CrossRef]
59. Da Costa, T.H.M.; Bluck, L.J. High lactation index is associated with insulin sensitivity. *J. Nutr. Biochem.* **2011**, *22*, 446–449. [CrossRef]
60. Ahima, R.S.; Osei, S.Y. Leptin signaling. *Physiol. Behav.* **2004**, *81*, 223–241. [CrossRef]
61. Johnstone, L.E.; Higuchi, T. Food intake and leptin during pregnancy and lactation. *Prog. Brain Res.* **2001**, *133*, 215–227.
62. Vernon, R.G.; Denis, R.G.P.; Sorensen, A.; Williams, G. Leptin and the adaptations of lactation in rodents and ruminants. *Horm. Metab. Res.* **2002**, *34*, 678–685. [CrossRef]
63. Casabiell, X.; Piñeiro, V.; Tomé, M.A.; Peinó, R.; Diéguez, C.; Casanueva, F.F. Presence of leptin in colostrum and/or breast milk from lactating mothers: A potential role in the regulation of neonatal food intake. *J. Clin. Endocrinol. Metab.* **1997**, *82*, 4270–4273. [CrossRef] [PubMed]
64. Palou, M.; Picó, C.; Palou, A. Leptin as a breast milk component for the prevention of obesity. *Nutr. Rev.* **2018**, *76*, 875–892. [CrossRef] [PubMed]
65. Picó, C.; Oliver, P.; Sánchez, J.; Miralles, O.; Caimari, A.; Priego, T.; Palou, A. The intake of physiological doses of leptin during lactation in rats prevents obesity in later life. *Int. J. Obes.* **2007**, *31*, 1199–1209. [CrossRef] [PubMed]
66. Koch, E.; Hue-Beauvais, C.; Galio, L.; Solomon, G.; Gertler, A.; Revillon, F.; Lhotellier, V.; Aujean, E.; Devinoy, E.; Charlier, M. Leptin gene in rabbit: Cloning and expression in mammary epithelial cells during pregnancy and lactation. *Physiol. Genomics* **2013**, *45*, 645–652. [CrossRef] [PubMed]
67. Ferrannini, E.; Camastra, S.; Coppack, S.; Fliser, D.; Golay, A.; Mitrakou, A. Insulin action and non-esterified fatty acids. *Proc. Nutr. Soc.* **1997**, *56*, 753–761. [CrossRef] [PubMed]

Article

Relationship between Body Chemical Composition and Reproductive Traits in Rabbit Does

Meriem Taghouti [1], Javier García [2], Miguel A. Ibáñez [3], Raúl E. Macchiavelli [4] and Nuria Nicodemus [2,*]

1 FeedInov CoLab, Integrated Production Systems, Quinta da Fonte Boa, 2005-048 Santarém, Portugal; myriam.taghouti@feedinov.com
2 Departamento de Producción Agraria, Escuela Técnica Superior de Ingeniería Agronómica, Agroalimentaria y de Biosistemas, Universidad Politécnica de Madrid, C/Senda del Rey 18, 28040 Madrid, Spain; javier.garcia@upm.es
3 Departamento de Economía Agraria, Estadística y Gestión de Empresas, Escuela Técnica Superior de Ingeniería Agronómica, Agroalimentaria y de Biosistemas, Universidad Politécnica de Madrid, C/Senda del Rey 18, 28040 Madrid, Spain; miguel.ibanez@upm.es
4 Colegio de Ciencias Agrícolas, Universidad de Puerto Rico, Mayagüez 00681-9000, Puerto Rico; raul.macchiavelli@upr.edu
* Correspondence: nuria.nicodemus@upm.es; Tel.: +34-9106-71072

Simple Summary: At the beginning of the productive life of rabbit does, there must be a balance between ensuring at least a minimal degree of bodily development to guarantee a successful reproductive life, and the minimization of the unproductive rearing period, but nowadays there is no clear recommendation about the optimal moment for the first artificial insemination (AI). A better body condition at the first AI (higher body protein, fat and energy), that indicates a higher degree of maturity of the rabbit doe, did not influence fertility at the first AI (that is usually very high), but improved it at the second AI (that is usually lower than the first one). The percentage of kits born alive at the first and at the second AI also were positively influenced by the body protein content at the first AI. We can conclude that the degree of maturity at the first AI is a key point to optimize the does reproductive success, with body fat and body protein content being relevant factors.

Abstract: The relationship among live weight, chemical body composition and energy content (at artificial insemination (AI) and three days before parturition), estimated by bioelectrical impedance with fertility rates and the percentage of kits born alive, was studied during the first three AI. The first AI was conducted at 16 weeks of age in 137 rabbit does that weighted 3.91 ± 0.46 kg. Their body chemical composition was 17.4 ± 0.50%, 16.1 ± 2.6%, 1067 ± 219 kJ/100 g body weight, for protein, fat and energy, respectively. An increase in body protein, fat and energy content at the first AI did not affect fertility at the first AI but improved it at the second AI ($p \leq 0.030$). Moreover, an increase in body fat and energy content at the second AI improved fertility at the second AI ($p \leq 0.001$). Fertility at the third AI was positively influenced by body protein at the third AI and the increase in body protein and fat between the second parturition and the third AI ($p \leq 0.030$). The percentage of kits born alive at the first and at the second AI improved with the increase in body protein at the first AI ($p \leq 0.040$). In conclusion, a minimal body protein and fat content is required at the first AI to optimize the reproductive performance in young does.

Keywords: body composition; fertility; kits born alive; rabbit does

Citation: Taghouti, M.; García, J.; Ibáñez, M.A.; Macchiavelli, R.E.; Nicodemus, N. Relationship between Body Chemical Composition and Reproductive Traits in Rabbit Does. *Animals* **2021**, *11*, 2299. https://doi.org/10.3390/ani11082299

Academic Editors: Rosa María García-García and Maria Arias Alvarez

Received: 29 May 2021
Accepted: 29 July 2021
Published: 4 August 2021

1. Introduction

In the last four decades, rabbit production underwent a noticeable change from a traditional and familiar organization to industrial and intensive systems. Consequently, genetic selection programs and new breeding management systems were established, improving the production of the new hybrid lines used. This has led to an increase in female nutritional requirements [1,2], health problems [3–7] and welfare necessity [8–11]. Nutritional strategies in reproductive females need to be global and consider both the short-term productive factors (litter size, milk production, or fertility) as well as the long-term factors (body condition or health). Therefore, the expected improvement in nutritional management should be based on an accurate analysis of the requirements of the doe, its evolution during successive reproductive cycles and the identification of crucial moments in the life of rabbit does to optimize productivity and longevity.

Two of the key points of the reproductive success of rabbit does are their birth weight and maturity at the first artificial insemination (AI). There is an optimal threshold for birth weight (>57 g) to optimize the initial reproductive performance, which is associated with an increase in the live weight and fat reserves at the first AI [12–14]. The latter are the traits used to define the maturity of rabbit does at the first AI, and both are also related to the nutritional rearing strategy and the time of AI [13,15]. However, there is no clear recommendation indicating the weight and/or body condition of the nulliparous rabbit doe at the onset of its reproductive life [16]. In the current management systems, does are inseminated at a fixed age with minor or no consideration of their weight and chemical body composition. As in other species, there must exist a balance between ensuring at least the minimal degree of body development needed to guarantee a successful reproductive life, and minimizing the unproductive rearing period. Accordingly, the study of the body chemical composition of rabbit does seems to be a useful tool not only to improve feeding, but also for general rabbit doe management [17]. The final aim is to extend the lifespan of rabbit does, which is limited by the relatively high early mortality and culling rate in intensive production systems [5].

A new non-destructive technique to estimate rabbit doe body chemical composition (moisture, ash, protein, fat, and energy content) based on the bioelectrical impedance measurement, live weight and physiological status of rabbit does, was developed by Pereda [18]. This method is easier and cheaper than TOBEC [19], and in both methods, the variations of gut contents are included in the error term. Furthermore, it allows the prediction of total fat and energy content (not only the perirenal fat content, as it is with the case with the ultra-sound technique [20]), as well as the body protein, which is not usually estimated with other methodologies to evaluate body condition.

The aim of this work was to establish the relationship between chemical body composition (at AI and parturition), determined using bioelectrical impedance, and both fertility and the percentage of kits born alive during the first three inseminations, and to identify the most important moments to record the body chemical composition.

2. Materials and Methods

2.1. Animal Husbandry and Management

Data of the estimated body chemical composition, the fertility and percentage of kits born alive recorded in two different farms in 2010 were used in order to obtain a wide variation in chemical body composition. In farm A, 106 crossbred (New Zealand White × Californian) rabbit does from the UPV hybrid line (genetic line selected for prolificacy in Polytechnic University of Valencia in Spain using the crossline A × Line V) were used. After the first parturition, does were submitted to three different breeding systems, defined by the parturition–AI and parturition–weaning intervals (4/32, 11/35 and 14/42, Table 1) to obtain a wide variability of body condition. In farm B, 37 crossbred (New Zealand White × Californian) rabbit does from the Hyplus hybrid line (prolific hybrid maternal line selected by Hypharm in France) were used and they were inseminated 11 d after parturition and litter weaned at 35 d of age. Rabbit does were inseminated for the first time between 16 and 18 weeks of age. Rearing period of rabbit does was not controlled and they were fed ad libitum after the first insemination. Non pregnant-non lactating rabbit does were restricted to around 150 g/d. Two commercial diets were used, one in farm A containing 17.5% crude protein, 32.0% neutral detergent fibre and 9.95 MJ digestible energy/kg (based on dehydrated alfalfa, wheat bran and barley), and another in farm B containing 16.5% crude protein, 30% neutral detergent fibre, and 10.8 MJ digestible energy/kg (based on dehydrated alfalfa, sugar beet pulp, sunflower meal, rye and wheat). All rabbit does were submitted to a cycle of 16 h light/8 h dark, and heating, cooling and forced ventilation systems allowed the building's temperature to be maintained between 18 and 24 °C. Animals were handled according to the principles for the care of animals in experimentation published by the Spanish Royal Decree 53/2013 [21] and favourably assessed retrospectively by the Ethics Committee of the Polytechnic University of Madrid.

Table 1. Description of the animals and breeding systems used.

Farm	Rhythm	Number and Strain	Artificial Insemination, d after Parturition	Weaning, d after Parturition
A	R4	37 UPV	4 d	32 d
A/B	R11	28 UPV/31 Hyplus	11 d	35 d
A	R14	41 UPV	14 d	42 d

The relationship between body chemical composition and fertility was studied during the first three cycles as the number of replicates decreased from the third parturition onwards (Table 2). The percentage of kits born alive was only studied at the first and second parturition (N = 121 and 83, respectively) due to the reduction in replicates in the third parturition. The range of variation in chemical body composition of these does at the moments of insemination and parturition is shown in Table 2.

Table 2. Estimated chemical body composition [1] of nulliparous and primiparous rabbit does during the first three reproductive cycles at artificial insemination and parturition.

	1st Artificial Insemination (N = 137)						1st Parturition (N = 137)					
	LW	Moist.	Prot.	Fat	Ash	Ener.	LW	Moist.	Prot.	Fat	Ash	Ener.
Mean	3906	60.1	17.4	16.1	2.9	1067	4323	60.2	17.3	15.4	2.9	1048
SD [2]	458	3.2	0.5	2.6	0.15	219	371	3.4	0.6	4.8	0.11	261
Min.	2720	53.4	15.0	8.7	2.7	373	3140	52.4	15.2	2.4	2.7	294
Max.	4836	76.4	18.6	22.1	3.4	1403	5321	77.3	19.0	28.5	3.2	1414
	2nd Artificial Insemination (N = 132)						**2nd Parturition (N = 96)**					
	LW	Moist.	Prot.	Fat	Ash	Ener.	LW	Moist.	Prot.	Fat	Ash	Ener.
Mean	4187	60.1	17.4	15.8	3.0	1077	4231	60.2	17.8	15.6	3.1	1125
SD	465	3.6	0.6	3.3	0.12	247	444	3.4	1.0	3.0	0.18	137
Min.	3055	51.6	14.9	6.2	2.7	293	3050	52.4	15.1	3.5	2.5	423
Max.	5280	72.4	19.1	23.1	3.3	1455	5397	77.3	19.8	22.0	3.6	1421
	3rd Artificial Insemination (N = 96)						**3rd Parturition (N = 84)**					
	LW	Moist.	Prot.	Fat	Ash	Ener.	LW	Moist.	Prot.	Fat	Ash	Ener.
Mean	4221	59.2	18.1	16.1	3.1	1159	4357	61.1	17.7	15.0	3.1	1091
SD	481	2.9	0.8	2.7	0.15	118	456	3.7	0.9	3.5	0.15	155
Min.	3219	49.8	15.7	8.4	2.6	848	3054	53.8	14.8	5.1	2.7	662
Max.	5680	66.8	19.5	26.0	3.4	1574	5126	71.6	19.7	22.5	3.6	1363

[1] LW: Live body weight (g); Moist. (moisture); Prot. (protein); fat and ash: % LW; Ener. (body energy): kJ/100 g body weight. [2] SD: Standard deviation; Min.: Minimal; Max.: Maximal.

Seminal doses with more than 20 million spermatozoa in 0.5 mL of a commercial diluent (Magapor S.L.) were made using a pool of fresh heterospermic semen from bucks selected for growth performance. In order to synchronize the oestrus in the second, third and fourth insemination, 48 h before insemination, the does were injected with 25 IU of eCG (Equine Chorionic Gonadotropin, Segiran, Lab. Ovejero, León) [22]. On the day of insemination, the does received an intramuscular injection of 10 µg of buserelin Suprefact®

(Hoechst Marison Roussel, S.A., Madrid). Buserelin is a Gonadotropin-releasing hormone agonist (GnRH agonist), which is used to induce ovulation in rabbit does [23].

2.2. Experimental Procedures and Data Recording

In order to determine the chemical body composition of the does, a bioelectrical impedance analysis (BIA) technique was used [18,24]. Measurements were taken with the four-terminal analyzer Quantum II (Model BIA-101, RJL Systems, Detroit, MI, USA). This device generates an alternating current of 425 μA of intensity at a frequency of 50 kHz. It is provided with 2 black electrodes to conduct the electrical current through the doe's body, and 2 red electrodes to register the resistance and reactance resulting from the passing current. When it is used with rabbit does, a needle (Terumo, 21G × 1 1/2″, 0.8 × 40 mm, nr 2) is inserted in the extremity of each electrode. The needles must pass through the skin of the rabbit doe at four reference points along the loin (two near the scapula and other in the distal part of the loin). Impedance is defined by the equation: Impedance = $(\text{Resistance}^2 + \text{Reactance}^2)^{1/2}$. These three parameters, in addition to the physiological status, live body weight and parturition order were used further to estimate the body composition of the rabbit does. Rabbit does were weighed and their body composition estimated on the days of artificial insemination and parturition (three days before parturition, both pregnant and non-pregnant does) during the first three reproductive cycles. The regression equations developed and validated by Nicodemus et al. [24] and Pereda [18] enabled the prediction of the doe's body content in moisture, protein, fat and ash expressed as percentages or as g/100 g of body weight. Furthermore, energetic body content expressed in MJ or in kJ/100 g of body weight could be also estimated.

2.3. Data Treatment and Statistical Analysis

In order to establish correlations among parameters of chemical body composition (moisture, fat, protein, ash and energy) at the moments of insemination and parturition during the first three parities, we used Pearson's correlation coefficient (CORR procedure of the SAS system) (SAS Inst., Cary, NC, USA). The GENMOD procedure of the SAS system [25] was used to study the relationship between chemical body composition, fertility during the first three parturitions and the percentage of kits born alive in the two first parturitions, as the logistic regression [26,27] is an adequate tool for modeling proportions [28]. In the generalized linear models used, the link function employed was 'logit' (Equation (1)) and we considered that both fertility and the percentage of kits born alive were proportions arising from a binomial distribution. In equation 1, 'p' is the mean of the proportion of fertility and kits born alive.

$$\text{Logit}(p) = \ln(p/1 - p);\ p \in (0,1) \tag{1}$$

Our first aim was to determine whether the variations in fertility and percentage of kits born alive (dependent variables) were significantly affected by the body chemical composition and its variation between insemination and parturitions (independent covariates). The model also included, as fixed effects, the farm and the breeding system (except for nulliparous does). Afterwards, only significant covariates were retained for the interpretation of coefficient estimates (β) and the linear predictor ($\eta = x'\beta$). In order to display the covariates' effects, three levels (low, medium and high), representative of the range of variation of each covariate, were fixed for each significant covariate (body constituent) based on the data used in this investigation. To predict the means of fertility and percentage of kits born alive expected for each level (the mean or predicted values, 'p') the inverse logit function was used (Equation (2)).

$$p = e^{\eta}/(1 + e^{\eta}) \tag{2}$$

Among the criteria of the goodness of fit, we used the likelihood ratio chi-square, 'Q_L', also called 'deviance', to establish whether models fitted appropriately [29]. The inverse function of the logit transformation was also used to calculate a confidence interval for the expected percentages of fertility and kits born alive for each level [29]. In fact, in the SAS output and for each estimator calculated, there are the values of lower (Lη) and upper (Uη) limits for its 95% confidence interval. In this case, the method consists in transforming the linear values of the limits to the log-link scale using Equation (3).

$$\text{Upper limit p} = e^{U\eta}/(1 + e^{U\eta});\ \text{Lower limit p} = e^{L\eta}/(1 + e^{L\eta}) \quad (3)$$

To calculate a standard error for the mean of predicted values, 'p', for each level of the covariates retained, we used the Delta method [30], which involves the standard error of the coefficient estimate (SEη) as well as the predicted value, as explained in Equation (4).

$$\text{SEp} \approx \text{p}(1 - \text{p})\ \text{SE}\eta \quad (4)$$

The results were transformed from the logit scale.

3. Results

3.1. Correlation among Does' Corporal Constituents during the First Three Parturitions

The initial live weight and chemical composition of rabbit does had a wide range of variation, which was more important for the fat than for the protein content (16 vs. 3%, respectively, for coefficient of variation (Table 2)). Rabbit does showed an important degree of growth between the first and the second AI (3906 vs. 4187 g, weight of does at 1st and 2nd AI, $p < 0.05$ (Table 2)). Live body weight was positively correlated ($p < 0.001$) with fat (r = 0.43 to 0.83) and energy body content (r = 0.43 to 0.71), and negatively with moisture (r = −0.35 to −0.72) (expressed on % and MJ/100 g body weight, respectively) from the first AI to the third parturition (Table 3). In this period, body fat was inversely and closely correlated with moisture (r = −0.95 to −0.99), and positively with body energy (r = 0.79 to 0.98), while body protein was positively correlated with ash content (r = 0.30 to 0.83). Regarding moisture, a negative correlation was observed with energy (r = −0.43 to −0.71).

Table 3. Correlation among estimated chemical body constituents [1] during the first three reproductive cycles at insemination and parturition [2].

	1st Artificial Insemination, N = 137					1st Parturition, N = 137					2nd Artificial Insemination, N = 132				
	LW	Prot.	Fat	Moist.	Ash	LW	Prot.	Fat	Moist.	Ash	LW	Prot.	Fat	Moist.	Ash
Prot.	−0.46 ***	1				NS	1				NS	1			
Fat	0.83 ***	−0.21 *	1			0.75 ***	0.30 ***	1			0.75 ***	0.25 **	1		
Moist.	−0.59 ***	−0.28 **	−0.95 ***	1		−0.59 ***	−0.57 ***	−0.97 ***	1		−0.72 ***	−0.35 ***	−0.99 ***	1	
Ash	−0.89 ***	0.64 ***	−0.66 ***	0.37 ***	1	NS	0.30 ***	NS	NS	1	−0.78 ***	0.45 ***	−0.62 ***	0.51 ***	1
Ener.	0.65 ***	NS	0.79 ***	−0.78 ***	−0.45 ***	0.63 ***	0.38 ***	0.83 ***	−0.85 ***	NS	0.71 ***	0.36 ***	0.85 ***	−0.89 ***	0.37 ***
	2nd Parturition, N = 96					**3rd Artificial Insemination, N = 96**					**3rd Parturition, N = 84**				
	LW	**Prot.**	**Fat**	**Moist.**	**Ash**	**LW**	**Prot.**	**Fat**	**Moist.**	**Ash**	**LW**	**Prot.**	**Fat**	**Moist.**	**Ash**
Protein	−0.62 ***	1				−0.52 ***	1				−0.59 ***	1			
Fat	0.58 ***	NS	1			0.54 ***	NS	1			0.43 ***	NS	1		
Moisture	−0.38 ***	−0.27 **	−0.93 ***	1		−0.42 ***	NS	−0.97 ***	1		−0.35 **	NS	−0.97 ***	1	
Ash	−0.83 ***	0.83 ***	−0.51 ***	0.19 †	1	−0.82 ***	0.80 ***	−0.56 ***	0.35 **	1	−0.77 ***	0.74 ***	−0.57 ***	0.40 ***	1
Ener.	0.48 ***	NS	0.96 ***	−0.98 ***	−0.3 **	0.51 ***	NS	0.98 ***	−0.98 ***	−0.47 **	0.43 ***	NS	0.98 ***	−0.99 ***	−0.48 ***

[1.] LW: Live body weight (g); Moist. (moisture); Prot. (protein); fat and ash are expressed in % LW; Ener. (energy): KJ/100 g body weight.
[2] ***: $p < 0.001$; **: $p < 0.01$; *: $p < 0.05$; †: $0.05 < p < 0.10$.

Protein content at the first AI was not correlated with protein content at any other moment of the cycle (Table 4). However, protein content was positively correlated at parturition and the subsequent AI (both at the first and second parturition; $p < 0.001$). In contrast, the body fat content was mainly positively correlated in the period between the first AI and the third parturition, and the same was observed for body energy content ($p < 0.05$ (Table 4)).

Table 4. Correlation among estimated protein, fat and energy body contents at artificial insemination (AI) and parturition during the first three cycles [1].

	Protein					Fat					Energy				
	1st AI	1st Partum	2nd AI	2nd Partum	3rd AI	1st AI	1st Partum	2nd AI	2nd Partum	3rd AI	1st AI	1st Partum	2nd AI	2nd Partum	3rd AI
1st partum	NS [2]	1				0.57 ***	1				0.82 ***	1			
2nd AI	NS	0.38 ***	1			0.33 ***	0.69 ***	1			0.74 ***	0.89 ***	1		
2nd partum	NS	−0.28 **	NS	1		0.37 **	0.48 ***	0.28 **	1		0.29 **	0.44 ***	0.23 *	1	
3rd AI	NS	−0.21 *	NS	0.73 ***	1	0.24 **	0.18 †	0.24 *	0.37 ***	1	NS	NS	0.21 *	0.35 **	1
3rd partum	NS	NS	NS	NS	NS	NS	0.32 **	NS	0.39 **	0.26 **	0.36 **	0.32 **	NS	0.31 **	0.26 **

[1] N = 137, 132, 96, 96 and 84 rabbit does for 1st partum, 2nd AI, 2nd partum, 3rd AI and 3rd partum, respectively. [2] NS: Not significant ($p > 0.05$). ***: $p < 0.001$; **: $p < 0.01$; *: $p < 0.05$; †: $0.05 < p < 0.10$.

3.2. Relationship between Body Condition and Fertility during the First Three Parturitions: Nulliparous, Primiparous and Multiparous Rabbit Does

The fertility rate registered in the first parturition was 93.4%. There were no differences between data from the two farms used for the analysis ($p > 0.05$). Neither live body weight, nor its chemical composition at the first AI, affected the fertility rate at this moment. In fact, there were no differences in the initial body composition and live weight at the first AI between pregnant and non-pregnant does (Table 5), although these results have to be taken with caution due to the small number of non-pregnant rabbits.

Table 5. Estimated chemical body composition of nulliparous pregnant and non-pregnant rabbit does at the first artificial insemination (AI) [1].

Parameter		Pregnant	Non-Pregnant	RMSE	*p*-Value
Live weight, kg	AI	3.81	3.70	0.43	0.44
	Parturition [2]	4.31	4.26	0.37	0.74
	Parturition–AI	0.49	0.56	0.29	0.47
Protein%	AI	17.5	17.7	0.48	0.29
	Parturition	17.3	18.8	0.60	<0.001
	Parturition–AI	−0.27	1.08	0.76	<0.001
Fat%	AI	15.6	15.7	2.91	0.95
	Parturition	15.7	19.4	3.00	<0.001
	Parturition–AI	0.15	3.78	2.48	<0.001
Moisture%	AI	60.8	60.3	4.41	0.71
	Parturition	61.2	55.7	4.07	<0.001
	Parturition–AI	0.35	−4.62	4.98	0.003
Ash%	AI	3.01	3.05	0.15	0.34
	Parturition	2.97	3.08	0.11	0.003
	Parturition–AI	−0.041	0.021	0.14	0.18
Energy, kJ/100 g	AI	1067.6	1161.4	214.2	0.19
	Parturition	1062.5	1336.4	249.8	0.001
	Parturition–AI	−0.77	177.0	136.3	<0.001

[1] N = 128 and 9 for pregnant and non-pregnant does, respectively. [2] Measured 3 d before parturition. RMSE: root mean square error.

As expected, in the second parturition, fertility was lower than in the first parturition (56.2 vs. 93.4%, respectively; $p < 0.05$). Breeding system and farm did not have any effect on fertility in the second parturition. Recorded fertility means were 54.0%, 50.8% and 65.8% for the studied breeding systems R4, R11, and R14, respectively. Live body weight at the moments of first AI, first parturition, and second AI did not affect the fertility rate in the second parturition. Fertility in the second AI was related to chemical body composition at the first AI, in contrast to that observed at the first parturition. Body protein content ($p = 0.007$), fat ($p = 0.030$), and energy ($p < 0.001$) at the first AI positively affected the fertility rate in the second AI (Table 6). The relationship between these parameters and fertility is lineal in the logit scale. Consequently, the model was fitted using three fixed levels of body protein, fat, energy, and protein/energy ratio. These levels were chosen to cover the data range used for the analysis. Fat and energy were positively correlated at the first AI (r = 0.79, $p < 0.001$), and with their values at the second AI (r = 0.33 and 0.74; $p < 0.001$ (Tables 3 and 4)), while at the first AI protein was negatively correlated with fat but did not have any correlation with energy, and was not related to protein at the second AI. The three levels determined for each constituent and used in the model are shown in Table 6. The increase in body protein from 16 to 18% and body fat from 10 to 20% increased fertility from 19.9 to 72.0% and from 31.3 to 69.5%, respectively. The results showed that does with higher energy content (1400 vs. 700 kJ/100 g) underwent improvements in fertility from 19.1 to 83.6%.

Table 6. Effect of estimated body composition at the first insemination on fertility in the second insemination [1].

Constituent	Levels	Fertility%	SE [2]	LL [3]	UL [3]	*p*-Value	Q_L [4]
Protein%						0.007	1.33
	16	19.9	0.09	9.1	45.7		
	17	44.5	0.06	33.2	56.2		
	18	72.0	0.06	57.2	83.0		
Fat%						0.030	1.36
	10	31.3	0.10	14.7	54.5		
	15	50.4	0.05	40.6	60.3		
	20	69.5	0.07	54.0	81.4		
Energy, kJ/100 g						<0.001	1.26
	700	19.1	0.07	8.5	37.5		
	1000	46.8	0.05	36.8	56.9		
	1400	83.6	0.06	69.0	92.0		
g protein/MJ energy						<0.001	1.27
	12	78.1	0.06	64.3	87.5		
	17	54.0	0.05	44.5	62.7		
	30	6.10	0.05	1.2	26.1		

[1] N = 132. [2] SE: Standard error. [3] LL and UL: lower and upper limits for a confidence interval (95%). [4] Q_L: Deviance of the models.

A higher body protein/energy ratio at the moment of the first AI negatively influenced the fertility rate in the second AI ($p < 0.001$ (Table 6)). This ratio reflects again the importance of the energetic content, since high ratios (30 g/MJ) corresponded to low corporal energy (800 kJ/100 g) rather than to high body protein content. In fact, the latter showed lower variability (2.9 vs. 20.5%, coefficient of variation of body protein and energy at the first AI, respectively (Table 2)).

Chemical body composition at the second AI was also related to fertility in the second AI, as detailed in Table 7. Fat and energy contents at the second AI were positively related to fertility at this moment ($p < 0.001$) in a similar way to their relation at the first AI. These two variables were positively correlated both at the first and at the second AI ($p < 0.001$ (Table 3)), and were also positively correlated with fat and energy content at the first AI ($p < 0.001$; Table 4), although their coefficient of variation increased at the second AI (Table 2).

Body protein at the second AI did not exert any effect on fertility at the second AI, in contrast with its observed effect at the first AI. It was positively correlated with fat and energy content at the second AI. Protein contents at the first and at the second AI were not correlated (Table 4), although their average values and variability were similar (Table 2), and were lower than body protein values observed from second parturition onwards. A higher body protein/energy ratio at the second AI again negatively affected the fertility rate at the second AI, reflecting the effect of energy content at the second AI.

Table 7. Effect of estimated body composition at the second AI on fertility in the second artificial insemination [1].

Constituent	Level	Fertility%	SE [2]	LL [3]	UL [3]	*p*-Value	Q_L [4]
Fat%						<0.001	1.26
	10	25.2	0.08	13.1	42.8		
	15	53.7	0.05	44.3	62.7		
	20	80.0	0.05	66.8	88.8		
Energy, kJ/100 g						<0.001	1.22
	800	29.3	0.07	17.4	45.0		
	1100	58.4	0.05	73.7	67.0		
	1300	76.1	0.05	64.8	84.4		
g protein/MJ energy						<0.001	1.23
	12	75.1	0.05	63.6	83.7		
	17	58.9	0.05	49.7	67.5		
	30	17.0	0.08	6.4	38.2		

[1] N = 132. [2] SE: Standard error. [3] LL and UL: lower and upper limits for a confidence interval (95%). [4] Q_L: Deviance of the models.

Moreover, to study the factors that influenced the fertility at the second AI, multiple linear regressions were fitted considering two independent variables: live weight or body composition at the first AI combined with their respective variations between the first and second AI (Table 8). Live body weight at the first AI and weight gain during the interval between the first AI and the first parturition as well as during the interval between the first and second AI were positively related to fertility at the second parturition.

Table 8. Effect of live weight at the first artificial insemination (AI) and its increment from the first insemination to the first parturition (model 1) or from the first to the second inseminations (model 2) on fertility at the second parturition.

	Model 1 (N = 137. QL1 = 1.33) [1]			Model 2 (N = 132. QL1 = 1.29)		
Weight at 1st AI, g	Increment-I2 [2]	Fertility%	SE	Increment-II3 [3]	Fertility%	SE [4]
3000	−2	9.98	0.08	−5	14.6	0.09
	10	21.7	0.10	15	35.2	0.10
	20	37.2	0.10	30	56.4	0.10
3500	−2	19.4	0.09	−5	25.7	0.08
	10	37.6	0.08	15	52.3	0.07
	20	56.4	0.07	30	72.3	0.09
4000	−2	34.4	0.08	−5	41.2	0.07
	10	56.7	0.05	15	68.9	0.05
	20	73.8	0.07	30	84	0.06
4500	−2	53.4	0.09	−5	58.5	0.08
	10	74.1	0.08	15	81.7	0.07
	20	85.9	0.07	30	91.4	0.05

[1] Q_L: Deviance of the models. [2] Increment-I: increment of body weight between the first AI and the first parturition expressed as a percentage of body weight at the first AI. [3] Increment-II: increment of body weight between the first and second AI expressed as a percentage of body weight at the first AI. [4] SE: standard error. Model 1: $p_{\text{-weight}} = 0.010$; $p_{\text{-increment1}} = 0.006$. Model 2: $p_{\text{-weight}} = 0.010$; $p_{\text{-increment2}} = 0.001$.

The fertility rate recorded at the third AI was 73.9%. No effect of breeding system or farm was found. Recorded fertility means were 56.2.0%, 90.3% and 75.8% for the studied breeding systems R4, R11, and R14, respectively, showing differences in the first two values ($p = 0.007$). The fertility rate recorded at the third AI was not affected by body composition (fat and energy) at the moments of the first and second AI. Unlike the fertility at the first and second AI, we registered a positive effect of protein content on fertility at the third AI ($p = 0.030$ (Table 9)). Body protein content at the third AI and second parturition were positively correlated (Table 4), and none of them were correlated with body fat or energy content (Table 3). Moreover, we studied the effect of the changes in chemical components between consecutive AI and parturitions. A positive effect of body protein gain between the first and third AI ($p = 0.020$), and between the second parturition and third AI ($p < 0.001$ (Table 9)), was observed. Similarly, body fat gain between the second parturition and third AI was also positively related to fertility rate at the third AI ($p = 0.020$ (Table 9)).

Table 9. Effect of estimated body protein content at the third artificial insemination (AI), its increments and fat increment on fertility at the third artificial insemination [1].

Constituent	Level	Fertility%	SE [2]	LL [3]	UL [3]	*p*-Value	Q_L [4]
Protein [5]%						0.030	1.11
	16	38.9	0.10	13.6	72.0		
	17	56.6	0.10	36.8	74.4		
	18	72.5	0.04	62.2	80.8		
Protein gain (AI3–AI1) [5]						0.021	1.09
	−5	56.8	0.10	39.3	72.9		
	5	69.1	0.05	58.1	78.4		
	15	79.2	0.07	68.0	87.2		
Protein gain (AI3–P2) [6]						<0.001	0.99
	−5	57.8	0.07	43.4	37.5		
	5	86.5	0.05	74.9	93.2		
	15	96.7	0.06	86.7	99.2		
Fat gain (AI3–P2) [7]						0.020	1.08
	−20	61.5	0.06	46.4	74.7		
	20	81.2	0.05	68.9	89.4		
	50	90.1	0.05	72.9	96.8		

[1] N = 96. [2] SE: standard error. [3] LL and UL: lower and upper limits for a confidence interval (95%). [4] Protein content at 3rd AI. [5] Protein increment between the moments of the first and third AI expressed as a percentage of the initial protein content.[6] Protein increment between the moments of third AI and second parturition expressed as a percentage of protein content at the second parturition. [7] Fat increment between the moments of the third AI and second parturition expressed as a percentage of fat content at the second parturition.

The effect of the previous reproductive success (or not) is reflected in the strong effects of both protein and fat gain before the third AI on the fertility at the third AI. In fact, pregnant does at the third AI showed lower fertility at the second AI, which allowed them a better body reserve recovery (especially protein) at the third AI compared to non-pregnant does (that showed higher fertility in the previous AI (Table 10)). This is confirmed by the positive relationship between fertility rates at the third and first AI ($p = 0.050$), and the negative relationship between fertility at the third and second AI ($p = 0.050$). We observed that does that gave birth at the second parturition had lower chances of becoming pregnant at the third AI ($p = 0.043$; Table 10).

Table 10. Effect of the change of live body weight and body composition, between the second parturition and the third artificial insemination (AI), and fertility at the second AI on the reproductive success at the third artificial insemination [1].

Change between 2nd Parturition and 3rd AI2 [2]	Pregnant at Third AI	Non-Pregnant at Third AI	RMSE [3]	*p*-Value
Live body weight	2.4	−4.7	8.08	<0.001
Protein	3.2	−3.3	7.08	<0.001
Fat	42.3	−5.1	210	NS [4]
Energy	6.8	−4.8	39.7	NS
Fertility at the second AI	62.0	84.0	-	0.043

[1] N = 96. [2] expressed as a percentage of the content at 2nd parturition. [3] RMSE: root mean square error. [4] NS: non-significant.

3.3. Relationship between Body Condition and Percentage of Kits Born Alive during the First Two Parturitions: Nulliparous and Primiparous Does

The percentage of kits born alive out of the total born in the first parturition was 93.4%, and the number of kits born alive per doe was 8.00 ± 2.97. Among chemical body constituents, body protein ($p = 0.040$) and energy contents ($p = 0.010$) at the first AI increased the percentage of kits born alive (Table 11). Both variables were not correlated at the first AI (Table 3). The relationship between these parameters and the percentage of kits born alive is linear in the logit scale. Consequently, the model with logit link was made using three fixed levels of body protein and energy contents. These levels were chosen to cover the data range used for the analysis. When protein content at the first AI increased from 16 to 18%, the percentage of kits born alive increased from 88.7% to 95.2%, respectively. Furthermore, when body energy at the first AI increased from 900 to 1300 kJ/100 g, the percentage of kits born alive increased from 92.4 to 95.0, respectively.

Table 11. Effect of body protein and energy content at the first insemination on the percentage of kits born alive at the first parturition [1].

Constituent	Level	Kits Born Alive%	SE [2]	LL [3]	UL [3]	*p*-Value	Q_L [4]
Protein%						0.040	2.94
	16	88.7	0.03	0.82	0.95		
	17	92.5	0.01	0.91	0.96		
	18	95.2	0.01	0.94	0.97		
Energy, kJ/100 g						0.010	2.94
	900	92.4	0.01	0.91	0.96		
	1100	93.8	0.01	0.93	0.98		
	1300	95.0	0.01	0.94	0.97		

[1] N = 121. [2] SE: standard error. [3] LL and UL: lower and upper limits for a confidence interval (95%). [4] Deviance of each model.

Body fat content was negatively correlated with body protein and positively correlated with energy content at the first AI (r = −0.21 and 0.79, respectively; $p < 0.05$ (Table 3)), and no effect on the percentage of kits born alive was observed. Live weight at the first AI was also correlated with body protein, energy and fat contents (r = −0.46, 0.65 and 0.83, respectively; $p < 0.001$ (Table 3)) but was not related to the percentage of kits born alive. Furthermore, a higher percentage of kits born alive was observed in farm B compared to farm A (98.1 and 88.7%, respectively; $p < 0.001$), but the number of kits born alive per doe was not different between the two farms.

The percentage of kits born alive recorded in the second parturition was 87.5% and the number of kits born alive per doe was 10.3 ± 4.46. This rate is lower than that observed in the first parturition (93.4%), but the number of kits born alive increased (8.00 ± 2.96 kits born alive/doe in the first parturition). No effect of the breeding system or farm on the percentage of kits born alive at the second parturition was reported. Weight and body composition at the first AI were not related to this trait. However, a higher body protein at the second AI ($p < 0.001$) was associated with an increase in the percentage of kits born at the second parturition (Table 12).

Table 12. Effect of body protein content at the second insemination on percentage of kits born alive at the second parturition [1].

Constituent	Levels	Kits Born Alive%	SE [2]	LL [3]	UL [3]	*p*-Value	Q_L [4]
Protein%						<0.001	5.12
	16	71.0	0.06	57.0	81.7		
	17	83.3	0.02	79.0	86.6		
	18	91.0	0.01	88.3	93.1		

[1] N = 83. [2] SE: standard error. [3] LL and UL: lower and upper limits for a confidence interval (95%). [4] Deviance of the model.

4. Discussion

A general problem observed in rabbit does is their low fertility rate in the second parturition [31]. It is usually explained by the energy deficiency observed before the second insemination [16]. In this period, it might be difficult for the rabbit doe to meet the requirements for both pregnancy and growth due to the limited feed intake [1,32–35]. However, when the energy supply was increased, it was dedicated to milk production with no limitation of reserve mobilization [36,37]. In this way, Pascual [13] hypothesized that this negative energetic balance would be a natural adaptation to optimize evolutionary success. In this context, it is interesting to study the relationship between the factors related to body chemical composition and their influence on fertility and kit survival at birth, in order to identify the threshold to be met by rabbit does at the beginning of their reproductive life.

The evolution of the live weight and chemical composition of rabbit does from their first AI onwards indicated that they were still growing when inseminated the first two times, which agreed with recent data [38]. Live body weight, body energy and fat were closely and positively correlated, which was similar to the correlation between perirenal fat thickness and body energy content reported previously [39,40]. In contrast, body protein had a minor or no correlation with the latter traits, but a negative one with live weight from the second parturition onwards. These results agree with rabbit doe maturation in this period, which would depend on maturity at the first AI and on reproductive success. Once maturity is reached (or nearly reached), the changes in body weight might be mainly associated with fat mobilization and/or deposition. This would agree with the moderated and positive correlation between live weight and body condition score [41].

The differences in chemical body composition and live weight between the first AI and the first parturition seemed to be related to the success in the first insemination, which was not influenced by body condition or live weight. This effect was reported by other authors [2,34,42]. It may be explained by the specific situation of non-pregnant does, which would use the entire intake for body protein and fat accretion, and accordingly, energy accretion, towards the completion of the final step to reach their maturity, where the fat deposition is much higher than the protein deposition (24 vs. 6% increment, respectively). Meanwhile, pregnant does have to supply gestation requirements that are especially important during the last 10 days of gestation, and which can impair not only fat content [17,43,44], but also protein balance [45], with respect to non-pregnant does. It must be taken into account that rabbit does inseminated later show a higher feed intake capacity [46] and lower growth requirements. Furthermore, rabbit does reduce feed intake in the days before parturition, contributing to the impairment of their nutrient balance [43].

The impairment of fertility in the second insemination reflects the specific situation of primiparous rabbit does, which suffer a negative energetic balance during their first pregnancy and lactation, compared to non-pregnant does, which seems to negatively influence reproductive performance and especially fertility [2,32,34], although this could be considered a natural adaptation, as commented before [13]. Live body weight at the moments of first AI, first parturition and second AI did not affect the fertility rate in the second parturition. This agrees with the findings of Rommers et al. [46], who did not observe any effect of body weight at the first AI on fertility rate in the first two parturitions. Nevertheless, the combination of a high live weight and high weight gain between the first two AI was also related to better fertility. Rabbit does that lose weight between the first two AI, regardless their initial body weight, were unable to present an acceptable reproductive performance in the second cycle. In this period, live weight was positively correlated with body fat and energy content, but not with protein, as does are finishing their protein accretion. The relationship between the fertility in the second AI and the chemical body composition at the first AI confirms the importance of the rearing management of rabbit does. Diets used [2,14,42,47] and the time of first insemination [15,48–50] influenced body composition, which consequently affected fertility. Therefore, at the end of the rearing period, reproductive does should reach an optimal body condition (minimal body protein, fat and energy content), assuring an adequate feed intake and body development, which enable high fertility rates during first parturitions [16]. In this sense, Pascual et al. [13] stated that body data at the first AI are a sign of doe soma and might be related to its productive potential. In this context, the supplementation of reproductive sows with certain daily amounts of amino acids enabled an adequate retention of nitrogen that led to an acceptable reproductive performance [51]. In this study, the threshold in the body composition at the first AI to avoid a sharp reduction in fertility in the second insemination might be set at 18% protein and 20% fat, but few does met this condition (12 and 7%, respectively). Other studies where the time of first insemination was delayed (to 18.4 or 19.5 weeks of age) rendered an increase in the body protein at that moment, as well as higher fertility values (83–87%) that were also associated with a higher fat content [38], although this was not always the case [52]. When the latter two studies were considered together, the does that were successful in the first five AI were lighter and had less body fat than the average (although their mean and range were similar to the current study), and the same body protein (although it was in the upper threshold that was previously mentioned: 17.9%) [53], suggesting the potential relevance of body protein at the beginning of the productive life. Another difference between these extraordinary does and the average population was their higher fat mobilization between the second AI and the first weaning (this was recovered between weaning and the third AI) [52]. These results partially differed from those of Theilgaard et al. [54], who indicated that there was no positive effect of perirenal fat at the first AI on reproductive life. In fact, a higher risk of culling was associated with high fat mobilization, although does that were too lean also seemed to increase their risk of culling. Similarly, Castellini et al. [50], using perirenal fat, found that does that were too fat and too lean (at AI) showed the poorest fertility. Recent results confirmed the negative effect of fatness at the first AI on the risk of being culled and litter size [14]. The disagreement among these studies and the current one might be related to the different fatness range (probably the absence of does that are too fat at the first AI in our work: maximal fat content at the first AI: 22.1%; Table 2), which might be associated with the time of the first AI (in the latter studies, nulliparous does were inseminated later than in the current one). Body fat and energy at the second AI was also related to fertility in the second AI, which might also reflect the observed positive influence of initial body condition on fertility at the second AI.

Quevedo et al. [55] suggested that the success of AI 11 days after parturition was conditioned by the rabbit doe's body condition at parturition rather than at insemination. This fact was not observed in this work, probably due to the fact that in our work, we measured body condition at parturition three days before birth (to avoid disturbing the doe) instead of after birth. This prevented the recording of an important proportion of fat mobilization, which was described by Savietto et al. [17]. Once confirmed, the BIA measurement did not alter the doe immediately after parturition (unpublished results); subsequent studies recorded it just after parturition of the doe [38,52].

The reproductive success at the beginning of reproductive life also influenced the fertility of the third AI. In fact, rabbits that are reproductively successful during the first two parturitions were more vulnerable, and consequently, their reproductive performance was impaired in the third AI if they did not have the opportunity to recover. In this sense, Castellini et al. [50,56] proposed the delaying of the second AI after weaning in order to allow rabbit does to recover properly from the first gestation lactation and continue their growth. Otherwise, reproductive success at the beginning of the reproductive life of rabbit does, when they cannot recover their body reserves, might worsen the rabbit doe's productivity and shorten its life span [13].

The absence of effects of the breeding system and farm may be explained by the synchronization of rabbit does at the moment of AI. Rebollar et al. [57] did not register a rhythm effect on fertility in the second parturition when using controlled lactation as the does' synchronization tool. However, in experiments without any synchronization method [56,58], it was concluded that the reproductive rhythm was related to fertility rate. Anyway, it must be stressed that the current study was not designed specifically to study the effect of breeding system on fertility rate. Besides, an influence of the breeding systems on the body condition could not be ruled out and further studies would be required to figure out the nature of their relationship.

There were also a positive influence of body protein and energy at the first AI on the percentage of kits born alive at the first parturition. Rommers et al. [15,59] also related a higher protein content at the first insemination, and lower fat content, with a trend towards increasing numbers of kits born alive and percentages of kits born alive at the first parturition. They observed that restricted nulliparous rabbit does inseminated at 17.5 weeks, compared with those fed ad libitum and inseminated at 14.5 weeks, showed higher protein and lower fat content with a similar live weight, and tended to increase the number of kits born alive and the percentage of kits born alive at the first parturition. In primiparous rabbit does, a better body condition (higher body protein, lipid, and energy contents) was also related with changes in metabolic signals (increase in serum protein and leptin concentrations) that might influence ovarian follicle and gamete quality and might be associated with an improved reproductive outcome [60]. Similarly, in sows, ovarian activity and oocyte quality were influenced by the body protein content [61,62]. Another explanation may involve body protein being related to fetal survival, as the increase in litter size has been related to a higher fetal survival, independently of ovulation rate [63].

Live weight and body fat content were also positively correlated with body energy content at the first AI, but negatively with body protein, and had no effect on the percentage of kits born alive. In contrast, Rommers et al. [46] reported that heavier females at the first AI (>4.0 kg) decreased the percentage of kits born alive, although this was combined with an improvement in the litter size at the first parturition. They related it to the development of the reproductive apparatus (larger uterine horns and more *corpora lutea* in the ovaries). Moreover, these results did not agree with those of Quevedo et al. [44], where rabbit does with higher perirenal fat thickness at 3 months of age tended to increase their percentage of kits born alive at the first parturition.

The reduction in the percentage of kits born alive in the second parturition, and the increase in the number of kits born alive, agreed with the results reported by Rommers et al. [46], who also observed, at the second parturition, a higher number of kits born alive and a lower percentage of kits born alive with respect to the first parturition. The different percentage of kits born alive observed in the two farms might be due to the different hybrids used and/or the different environmental management conditions in each farm. No effect of the breeding system or farm on percentage of kits born alive at the second parturition was detected. Weight and body composition at the first AI were not related to this trait. However, higher body protein at the second AI increased the percentage of kits born alive. This result is similar to that recorded at the first AI (on the percentage of kits born alive at the first parturition) and again suggests a positive role of nitrogen content on conception success and fetus viability.

5. Conclusions

An adequate body chemical composition at the first AI (around 18% protein and 20% fat) allowed a better fertility at the second AI. However, the consecutive reproductive success at the first and second AI did not allow rabbit does to recover body reserves and impaired fertility at the third AI. Body composition also affected the percentage of kits born alive at the first AI and at the second AI that increased with body protein content.

Consequently, rearing management (e.g., time of first AI) is key to avoiding low fertility rates in primiparous rabbit does. Furthermore, when reproductive success is reached, rabbit does may require alternative management strategies to recover their body condition. Finally, determining body composition at moments of AI may be an adequate tool to anticipate the reproductive success possibilities of rabbit does. Further studies considering different ages at the first AI and breeding systems are warranted to confirm these conclusions and to clarify the relationship between breeding systems and body conditions.

Author Contributions: N.N., M.T. and J.G. contributed to the concept and design of the experiment. M.T., N.N., and J.G. conducted the data processing, with the help of R.E.M. and M.A.I. in the statistical analysis and modelling. All authors contributed to the interpretation of the results. M.T., J.G. and N.N. wrote the manuscript. All authors have read and agreed to the published version of the manuscript.

Funding: This research was financed by the project MINECO-FEDER (AGL2008-00627) and the grant from IAMZ-CIHEAM obtained by Myriam Taghouti to develop her MSc. thesis.

Institutional Review Board Statement: The study was conducted according to the guidelines of the Declaration of Helsinki, and approved by the Institutional Review Board (or Ethics Committee) of Universidad Politécnica de Madrid (approved retrospectively on 28 July 2021).

Data Availability Statement: The data presented in this study are available on request from the corresponding author. The data are not publicly available due to confidentiality requirements agreed with one of the farms.

Acknowledgments: This work was promoted by Jhonny Demey (RIP, formerly of the Universidad Central de Venezuela, Maracay, Venezuela), who we all have in our memories.

Conflicts of Interest: The authors declare no conflict of interest.

References

1. Xiccato, G. Nutrition of lactating does. In Proceedings of the 6th World Rabbit Congress; Association Française de Cuniculture: Toulouse, France, 1996; Volume 1, pp. 29–47.
2. Xiccato, G.; Trocino, A. Energy and protein metabolism and requirements. In *Nutrition of the Rabbit*, 3rd ed.; CABI Publishing CAB International: Wallingford, UK, 2020; pp. 89–125.
3. Lebas, F. Biología (Biology). In *Enfermedades del Conejo*; Rosell, J.M., Ed.; Mundi-Prensa Libros S.A.: Madrid, Spain, 2000; Volume 1, Chapter 1, pp. 55–126, ISBN 8471149087.
4. Quevedo, F.; Pascual, J.J.; Blas, E.; Cervera, C. Influencia de la madre sobre el crecimiento y la mortalidad de los gazapos en cebo. In *Proc. XXVIII Symposium de Cunicultura*; Diputación General de Aragón: Teruel, Spain, 2003; pp. 115–122.
5. Rosell, J.M.; De La Fuente, L.F. Culling and mortality in breeding rabbits. *Prev. Vet. Med.* **2009**, *88*, 120–127. [CrossRef]
6. Rosell, J.M.; De La Fuente, L.F. Mastitis on Rabbit Farms: Prevalence and Risk Factors. *Animals* **2018**, *8*, 98. [CrossRef]
7. Rosell, J.M.; De La Fuente, L.F.; Carbajo, M.T.; Fernández, X.M. Reproductive Diseases in Farmed Rabbit Does. *Animals* **2020**, *10*, 1873. [CrossRef]
8. Hoy, S.; Verga, M. Welfare indicators. In *Recent Advances in Rabbit Sciences*; Maertens, L., Coudert, P., Eds.; ILVO: Melle, Belgium, 2006; pp. 71–74, ISBN 92-898-0030.EPS.
9. Dixon, L.M.; Hardiman, J.R.; Cooper, J.J. The effects of spatial restriction on the behavior of rabbits (Oryctolagus cuniculus). *J. Vet. Behav.* **2010**, *5*, 302–308. [CrossRef]
10. Szendrő, Z.; Trocino, A.; Hoy, S.; Xiccato, G.; Villagrá, A.; Maertens, L. A review of recent research outcomes on the housing of farmed domestic rabbits: Reproducing does. *World Rabbit Sci.* **2019**, *27*. [CrossRef]
11. López, M.; Cervera, C.; Pascual, J. Bienestar y resultados zootécnicos en conejas de aptitud carne. Revisión bibliográfica. *Inf. Tec. Econ. Agrar.* **2020**, 116. [CrossRef]
12. Szendrő, Z.; Gyovai, M.; Maertens, L.; Biró-Németh, E.; Radnai, I.; Matics, Z.; Princz, Z.; Gerencsér, Z.; Horn, P. Influence of birth weight and nutrient supply before and after weaning on the performance of rabbit does to age of the first mating. *Livest. Sci.* **2006**, *103*, 54–64. [CrossRef]
13. Pascual, J.J.; Savietto, D.; Cervera, C.; Baselga, M. Resources allocation in reproductive rabbit does: A review of feeding and genetic strategies for suitable performance. *World Rabbit Sci.* **2013**, *21*. [CrossRef]
14. Martínez-Paredes, E.; Ródenas, L.; Pascual, J.J.; Savietto, D. Early development and reproductive lifespan of rabbit females: Implications of growth rate, rearing diet and body condition at first mating. *Animals* **2018**, *12*, 2347–2355. [CrossRef]
15. Rommers, J.M.; Meijerhof, R.; Noordhuizen, J.P.; Kemp, B. Effect of feeding program during rearing and age at first insemination on performances during subsequent reproduction in young rabbit does. *Reprod. Nutr. Dev.* **2004**, *44*, 321–332. [CrossRef]
16. Castellini, C.; Bosco, A.D.; Alvarez, M.A.; Lorenzo, P.L.; Cardinali, R.; Rebollar, P.G. The main factors affecting the reproductive performance of rabbit does: A review. *Anim. Reprod. Sci.* **2010**, *122*, 174–182. [CrossRef] [PubMed]
17. Savietto, D.; Marono, S.; Martinez, I.; Martínez-Paredes, E.; Ródenas, L.; Cervera, C.; Pascual, J. Patterns of body condition use and its impact on fertility. *World Rabbit Sci.* **2016**, *24*, 39. [CrossRef]
18. Pereda, N. Evaluación de la Técnica del Análisis de Impedancia Bioeléctrica Para Predecir la Composición Corporal: Aplicación en Conejas Sometidas a Diferentes Sistemas de Alimentación Durante la Recría. Ph.D. Thesis, Universidad Politécnica de Madrid, Madrid, Spain, 2010; p. 194.
19. Fortun-Lamothe, L.; Lamboley-Gaüzère, B.; Bannelier, C. Prediction of body composition in rabbit females using total body electrical conductivity (TOBEC). *Livest. Prod. Sci.* **2002**, *78*, 133–142. [CrossRef]

20. Pascual, J.; Cervera, C.; Fernández-Carmona, J. The effect of dietary fat on the performance and body composition of rabbits in their second lactation. *Anim. Feed. Sci. Technol.* **2000**, *86*, 191–203. [CrossRef]
21. BOE.es—Documento BOE-A-2013-1337. Available online: https://www.boe.es/diario_boe/txt.php?id=BOE-A-2013-1337 (accessed on 28 July 2021).
22. Rebollar, P.; Milanés, A.; Pereda, N.; Millan, P.; Cano, P.; Esquifino, A.; Villarroel, M.; Silván, G.; Lorenzo, P.L. Oestrus synchronisation of rabbit does at early post-partum by doe–litter separation or ECG injection: Reproductive parameters and endocrine profiles. *Anim. Reprod. Sci.* **2006**, *93*, 218–230. [CrossRef]
23. Quintela, L.A.; Peña, A.I.; Vega, M.D.; Gullón, J.; Prieto, M.C.; Barrio, M.; Becerra, J.J.; Maseda, F.; Herradón, P.G. Ovulation induction in rabbit does submitted to artificial insemination by adding buserelin to the seminal dose. *Reprod. Nutr. Dev.* **2004**, *44*, 79–88. [CrossRef]
24. Nicodemus, N.; Pereda, N.; Romero, C.; Rebollar, P.G. Évaluatuion de la technique d´impédance bioélectrique (IBE) puor estimer la composition corporelle de lapines reproductrices. In Proceedings of the 13émes Jornées de la Recherche Cunicole (INRA/ITAVI), Le Mans, France, 17–18 November 2009; pp. 109–112.
25. Allison, P.D. *Logistic Regression Using the SAS® System: Theory and Application*; SAS Institute Inc.: Cary, NC, USA, 1999; 287p, ISBN 978-0-471-22175-3.
26. McCullagh, P.; Nelder, J.A. *Generalized Linear Models*, 2nd ed.; Chapman and Hall: New York, NY, USA, 1989; 506p, ISBN 0-412-31760-5.
27. Agresti, A. *Categorical Data Analysis*; John Wiley and Sons Inc.: New York, NY, USA, 1986; 373p, ISBN 978-0-471-22618-5.
28. Orelien, J.G. Model Fitting in PROC GENMOD. In Proceedings of the 2001 SUGI, Long Beach, CA, USA, 22–25 April 2001.
29. Ziegel, E.R.; Stokes, M.; Davis, C.; Koch, G. Categorical Data Analysis Using the SAS System. *Technometrics* **1996**, *38*, 413. [CrossRef]
30. Pasta, D.J.; Cisternas, M.G. Estimating Standard Errors for CLASS Variables in Generalized Linear Models Using PROC IML. In Proceedings of the 2003 SUGI, Seattle, WA, USA, 30 March–2 April 2003; pp. 264–282.
31. Theau-Clément, M. Preparation of the rabbit doe to insemination: A review. *World Rabbit Sci.* **2010**, *15*. [CrossRef]
32. Pascual, J.J.; Cervera, C.; Blas, E.; Fernández-Carmona, J. High energy diets for reproductive rabbit does: Effect of energy source. *Nutr. Abs. Revs.* **2003**, *73*, 27–39.
33. Fortun-Lamothe, L.; Prunier, A. Effects of lactation, energetic deficit and remating interval on reproductive performance of primiparous rabbit does. *Anim. Reprod. Sci.* **1999**, *55*, 289–298. [CrossRef]
34. Fortun-Lamothe, L. Energy balance and reproductive performance in rabbit does. *Anim. Reprod. Sci.* **2006**, *93*, 1–15. [CrossRef] [PubMed]
35. Alvarez, M.A.; Garcia, R.; Rebollar, P.; Revuelta, L.; Millán, P.; Lorenzo, P.L. Influence of metabolic status on oocyte quality and follicular characteristics at different postpartum periods in primiparous rabbit does. *Theriogenology* **2009**, *72*, 612–623. [CrossRef]
36. Xiccato, G.; Parigi-Bini, R.; Zotte, A.D.; Carazzolo, A.; Cossu, M.E. Effect of dietary energy level, addition of fat and physiological state on performance and energy balance of lactating and pregnant rabbit does. *Anim. Sci.* **1995**, *61*, 387–398. [CrossRef]
37. Parigi Bini, R.; Xiccato, G.; Dalle Zotte, A.; Castellini, C.; Stradaioli, G. Effect of remating interval and diet on the performance and energy balance of rabbit does. In Proceedings of the 6th World Rabbit Congress, Toulouse, France, 9–12 July 1996; pp. 253–258.
38. Delgado, R.; Nicodemus, N.; Abad-Guamán, R.; Sastre, J.; Menoyo, D.; Carabaño, R.; García, J. Effect of dietary soluble fibre and n-6/n-3 fatty acid ratio on growth performance and nitrogen and energy retention efficiency in growing rabbits. *Anim. Feed. Sci. Technol.* **2018**, *239*, 44–54. [CrossRef]
39. Pascual, J.J.; Castella, F.; Cervera, C.; Blas, E.; Fernández-Carmona, J. The use of ultrasound measurement of perirenal fat thickness to estimate changes in body condition of young female rabbits. *Anim. Sci.* **2000**, *70*, 435–442. [CrossRef]
40. Pascual, J.J.; Blanco, J.; Piquer, O.; Quevedo, F.; Cervera, C. Ultrasound measurements of perirenal fat thickness to estimate the body condition of reproducing rabbit does in different physiological states. *World Rabbit Sci.* **2010**, *12*, 07–21. [CrossRef]
41. De La Fuente, L.; Rosell, J. Body weight and body condition of breeding rabbits in commercial units1. *J. Anim. Sci.* **2012**, *90*, 3252–3258. [CrossRef] [PubMed]
42. Rebollar, P.; Pereda, N.; Schwarz, B.; Millan, P.; Lorenzo, P.L.; Nicodemus, N. Effect of feed restriction or feeding high-fibre diet during the rearing period on body composition, serum parameters and productive performance of rabbit does. *Anim. Feed. Sci. Technol.* **2011**, *163*, 67–76. [CrossRef]
43. Pascual, J.J.; Motta, W.; Cervera, C.; Quevedo, F.; Blas, E.; Fernández-Carmona, J. Effect of dietary energy source on the performance and perirenal fat thickness evolution of primiparous rabbit does. *Anim. Sci.* **2002**, *75*, 267–279. [CrossRef]
44. Quevedo, F.; Cervera, C.; Blas, E.; Baselga, M.; Costa, C.; Pascual, J.J. Effect of selection for litter size and feeding programme on the performance of young rabbit females during rearing and first pregnancy. *Anim. Sci.* **2005**, *80*, 161–168. [CrossRef]
45. Parigi Bini, R.; Xiccato, G.; Cinetto, M. Utilizzazione e ripartizione dell'energia e della proteina digeribile in coniglie non gravide durante la prima lattazione. *Zootec. E Nutr. Anim.* **1991**, *17*, 107–120.
46. Rommers, J.M.; Meijerhof, R.; Noordhuizen, J.P.T.M.; Kemp, B. Relationships between body weight at first mating and subsequent body development, feed intake, and reproductive performance of rabbit does. *J. Anim. Sci.* **2002**, *80*, 2036–2042. [CrossRef] [PubMed]
47. Fortun-Lamothe, L. Effects of pre-mating energy intake on reproductive performance of rabbit does. *Anim. Sci.* **1998**, *66*, 263–269. [CrossRef]

48. Bonanno, A.; Mazza, F.; Di Grigoli, A.; Alicata, M.L. Assessement of a method for evaluating the body condition of lactating rabbit does: Preliminary results. In Proceedings of the ASPA 16th Congress, Torino, Italy, 28–30 June 2005; Volume 4, p. 560.
49. Bonanno, A.; Mazza, F.; Di Grigoli, A.; Alicata, M.L. Body condition score and related productive responses in rabbit does. In Proceedings of the 9th World Rabbit Congress, Verona, Italy, 10–13 June 2008; pp. 297–301.
50. Castellini, C.; Bosco, A.D.; Cardinali, R. Long term effect of post-weaning rhythm on the body fat and performance of rabbit doe. *Reprod. Nutr. Dev.* **2006**, *46*, 195–204. [CrossRef] [PubMed]
51. Everts, H.; Dekker, R.A. Effect of nitrogen supply on the retention and excretion of nitrogen and on energy metabolism of pregnant sows. *Anim. Sci.* **1994**, *59*, 293–301. [CrossRef]
52. Delgado, R.; Abad-Guamán, R.; De La Mata, E.; Menoyo, D.; Nicodemus, N.; García, J.; Carabaño, R. Effect of dietary supplementation with arginine and glutamine on the performance of rabbit does and their litters during the first three lactations. *Anim. Feed. Sci. Technol.* **2017**, *227*, 84–94. [CrossRef]
53. Delgado, R.; Abad-Guamán, R.; Carabaño, R.; García, J.; Nicodemus, N. La movilización de grasa entre la segunda insemi-nación y el destete está positivamente relacionada con el éxito reproductivo en conejas primíparas. In Proceedings of the XLII Symposium de Cunicultura de ASESCU, Murcia, Spain, 29–30 April 2017; pp. 56–59, ISBN 978-84-92928-77-4.
54. Theilgaard, P.; Sanchez, J.; Pascual, J.; Friggens, N.; Baselga, M. Effect of body fatness and selection for prolificacy on survival of rabbit does assessed using a cryopreserved control population. *Livest. Sci.* **2006**, *103*, 65–73. [CrossRef]
55. Quevedo, F.; Cervera, C.; Blaß, E.; Baselga, M.; Pascual, J.J. Long-term effect of selection for litter size and feeding programme on the performance of reproductive rabbit does 2. Lactation and growing period. *Anim. Sci.* **2006**, *82*, 751–762. [CrossRef]
56. Castellini, C.; Bosco, A.D.; Mugnai, C. Comparison of different reproduction protocols for rabbit does: Effect of litter size and mating interval. *Livest. Prod. Sci.* **2003**, *83*, 131–139. [CrossRef]
57. Rebollar, P.; Pérez-Cabal, M.; Pereda, N.; Lorenzo, P.; Arias-Álvarez, M. Effects of parity order and reproductive management on the efficiency of rabbit productive systems. *Livest. Sci.* **2009**, *121*, 227–233. [CrossRef]
58. Xiccato, G.; Trocino, A.; Boiti, C.; Brecchia, G. Reproductive rhythm and litter weaning age as they affect rabbit doe performance and body energy balance. *Anim. Sci.* **2005**, *81*, 289–296. [CrossRef]
59. Rommers, J.M.; Meijerhof, R.; Noordhuizen, J.P.T.M.; Kemp, B. Effect of different feeding levels during rearing and age at first insemination on body development, body composition, and puberty characteristics of rabbit does. *World Rabbit Sci.* **2001**, *9*, 101–108. [CrossRef]
60. Arias-Alvarez, M.; García-García, R.M.; Rebollar, P.G.; Nicodemus, N.; Millán, P.; Revuelta, L.; Lorenzo, P.L. Follicular, Oocyte and Embryo Features Related to Metabolic Status in Primiparous Lactating does Fed with High-Fibre Rearing Diets. *Reprod. Domest. Anim.* **2009**, *45*, e91–e100. [CrossRef]
61. Clowes, E.J.; Aherne, F.X.; Foxcroft, G.R.; Baracos, V.E. Selective protein loss in lactating sows is associated with reduced litter growth and ovarian function. *J. Anim. Sci.* **2003**, *81*, 753–764. [CrossRef] [PubMed]
62. Clowes, E.J.; Aherne, F.X.; Schaefer, A.L.; Foxcroft, G.R.; Baracos, V.E. Parturition body size and body protein loss during lactation influence performance during lactation and ovarian function at weaning in first-parity sows1. *J. Anim. Sci.* **2003**, *81*, 1517–1528. [CrossRef]
63. García, M.L.; Baselga, M. Genetic response to selection for reproductive performance in a maternal line of rabbits. *World Rabbit Sci.* **2010**, *10*, 71–76. [CrossRef]

 *animals* 

Article

Genotype Imputation to Improve the Cost-Efficiency of Genomic Selection in Rabbits

Enrico Mancin [1,†], Bolívar Samuel Sosa-Madrid [2,*,†], Agustín Blasco [2] and Noelia Ibáñez-Escriche [2,*]

1 Department of Agronomy, Food, Natural Resources, Animals and Environment (DAFNAE), University of Padova, viale dell'Università 16, 35020 Legnaro, PD, Italy; enrico.mancin@phd.unipd.it
2 Institute for Animal Science and Technology, Universitat Politècnica de València, 46022 Valencia, Spain; ablasco@dca.upv.es
* Correspondence: bosomad@posgrado.upv.es (B.S.S.-M.); noeibes@dca.upv.es (N.I.-E.); Tel.: +34-963-877-438 (N.I.-E.)
† E.M. and B.S.S.-M. share the first author position.

Simple Summary: Genotyping costs are still the major limitation for the uptake of genomic selection by the rabbit meat industry, as a large number of genetic markers are needed for improving the prediction of breeding values by genomic data. In this study, several genotyping strategies were examined through simulation scenarios to disentangle the best feasible options of implementing genomic selection in rabbit breeding programs. Most scenarios emphasized the genotyping of candidate animals with a low Single Nucleotide Polymorphism (SNP) density platform. Imputation accuracies were high for the scenarios with ancestors genotyped at high or medium SNP-densities. However, the scenario with male ancestors genotyped at high SNP-density and only dams genotyped at medium SNP-density showed the best economically feasible strategy, taking into account the trade-off among genotyping costs, the accuracy of breeding values and response to selection. The results confirmed that by combining the imputation technique with a mindful selection of the animals to be genotyped, it is possible to achieve better performance than Best Linear Unbiased Prediction (BLUP), reducing genotyping cost at the same time.

Citation: Mancin, E.; Sosa-Madrid, B.S.; Blasco, A.; Ibáñez-Escriche, N. Genotype Imputation to Improve the Cost-Efficiency of Genomic Selection in Rabbits. *Animals* **2021**, *11*, 803. https://doi.org/10.3390/ani11030803

Academic Editors: Rosa M. García-Garcia and Maria Arias Alvarez

Received: 2 February 2021
Accepted: 5 March 2021
Published: 13 March 2021

Abstract: Genomic selection uses genetic marker information to predict genomic breeding values (gEBVs), and can be a suitable tool for selecting low-hereditability traits such as litter size in rabbits. However, genotyping costs in rabbits are still too high to enable genomic prediction in selective breeding programs. One method for decreasing genotyping costs is the genotype imputation, where parents are genotyped at high SNP-density (HD) and the progeny are genotyped at lower SNP-density, followed by imputation to HD. The aim of this study was to disentangle the best imputation strategies with a trade-off between genotyping costs and the accuracy of breeding values for litter size. A selection process, mimicking a commercial breeding rabbit selection program for litter size, was simulated. Two different Quantitative Trait Nucleotide (QTN) models (QTN_5 and QTN_44) were generated 36 times each. From these simulations, seven different scenarios (S1–S7) and a further replicate of the third scenario (S3_A) were created. Scenarios consist of a different combination of genotyping strategies. In these scenarios, ancestors and progeny were genotyped with a mix of three different platforms, containing 200,000, 60,000, and 600 SNPs under a cost of EUR 100, 50 and 11 per animal, respectively. Imputation accuracy (IA) was measured as a Pearson's correlation between true genotype and imputed genotype, whilst the accuracy of gEBVs was the correlation between true breeding value and the estimated one. The relationships between IA, the accuracy of gEBVs, genotyping costs, and response to selection were examined under each QTN model. QTN_44 presented better performance, according to the results of genomic prediction, but the same ranks between scenarios remained in both QTN models. The highest IA (0.99) and the accuracy of gEBVs (0.26; QTN_44, and 0.228; QTN_5) were observed in S1 where all ancestors were genotyped at HD and progeny at medium SNP-density (MD). Nevertheless, this was the most expensive scenario compared to the others in which the progenies were genotyped at low SNP-density (LD). Scenarios with low average costs presented low IA, particularly when female ancestors were genotyped at LD (S5) or non-genotyped (S7). The S3_A, imputing whole-genomes, had the lowest accuracy of gEBVs (0.09), even worse than Best Linear Unbiased Prediction (BLUP). The best trade-off between genotyping costs and the accuracy of gEBVs (0.234; QTN_44 and 0.199) was in S6, in which dams were genotyped with MD whilst grand-dams were non-genotyped. However, this relationship would depend mainly on the distribution of QTN and SNP across the genome, suggesting further studies on the characterization of the rabbit genome in the Spanish lines. In summary, genomic selection with genotype imputation is feasible in the rabbit industry, considering only genotyping strategies with suitable IA, accuracy of gEBVs, genotyping costs, and response to selection.

Keywords: genomic selection; imputation; litter size; rabbits; genomic simulation

1. Introduction

The rabbit industry still plays an important role throughout the agricultural sector in some European countries, such as Spain, Italy, France, Hungary, Portugal, Germany, Belgium, Poland and Malta [1,2]. In recent years, the rabbit industry is currently facing a critical period, mainly due to the increase in feeding, management costs and a constant decline in rabbit meat consumption. Hence, farmers and researchers have been looking for promising strategies to improve the current situation, one of them being the optimization of genetic selection by using genomic information. Genetic selection can improve productive and reproductive traits, such as meat characteristics, reducing feeding and management costs, which makes rabbit more appealing for consumers, and hence, will aid the rabbit industry [1]. Reproductive traits, especially litter size, are those with relevant economic weight in the rabbit industry [3]. However, the response to selection for this trait has been relatively low using traditional selection by Best Linear Unbiased Prediction (BLUP), mainly due to its low-heritability [4–6]. Selection using genomic information can be an efficient tool to guarantee a higher genetic gain for traits with low heritability and measured in only one sex such as litter size. [7,8]. Genomic selection has generally showed better accuracy for the predicted breeding value (BV) [9] due to a potentially more accurate kindship estimation between animals. This method has produced positive selection results for traits in dairy cattle [10,11], poultry [12] and pigs [13–15]. However, in rabbits, a high-density commercial Single Nucleotide Polymorphism (SNP) platform (~200K SNPs) was not available until 2015, which has delayed genomic selection application [16]. Further, additional issues such as the small economic value of paternal rabbits, the costs of the commercial SNP platform, and the short generation interval are still limiting genomic selection as an evaluating method [17]. Strategies allowing us to diminish high genotyping costs are vital in the rabbit industry. The imputation of a low SNP-density platform using a high SNP-density platform has been carried out in other species, keeping or improving the genetic progress. The technique normally consists of genotyping ancestors (grandparents and parents) at high SNP-density in order to assign (impute) the most likely SNP allele to missing genotypes of young candidate animals genotyped at lower SNP-density [10,18,19]. This approach depends on multiple factors concerning the level of SNP density, methods, the structure and size of the reference population, the minor allele frequency of missing or untyped SNPs, the genetic architecture of traits and the particular breeding scheme of a livestock species [20–22].

The aim of this study was to disentangle the most appropriate imputation strategies for implementing genomic selection in maternal rabbit breeding schemes. Under this aim, the imputation accuracy was evaluated from low to moderate SNP-density platforms considering the cost-effectiveness of each strategy. In addition, we investigated how the imputation strategies influence the estimation of breeding values and consequently the response to selection.

2. Materials and Methods

2.1. Simulation Structure

We used stochastic simulation to create the populations for developing imputation analyses. The structure of simulations is shown in Figure 1. The simulations were performed by the AlphaSim program [23], available at https://alphagenes.roslin.ed.ac.uk/wp/software-2/alphasimr/ (accessed on 10 June 2020). The objective was to simulate a rabbit population mimicking a single maternal line under the common rabbit breeding scheme of a small nucleus breeding. The simulated trait was litter size at birth. First, founder genomes were generated. The rabbit genome was simulated by sampling 2000 haplotype sequences for each of 20 chromosomes using the Markovian Coalescent Simulator (*MaCS*) [24]. Each chromosome was 100 cM long, as genetic distance, and included 124.43×10^6 base pairs, as physical distance. Chromosomes were simulated according to rabbit population history defined by the *MaCS* program using a per-site recombination rate 8.57×10^{-9}, a per-site mutation rate 1.74×10^{-9}, and an effective population size varying over time (according to "Internal rabbit" as an option of population history in *MaCS*). Later, in the first generation of the foundation of the maternal line (initial population), quantitative trait nucleotides (QTNs) were chosen randomly from the segregating sequence variants and an equal number of QTNs were assigned to each chromosome. The QTNs had additive effects sampled from a Gamma distribution with a shape of 0.60 and a scale of 0.80. These parameters were chosen after exploratory analysis evaluating various values of Gamma distribution parameters against selection response after 20 generations (analyses not shown). The heritability and residual variance were calculated relative to the additive variance in the initial population. The heritability was 0.113 and the trait genetic variance was 0.675 in the base population [5,25].

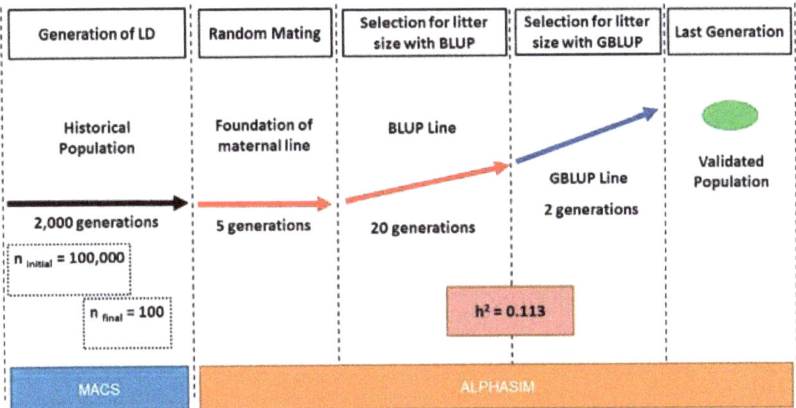

Figure 1. Design of simulations. A first period, generating the linkage disequilibrium (LD) across the rabbit genome using *MaCS* program. The remaining periods are carried out using *AlphaSim*: foundation of maternal line, selection for litter size with Best Linear Unbiased Prediction (BLUP), and with Genomic Best Linear Unbiased Prediction (GBLUP).

After five generations of random mating, for details as in [26], a base population was established using 138 does and 77 sires. The selection was carried out to select 70 does and 35 sires per generation, involving 20 generations with the BLUP method (see the parameters in Table S1). Litter size was assumed as a trait of sex-limited expression, hence, only does presented phenotypic records. In every generation, the 70 does from the previous generation with the highest estimated breeding values (EBVs) were selected to produce the next generation. A total of 35 males were also selected, which stand for 25 principal and 10 surrogate sires used in practice. In addition, two further generations evaluated with the Genomic Best Linear Unbiased Prediction (GBLUP) method were generated to obtain the last population, which consisted of a progeny with 1500 does and 1500 males. The two selection methods used in these simulations (BLUP and GBLUP) are directly implemented in the *AlphaSim* program.

The high SNP-density platform (HD; 200,000 SNPs) used in genomic selection was generated by AlphaSim. Another two SNP platforms were set up according to SNP-density size: medium SNP-density (MD; 6000 SNPs) and low SNP-density (LD; 600 SNPs) platforms. All SNPs were assigned proportionally per chromosome. Genetic markers for the MD platform were selected at random from the HD platform and, consequently, genetic markers for the LD platform were selected at random from the MD platform (see the parameters in Table S1). Regarding costs for implementing genomic selection, we only considered the costs of genotyping based on the SNP platforms in the last three generations; therefore, other indirect costs were ignored. Each platform contains 96 "BeadChips" (genotyping proof). We assumed that the cost of the HD platform is EUR 9600. For the other platforms, the costs are around EUR 4800 and EUR 1056 for MD and LD, respectively. This information came from Thermo Fisher Scientific Inc. supplied by genotyping budgets for previous genomic experiments in the Animal Breeding Group at the Institute for Animal Science and Technology in the Universitat Politècnica de València.

An underlying polygenic nature was assumed for litter size, representing the genetic architecture of complex traits [27]. We considered two QTN models, depicting: (1) a polygenic trait controlled by many QTNs with a total of 880 (44 per chromosome; QTN_44) and (2) a polygenic trait controlled by a small number of 100 QTNs (5 per chromosome; QTN_5). Simulated data were generated from 36 replicates for every QTN model. The results were summarized over 36 replicates within the QTN model, and presented graphically. The graphics were obtained by the R program [28].

2.2. Imputation Strategies

All imputation analyses were developed by AlphaImpute program *v1.96*. This program is available at https://alphagenes.roslin.ed.ac.uk/wp/software-2/alphaimpute/ (accessed on 15 August 2020). AlphaImpute uses a hybrid imputation algorithm in which the first step consists of a long-range phasing process, followed up haplotypes library construction, and finally, pedigree-based imputations are performed [29].

Imputation accuracy (IA) was measured by the Pearson's correlation between the imputed allele and the true genotype at untyped SNP markers. The correlation was computed one individual at a time and averaged over individuals [22]. This parameter stands for the genotype probability for reasons sketched out in [30]. Genotype yield was also computed for each imputation strategy below, as the percentage of the SNP allele calls at untyped SNP markers after the imputation process.

To assess the trade-off between IA, genomic prediction accuracy and genotyping cost, a number of hypothetical test scenarios were established, as outlined below and represented in Table 1. We analyzed the simulated data in seven sets of hypothetical scenarios with a replicate of the third scenario (S3_A). All grand-sires and sires were genotyped at HD platforms, whilst grand-dams and dams were genotyped according to the imputation strategy. The progenies were genotyped at the LD platform except the first scenario (S1), as this scenario used MD platforms to genotype progeny and HD platforms for the genotyping of grand-dams and dams. The second scenario (S2) was like the first scenario, but LD platforms were used for progeny genotyping. The third scenario (S3) used MD platforms for the genotyping of grand-dams, and only half of the progeny was genotyped. S3_A included the further half of the progeny, having their imputed whole-genomes. The fourth scenario (S4) used MD platforms for the genotyping of grand-dams and dams, whilst the fifth scenario (S5) used LD platforms. The sixth scenario (S6) used MD platforms for genotyping dams, but the grand-dams were non-genotyped. The seventh scenario (S7) had non-genotyped grand-dams and dams. These above-mentioned scenarios summarized all exploratory analyses concerning imputation strategies.

Table 1. Structure of imputation strategies and number of genotyped animals for implementing genomic selection in rabbits.

Imputation Strategy	Training Populations				For Imputation	Validated Population (Genomic Prediction)
	26th Generation		27th Generation		28th Generation	
	Grand-Dams (150)	Grand-Sires (35)	Dams (150)	Sires (35)	Progeny (1500)	Progeny (1500)
S1	HD	HD	HD	HD	MD	*i*-HD
S2	HD	HD	HD	HD	LD	*i*-HD
S3	MD	HD	HD	HD	$\frac{1}{2}$ LD	$\frac{1}{2}$ *i*-HD + $\frac{1}{2}$ NG
S4	MD	HD	MD	HD	LD	*i*-HD
S5	LD	HD	LD	HD	LD	*i*-HD
S6	NG	HD	MD	HD	LD	*i*-HD
S7	NG	HD	NG	HD	LD	*i*-HD
S3_A	MD	HD	HD	HD	$\frac{1}{2}$ LD	$\frac{1}{2}$ *i*-HD + $\frac{1}{2}$ *i*-WG

The "S" stands for scenarios. Animals of every scenario were genotyped according to different Single Nucleotide Polymorphism (SNP)-density platforms. HD: high SNP-density platform (200,000 SNPs), MD: medium SNP-density platform (6000 SNPs), LD: low SNP-density platform (600 SNPs). *i*-HD: animals with imputed genotypes to high SNP-density. *i*-WG: animals with imputed whole-genome. NG: non-genotyped animals.

2.3. Estimating Breeding Values and Response to Genomic Selection

Typically, genomic selection entails a training population or reference population (with genotyped and phenotyped individuals) and an evaluated population of young candidates (with only genotyped individuals), using pseudo-phenotypes for sex-limited traits [7,31]. In this study, the rabbit reference population comprised up to 300 females (Table 2). As the reference population was small, genomic prediction was estimated using the Single-Step GBLUP (ssGBLUP) algorithm. It allows us to evaluate jointly non-genotyped and genotyped animals, combining pedigree and markers information into one matrix [32,33]. Otherwise, prediction by GBLUP hinders a greater rate of genetic progress compared to BLUP selection, with numerically small reference populations [11,34].

Table 2. Number of data used in the genomic analyses per generation.

Generation	Pedigree	Phenotypic [1]	Genomic [2]
23th	300	150	0:0
24th	300	150	0:0
25th	300	150	0:0
26th	300	150	35:150
27th	300	150	35:150
28th	1500	0	0:1500

[1] Selected does are 47% of total females (150). Each female contributes to the progenies with the same proportion of males and females (1:1). [2] Number of males: females with genotypes in each generation. The number varies according to the imputation strategy given standard *BLUPF90* parameters of quality control.

In each simulation, the accuracy of genomic breeding values (gEBVs) was estimated using the imputed genotypes of the eight scenarios described above. In addition, the EBVs of candidate animals were also estimated using only pedigree information (BLUP scenarios). The model was the same for BLUP and ssGBLUP:

$$y = \mathbf{1}\mu + \mathbf{Z}\mathbf{a} + \mathbf{e} \tag{1}$$

where $\boldsymbol{y}$ is the vector of phenotypes, μ is the population's mean, $\boldsymbol{a}$ is the vector of additive genetic effects of animals, $\boldsymbol{e}$ is the vector of residuals, and $\boldsymbol{Z}$ is the incidence matrix for additive genetic effects. Residual effects are sampled from distribution $N(\mathbf{0},\boldsymbol{I}\sigma_e^2)$. In BLUP, random additive genetic effects are sampled from distribution $N(\mathbf{0},\boldsymbol{A}\sigma_a^2)$; σ_a^2 is the genetic additive variance and $\boldsymbol{A}$ is the identity by descent (IBD) relationship matrix constructed from pedigree information. In ssGBLUP, additive genetic effects are sampled from distribution with $\boldsymbol{a} \sim N(\boldsymbol{0},\boldsymbol{H}\sigma_a^2)$. $\boldsymbol{H}$ matrix can be interpreted as the (co)variances of multivariate normal distributions of additive effects of both genotyped and non-genotyped animals [32,33]. In ssGBLUP, the inverse of the (co)variance structure of random effect was replaced by $\boldsymbol{H}^{-1}$, described as:

$$\boldsymbol{H}^{-1} = \boldsymbol{A}^{-1} + \begin{pmatrix} 0 & 0 \\ 0 & \boldsymbol{G}^{-1} - \boldsymbol{A}_{22}^{-1} \end{pmatrix} \quad (2)$$

where $\boldsymbol{A}^{-1}$ is the inverse of pedigree matrix and $\boldsymbol{A}_{22}^{-1}$ is the inverse of the sub-covariance structure containing only genotyped animals. $\boldsymbol{G}^{-1}$ is the inverse of the matrix built as described in [35]. According to Garcia-Baccino et al. [36] and Aguilar et al. [37], $\boldsymbol{A}^{-1}$ was computed accounting for inbreeding in order to avoid inflation (bias), particularly for BLUP scenarios, and to minimize blending problems between genomic and pedigree matrices.

All analyses were performed using the *BLUPF90* suite of programs under their standard parameters of quality control for genomic database [38]. Pedigree information from the 23rd to 28th generations was retained (Table 2). Phenotypic data comprised does from the 23rd to 27th generations. Phenotypic records of progenies at the last generation (28th generation) were not considered because they represented young candidate animals. Although all scenarios (S1–S7) used genotypes of grand-sires and sires (26th and 27th generations) for genomic prediction, a few scenarios (S1 and S2) also used all genotypes of grand-dam and dams. In the other scenarios (S3–S7), genotypes from grand-dams and dams were discarded due to low IA and the large number of missing SNPs [39]. On the other hand, in S3, EBVs of non-genotyped progeny were estimated as means of their parents' EBVs (Table 1).

The accuracy of gEBVs was measured by the Pearson's correlation between predicted and true breeding values (TBVs) on the progeny at the validated population, animals belonging to the 28th generation. To assess the gain of accuracies when genomic information was introduced, the mathematical differences of accuracies between each scenario (S1–S7) and BLUP were calculated within each simulation.

This study emphasizes the genetic improvement via doe selection. Hence, response to selection was calculated by subtracting the TBV mean of young candidates (1500 does), within each imputation strategy, from the TBV mean of the 150 selected individuals. We also computed the percentage of candidate animals correctly selected. Results within each scenario are exposed, comparing IA, the accuracy of gEBVs, the selection response and genotyping cost.

3. Results

3.1. Simulation Outcomes

The outcomes from the 36 simulations were similar with regards to the response to the selection of experiments using Spanish rabbit lines and the BLUP method for genetic evaluations: an average of 2.53 kits after 20 generations of selection [4,5]. For genomic selection, the average response to selection across QTN models computed for the first two generations with HD platforms was 0.15 kits (0.13 and 0.17 for QTN_5 and QTN_44 models, respectively), which corresponds to a response per generation of 0.08 (ranging between 0.002 and 0.20) across QTN models. These results are in line with outcomes of pig genomic selection experiments [40], as hitherto there have been no empirical genomic selection experiments in rabbits. As expected, the additive genetic variance was smaller in QTN_5 than in QTN_44. As expected, scenarios with 44 QTNs per chromosome presented higher additive genetic variance than QTN_5 ones. This is strictly correlated with the number of QTNs that were fixed during haplotype creation. On the other hand, this discrepancy did not have a clear effect on the response to selection as reported on Figure 2.

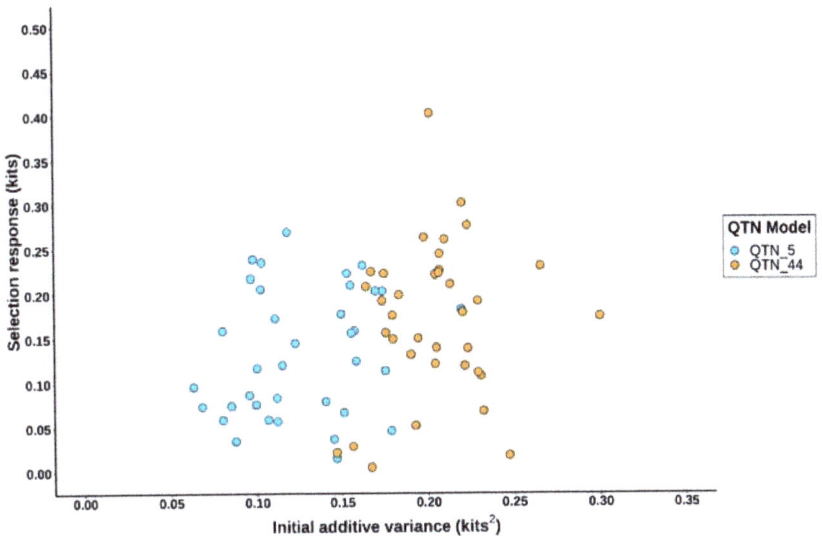

Figure 2. Relationship between initial additive variance and the response to genomic selection. The initial additive variance is the additive genetic variance at the 25th generation in which the genomic evaluations begin. QTN_5 stands for the QTN model with 5 QTNs per chromosome and QTN_44 stands for the QTN model with 44 QTNs per chromosome. The simulated rabbit genome comprised 20 chromosomes.

3.2. Performance of Imputation Strategies

The average of IA for the animals in the validated population (28th generation) was computed for each scenario. The results are shown in Figure 3. No difference, in terms of IA, was shown between two different QTN models. The greatest IA (0.99) was achieved in the S1. The IA decreased to 0.941 (S2) when the LD platform was used. A great decline in IA was found in S3_A (0.797) when the half validated population had imputed the whole-genome, unlike S3 (0.935). IA decreased to 0.918 when dams of the training population were genotyped with the MD platform (S4). S6 also presented an intermediate value (0.902) between all scenarios. Using LD platforms on does, IA decreases to 0.858 (S5); whereas IA declines to 0.811 with non-genotyped does (S7). In addition, lower standard deviation was obtained in S1 (0.0003), S2 (0.002) and S3 (0.002). Conversely, S7 and S5 presented the highest values of standard deviation (up to 0.005 and 0.004, respectively).

In the same way as IA, genotype yields across QTN models were higher in S1 (0.999) and S2 (0.984). The values were intermediate for S3 (0.952), S3_A (0.951) and S4 (0.945). Lower genotype yields were presented in S6 (0.899), S5 (0.893) and S7 (0.866).

3.3. Gemomic Prediction vs. Pedigree-Based Analyses

The accuracy of gEBVs presented, on average, a higher accuracy of prediction for QTN_44 than for QTN_5 (Figure 4). The accuracies of gEBVs of S1 were 0.26 ± 0.015 (QTN_44) and 0.228 ± 0.014 (QTN_5), representing mean ± standard error. S2 presented lower accuracy for both QTN models, 0.237 ± 0.014 (QTN_44) and 0.205 ± 0.013 (QTN_5). S3 presented similar values compared to S2, 0.237 ± 0.016 (QTN_44) and 0.193 ± 0.015 (QTN_5). S4 and S6 were very similar to S3 in QTN_44, with 0.232 ± 0.016 (S4) and 0.234 ± 0.016 (S6), and had slightly higher values than S3 in QTN_5, with 0.197 ± 0.015 (S4) and 0.199 ± 0.016 (S6). Conversely, lower accuracy values were found for S5 and S7, with 0.223 ± 0.016 (S5) and 0.22 ± 0.015 (S7) for QTN_44, and 0.185 ± 0.014 (S5) and 0.18 ± 0.015 (S7) for QTN_5. The accuracy of gEBVs drastically declined in S3_A, 0.09 ± 0.016 (QTN_44) and 0.09 ± 0.013 (QTN_5), because of low IA presented in progeny with its imputed whole-genome. The S3_A values were even worse than BLUP accuracies, which showed 0.202 ± 0.014 (QTN_44) and 0.166 ± 0.015 (QTN_5).

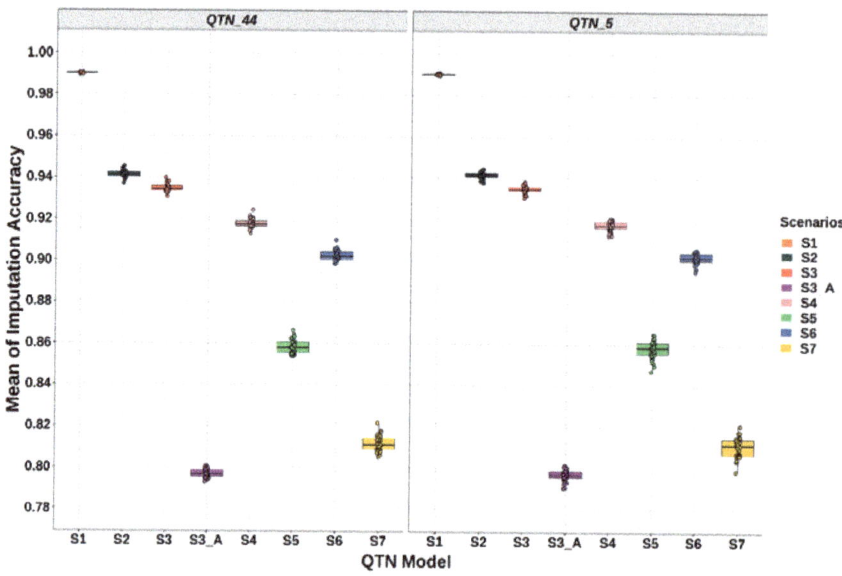

Figure 3. Imputation accuracy for each imputation strategy within QTN models. "S" stands for scenario. QTN_5 stands for the QTN model with 5 QTNs per chromosome and QTN_44 stands for the QTN model with 44 QTNs per chromosome. The simulated rabbit genome comprised 20 chromosomes. Each dot stands for a simulation value within each scenario.

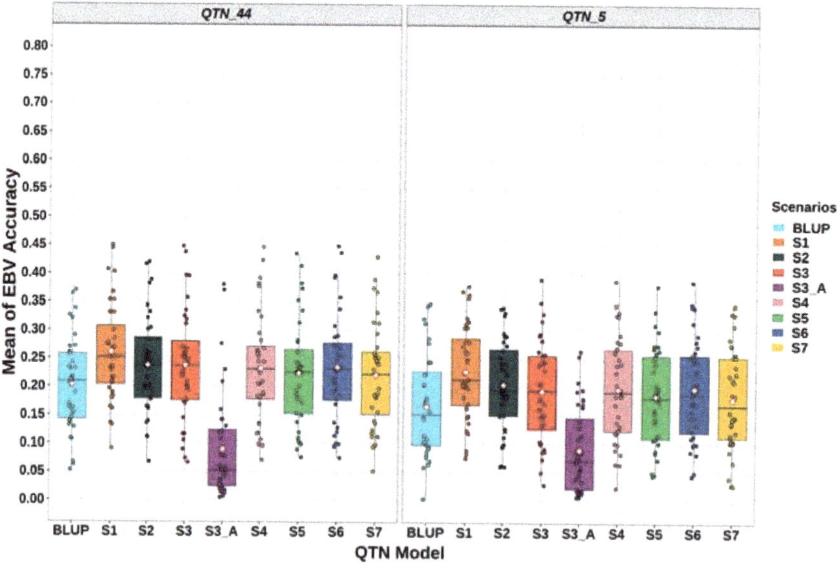

Figure 4. Accuracy of the estimated breeding values (EBVs) for each scenario within QTN models. "S" stands for scenario. QTN_5 stands for the QTN model with 5 QTNs per chromosome and QTN_44 stands for the QTN model with 44 QTNs per chromosome. The simulated rabbit genome comprised 20 chromosomes. Each dot stands for a simulation value within each scenario. White rhombus stands for the mean by scenario.

The results presented a great variability on simulation-based analysis (on average S.D. of 0.08). Figure 5 shows the mathematical differences of accuracy between all genomic scenarios (S1–S7) and BLUP scenarios for each simulation. S1 presented better results than BLUP over 36 simulations for QTN_5, whilst it had better results in 95% of simulations for QTN_44. S3 outperformed in 82% of simulations, whilst S2 outperformed in 75% for QTN_5. By contrast, S3 was in the 75% of simulations better than BLUP for QTN_44, whilst S2 was in the 80%. S4 and S6 performed in a rank of 70–80%, whereas S5 and S7 performed in a rank of only 50–60%. S3_A was better than the BLUP scenario in only 25% of simulations.

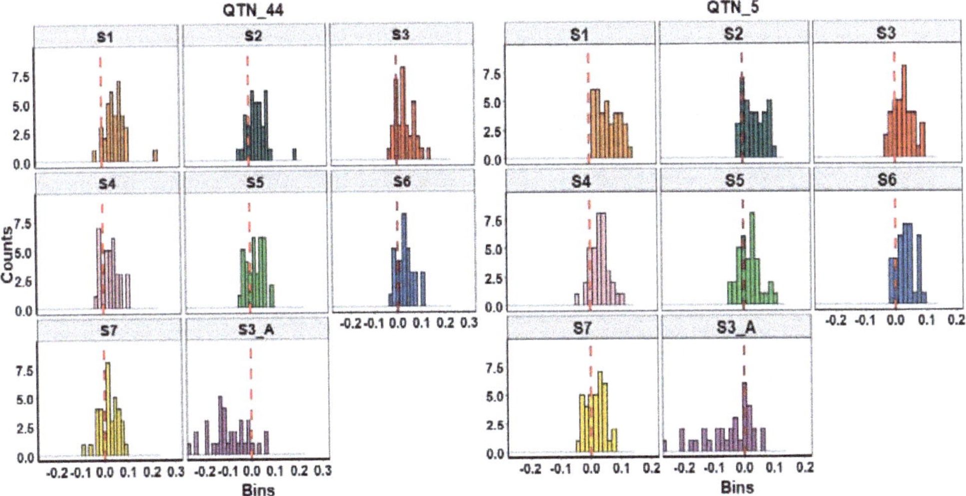

Figure 5. Histograms of each scenario representing the differences between accuracy from genomic predictions (ssGBLUP) versus genetic prediction (BLUP) per simulation. Right-sides of red dashed lines represent the number of simulations in which genomic selection outperformed BLUP.

The response to genomic selection and the percentages of correctly selected candidates are showed in Table 3. These parameters are strictly correlated with the accuracy of gEBVs in prolific livestock; however, they represent more pragmatic methods of comparison and dissemination for farmers. The correlations were 0.90 and 0.88 between the accuracy of gEBVs and response to genomic selection, and the first one and percentages of animals correctly selected, respectively. A response to selection of 0.105 was found when only pedigree information was used. In S1, the number of kits per generation increased by 22% with respect to BLUP. A slight decline was observed in scenarios S2, S3, S4 and S6 with a value of 0.120, 0.117, 0.114 and 0.116, respectively. S5 and S7 did not show any significant augmentation with respect to the BLUP scenario. As expected, a lower selection response than BLUP was noticed in S3_A (0.046). The percentages of correctly selected animals were similar to the trend of selection response, thus best performance was obtained in S1 with a value of 30.54%, followed by S2 and S3 with a percentage of 29.45% and 29.36%, respectively. S3 to S7 presented a range between 29.06% and 28.42%. BLUP presents a percentage of correctly selected animals of 27.81%. Even for this parameter, the values in S3_A were lower than in the BLUP scenario.

Table 3. Response to genomic selection and percentages of correctly selected animals across QTN models for each scenario.

Scenario	Selection Response [1]	SE-1 [2]	Percentage of ACS [3]	SE-2 [4]
BLUP	0.105	0.007	27.81	0.56
S1	0.129	0.007	30.54	0.56
S2	0.120	0.007	29.45	0.54
S3	0.117	0.008	29.36	0.60
S3_A	0.046	0.008	23.62	0.65
S4	0.114	0.007	28.93	0.54
S5	0.108	0.007	28.42	0.53
S6	0.116	0.007	29.06	0.55
S7	0.109	0.007	28.26	0.56

[1] Mean of response to genomic selection (kits). [2] Standard error of response to genomic selection. [3] Percentage of animals correctly selected. [4] Standard error of percentage of animals correctly selected.

3.4. Genotyping Costs

Figure 6 shows the relationships between the price of genotyping and IA, and the accuracy of gEBVs. As expected, the more investments, in terms of genotyping, the more IA and accuracy of gEBVs were found. However, that is not a strictly linear correlation, especially for IA. S1 showed the most expensive investment (EUR 112,000) due to progeny genotyping at the HD platform. The second ranking position was S2 with a genotyping cost of EUR 53,500. S3, S4 and S6 presented a lower cost compared to S2 from EUR 38,500 to EUR 31,000. S5 and S7 are the cheapest scenarios, with costs of EUR 26,800 and EUR 23,500, respectively.

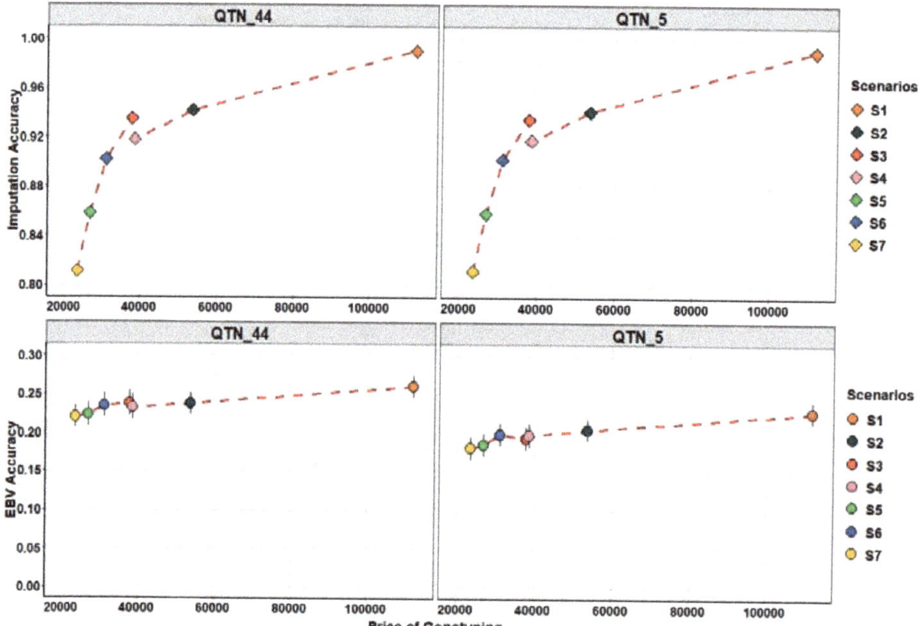

Figure 6. Plots of relationship between price of genotyping and imputation accuracy (top figure), and price of genotyping and EBV accuracy (bottom figure). Vertical lines are the mean ± standard error.

4. Discussion

Our results highlight four main points for discussion: (1) imputation strategies; (2) causes that affect genomic prediction; (3) comparison with other studies; and (4) searching for trade-off: cost and genetic accuracy trends.

4.1. Imputation Strategies

Genotype imputation was introduced as a tool to increase detection power on association studies for linking results across studies that rely on different genotype platforms [41]. On the other hand, imputation can be a suitable tool to reduce the cost of genotyping in both plant breeding [30,42] and animal breeding programs [18,43,44]. In this case, close ancestors are typically genotyped at HD platforms, whilst progeny are genotyped at lower SNP density. When the marker position of different platforms perfectly overlaps and pedigree information is present, imputation accuracies (IAs) are usually high [29]. In our situation, in which all close ancestors are genotyped with the HD platform, this approach may not produce the same benefit as in the other species due to the prohibitive cost of HD platforms. For this reason, we examined how IAs are affected by the different genotype information of the ancestors. There are several factors influencing IA, highlighting the missing rate of LD platforms as one of the main factors [22,30]. Missing rate is the percentage of SNPs present at an HD platform that are untyped (not covered) at LD platforms. The missing rates were 70% and 99.7% for MD and LD platforms in the current study. Although the first imputation studies suggested values between 50% and 75% [29,45], the values of the current study are in line with other studies that presented very high IA for both plant [42] and animal breeding programs [18,43,46]. The missing rate can be even higher when pedigree information is available for the imputation process. AlphaImpute, which uses both genotype and pedigree information, has been demonstrated to have high imputation performance [18,29,43], even in populations with low levels of linkage disequilibrium [47]. Thus, software and missing rates seem appropriate for imputation in rabbit breeding programs.

The number of animals in the reference population, as a second factor, influences the resolution of haplotypes during the phasing process. A large number of individuals in the reference population ensures high IA in validated populations. However, it can be reduced considerably if the animals from both populations are close relatives, sharing the structure of linkage disequilibrium and haplotypes across wide chromosome segments [22,29]. S7 showed a high IA using only 70 male ancestors, which is explained by the close relationship between them, the female ancestors and progeny under ongoing selection. As expected, S1 presented the highest IA and genotype yields due to the larger number of animals in the reference population (training) and lower missing rate. In general, many studies showed that imputation using HD platforms on ancestors and MD platforms on the progenies produces high levels of IA and concordance rate in dairy cattle [46,48,49], pigs [13,18,19], poultry [12,50], sheep [20] and farmed Atlantic salmon [43]. When the SNP densities of MD (S1) platforms were reduced to LD (S2) in the validated population, IAs decreased only five percentage points. Hence, vast haplotype information keeps retrieving, using LD platforms, because of the high relatedness of rabbits. The IA results of S1 and S2 were similar to those reported in a pig imputation study, especially for Landrace and Yorkshire breeds [19]. In the current study, fewer differences in IA were found when grand-dams were genotyped at MD platforms (less than two percentage points, S3 and S4) compared to S2. This demonstrated that dams at MD platforms were enough to retrieve female haplotypes and to keep high IAs, being noticeable when S6 presented better IA values than S5. The female haplotype resolution is better when dams are genotyped at a higher SNP density than any level of SNP density on grand-dams. On the other hand, the imputation of whole genomes (S3_A) seems to not be a feasible technique for rabbit breeding programs. IA declined up to 0.797, probably due to the small number of training populations and the higher error rate of imputed SNPs associated with QTNs—very important if they have low minor allele frequency (MAF). Many animals with imputed whole-genomes presented very low IA—less than 60%. Conversely, some studies showed the benefits of strategies based on imputed whole-genomes in part of training populations [47,51] and evaluated populations using a larger number of genotyped individuals [51].

4.2. Causes That Affect Genomic Prediction

In this study, ssGBLUP was used as a method for genomic evaluations. This method includes a different sort of information: phenotypes, pedigree, and genotypes. Previous studies demonstrated that ssGBLUP gives potentially more accurate and less biased genomic gEBVs than multistep methods, especially in the presence of small populations and sex-limited traits [31,52].

The accuracy of the EBVs varies greatly within each scenario with respect to IA (Figures 3 and 4). Conversely to the IA, which is mainly influenced by the genotyping strategies, the accuracy of gEBVs is affected by several factors. Some of these are related to the genetic architecture of the traits such as QTNs distribution and their allele frequency [22,53]. The number of SNPs in LD with these QTNs and the allele frequency of those SNPs also played an essential role in the accuracy of genomic prediction [54]. In addition, working with imputed genotypes, it is important to consider the influence of the imputation error of those SNPs and therefore IA [22,53]. In these studies, scenarios with a higher IA value also presented higher accuracy on genomic prediction and the opposite. However, a strong correlation between these two-type accuracies cannot be defined due to the high variability of gEBV accuracy within the scenarios. Additionally, with S3_A and S7, it was confirmed that low IA values (under 85%) can be deleterious for genomic prediction. For that reason, conservative thresholds for genotype imputation and quality control before genomic selection must be adopted. Furthermore, Cleveland and Hickey [18] showed that differences in gEBVs may be due to both different values of IA but also to the intrinsic genotyping structure of each scenario. There may be animals that have a marginal influence on IA but can significantly affect gEBVs.

As mentioned previously, variability in these scenarios was also caused by different allele frequencies of SNPs and QTNs. High heterogeneity of these factors was observed between simulations due to the random events that occurred during the 28th generation of mating. In some simulations, a larger number of QTNs were fixed, and in some, many SNPs were not associated with the rest of the QTNs. This would also explain the variability between simulations, and why few simulations presented a lower accuracy of gEBVs for genomic scenarios compared to BLUP, even in the scenarios with high IA as in S2 and S3.

Regarding QTN distributions, QTN_44 presented better performance than QTN_5. This trend agrees with the study of Zang et al. [53], in which ssGBLUP outperformed when the phenotype was controlled by several genes of equal effect sizes. A method that assumes unequal variances for each marker could suit for the genomic prediction of QTN_5. Modeling SNP's effect and its variance can potentially give better results in short term selection for this simulation. Iterative ssGBLUP and/or nonlinear weight **A** can easily be implemented and can potentially lead to an increase in prediction accuracy [55,56]. Fernando et al. [57] also propose a single-step Bayesian regression in which it is possible to model the distribution of marker effects in many forms such as *t* distribution, variable selection model, and mixture distributions. Despite this, the method of estimation and of building **G** matrices was kept the same in all simulations for the sake of simplicity and for comparison with other studies.

4.3. Comparison with Other Studies

S1 represented the typical genotyping strategy used in all livestock species in which parents are genotyped with HD platforms and progeny are genotyped at LD platforms. As expected, S1 exhibits the best accuracies of gEBVs, and the accuracy of genomic prediction presented in S1 is close to that presented when the same candidate animals are genotyped with HD platforms; correlation is almost one in all simulations for both QTN models (data not shown). A similar correlation was found in pigs [18,58] and cattle [21,22,53], even if the proportion between HD and LD animals was different due to distinct breeding schemes present in these species. Previous imputation studies conducted in commercial pig breeding programs are a good comparison method due to a similar mating system, although the number of animals is lower in rabbits. Scenarios comparable to those present in this study have been reported in Cleveland and Hickey [18]. A similar sharp drop in IA and the accuracy of gEBVs was observed when animals of validated populations were genotyped with MD to LD platforms. In addition, an analogous decline in gEBVs was observed when grand-dams and dams were genotyped with LD or MD, although in both studies IA was quite high for these scenarios. However, Cleveland and Hickey [18] demonstrated that genotyping at HD platforms for animals that are not related with LD animals has little impact concerning IA, but it can significantly affect the accuracy of gEBVs. Nevertheless, this scenario was not included in our study because the limited number of animals would not guarantee the same gap of accuracy presented in pigs. The same consideration made on Grossi et al. [19] can be also made for our study; genotyping reference animals with the LD platform can ensure good accuracy levels of IA and gEBVs when parents are genotyped with HD, if not low levels of prediction have been observed, i.e., S5 or S7.

4.4. *Searching for Trade-Off: Cost and Genetic Accuracy Trends*

Cost evaluations are usually performed by comparing imputation strategies against an idealistic genomic selection in which all candidate animals are genotyped at HD platforms. Under this condition, imputation strategies are always a cost-effective technique [18,42,43]. Here, we show only genotyping costs to evaluate investment for moving from traditional genetic evaluation (BLUP) to genomic selection (ssGBLUP). Imputation was further demonstrated as a reliable tool for reducing genotyping investment, with an accuracy higher than BLUP in the majority of cases.

However, some scenarios present a cost-benefit inefficiency when compared with the other scenarios. Clearly, S3_A cannot be taken as an option due to the low performance of prediction—even worse than BLUP. Results from this scenario suggest that a method based on high thresholds of individual-specific IA must be considered, especially when genomic characterization is available [22], and for strategies with non-genotyped animals [51]. Using animals with imputed whole-genomes for genomic selection is still a challenge due to the imputation error rate [51,59]. In this sense, high error rates of imputed SNPs associated with QTNs are clear in S3_A, presenting 75% of simulations lower than BLUP scenarios. BLUPF90 reported genotypes duplicated for S3_A, as *AlphaImpute* copied whole-genomes of full-sibs for some progenies.

The same reason can be applied to S7 and S5 as, even if lower genotype investment was present in these scenarios, these strategies cannot be considered one of the best scenarios because the number of simulations in which genomic selection outperformed BLUP is limited—about half of the cases. Conversely, S3, S6, and S4 presented a considerable increase in accuracy and response to selection with a marginal increment investment.

The opposite situation was presented in S1, and even if a significant reduction in accuracy was observed switching from MD platforms to LD platforms (S2), the big drop in price observed can potentially justify this decline in accuracy. No significant differences in terms of percentage of correctly selected animals and response to selection were observed between S2 with S3 and S6, thus these two scenarios are preferable over S2 due to the lower cost. In addition, S4 can be discarded among the best scenarios, as it presents a lower selection response than S6 and S3 but at a higher price. In conclusion, the best scenarios can be identified as S3 and S6. Genotyping only half of the animals per litter (S3) is an effective strategy to reduce the impact of the cost of genotyping, especially in prolific species such as rabbit and pig. This approach is commonly adopted in the pig industry selection scheme [15,17,18]. On the other hand, S6 demonstrated that genotyping grand-dams (S4) leads to a negligible difference in accuracies when sires and dams are genotyped, with S6 being cheaper than S3 (+ EUR 6750). Thus, we considered S6 as the strategy with the best trade-off for implementing genomic selection in rabbits.

5. Conclusions

Imputation strategies are feasible in rabbit breeding programs as in other species, as IAs were high, particularly in scenarios with parents' genotype at HD platforms. Furthermore, the slight positive correlation between IA and the accuracy of gEBVs was also demonstrated, and scenarios with IA also have a high gEBV accuracy. The results of the response to genomic selection and the percentage of correctly selected candidates are in the same line as genetic accuracy values. Despite this, gEBV accuracy showed great variance among simulations. Therefore, we must be cautious with the results from simulated data as they are conditioned to several factors of the small reference population (e.g., size, relationships between individuals, inbreeding level, and update). Another comparison between genomic selection and BLUP could be made by enlarging the reference population using dams of the nucleus farms and the multipliers, or even crossbreeding from commercial farms. In conclusion, the adoption of imputation strategies can be an effective strategy for drastically reducing the genotyping cost in rabbits, maintaining an accuracy slightly lower than that with all HD animals. Hence, the best trade-off scenario can be identified in S6; although, as previously stated, these results may change under different conditions.

Supplementary Materials: The following are available online at https://www.mdpi.com/2076-2615/11/3/803/s1, Table S1: Parameters and info to run simulations in *AlphaSim* and *MaCS*.

Author Contributions: Conceptualization, N.I.-E. and A.B.; methodology, B.S.S.-M. and N.I.-E.; software, B.S.S.-M. and E.M.; validation, B.S.S.-M., E.M. and N.I.-E.; formal analysis, E.M. and B.S.S.-M.; investigation, B.S.S.-M.; resources, A.B. and N.I.-E.; data curation, B.S.S.-M. and E.M.; writing—original draft preparation, E.M. and B.S.S.-M.; writing—review and editing, E.M., B.S.S.-M., N.I.-E. and A.B.; visualization, E.M. and B.S.S.-M.; supervision, N.I.-E.; project administration, B.S.S.-M.; funding acquisition, N.I.-E. and A.B. All authors have read and agreed to the published version of the manuscript.

Funding: This research study was funded by AGL2017-86083-C2-P1 from "Plan Nacional de Investigación Científica" of Spain-Project I+D. B. Samuel Sosa Madrid was supported by FPI grant, number BES-2015-074194, from "Ministerio de Ciencia e Innovación".

Data Availability Statement: The databases used and analyzed in the current study are available from The Figshare Repository. Some scripts are also included in the repository. The link addresses are https://doi.org/10.6084/m9.figshare.13350443 or https://figshare.com/s/2591beabd3ef5bc6d220 (accessed on 9 March 2021).

Acknowledgments: We express our profound thanks to Francisco Rosich of "Área de Sistemas de Información y Comunicaciones" (ASIC) for his contributions to pipelines of simulations and imputation analyses.

Conflicts of Interest: The authors declare no conflict of interest.

References

1. Cullere, M.; Dalle Zotte, A. Rabbit meat production and consumption: State of knowledge and future perspectives. *Meat Sci.* **2018**, *143*, 137–146. [CrossRef]
2. European Commission. Overview Report: Commercial Farming of Rabbits in the European Union. *Luxemb. Publ. Off. Eur. Union* **2017**, 16. [CrossRef]
3. Cartuche, L.; Pascual, M.; Gómez, E.A.; Blasco, A. Economic weights in rabbit meat production. *World Rabbit Sci.* **2014**, *22*, 165–177. [CrossRef]
4. Baselga, M. Genetic Improvement of Meat Rabbits. Programmes and Diffusion. In Proceedings of the 8th World Rabbit Congress, Puebla, Mexico, 7–14 September 2004; pp. 1–13.
5. Fernández, E.N.; Sánchez, J.P.; Martínez, R.; Legarra, A.; Baselga, M. Role of inbreeding depression, non-inbred dominance deviations and random year-season effect in genetic trends for prolificacy in closed rabbit lines. *J. Anim. Breed. Genet.* **2017**, *134*, 441–452. [CrossRef] [PubMed]

6. Ragab, M.; Baselga, M. A comparison of reproductive traits of four maternal lines of rabbits selected for litter size at weaning and founded on different criteria. *Livest. Sci.* **2011**, *136*, 201–206. [CrossRef]
7. Meuwissen, T.; Hayes, B.; Goddard, M. Genomic selection: A paradigm shift in animal breeding. *Anim. Front.* **2016**, *6*, 6. [CrossRef]
8. da Cruz, V.A.R.; Brito, L.F.; Schenkel, F.S.; de Oliveira, H.R.; Jafarikia, M.; Feng, Z. Strategies for within-litter selection of piglets using ultra-low density SNP panels. *Livest. Sci.* **2019**, *220*, 173–179. [CrossRef]
9. Habier, D.; Fernando, R.L.; Dekkers, J.C.M. The impact of genetic relationship information on genome-assisted breeding values. *Genetics* **2007**, *177*, 2389–2397. [CrossRef]
10. Jattawa, D.; Elzo, M.A.; Koonawootrittriron, S.; Suwanasopee, T. Imputation Accuracy from Low to Moderate Density Single Nucleotide Polymorphism Chips in a Thai Multibreed Dairy Cattle Population. *Asian-Australas. J. Anim. Sci.* **2016**, *29*, 464–470. [CrossRef] [PubMed]
11. Schöpke, K.; Swalve, H.H. Review: Opportunities and challenges for small populations of dairy cattle in the era of genomics. *Animal* **2016**, *10*, 1050–1060. [CrossRef]
12. Wolc, A.; Kranis, A.; Arango, J.; Settar, P.; Fulton, J.E.; O'Sullivan, N.P.; Avendano, A.; Watson, K.A.; Hickey, J.M.; de los Campos, G.; et al. Implementation of genomic selection in the poultry industry. *Anim. Front.* **2016**, *6*, 23–31. [CrossRef]
13. Knol, E.F.; Nielsen, B.; Knap, P.W. Genomic selection in commercial pig breeding. *Anim. Front.* **2016**, *6*, 15–22. [CrossRef]
14. Garrick, D.J. The role of genomics in pig improvement. *Anim. Prod. Sci.* **2017**, *57*, 2360–2365. [CrossRef]
15. Ibáñez-Escriche, N.; Forni, S.; Noguera, J.L.; Varona, L. Genomic information in pig breeding: Science meets industry needs. *Livest. Sci.* **2014**, *166*, 94–100. [CrossRef]
16. Blasco, A.; Pena, R.N. Current Status of Genomic Maps: Genomic Selection/GBV in Livestock. In *Animal Biotechnology 2: Emerging Breeding Technologies*; Niemann, H., Wrenzycki, C., Eds.; Springer International Publishing: Cham, Germany, 2018; pp. 61–80. ISBN 978-3-319-92348-2.
17. Ibáñez-Escriche, N.; González-Recio, O. Review. Promises, pitfalls and challenges of genomic selection in breeding programs. *Span. J. Agric. Res.* **2011**, *9*, 404–413. [CrossRef]
18. Cleveland, M.A.; Hickey, J.M. Practical implementation of cost-effective genomic selection in commercial pig breeding using imputation1. *J. Anim. Sci.* **2013**, *91*, 3583–3592. [CrossRef]
19. Grossi, D.A.; Brito, L.F.; Jafarikia, M.; Schenkel, F.S.; Feng, Z. Genotype imputation from various low-density SNP panels and its impact on accuracy of genomic breeding values in pigs. *Animal* **2018**, *12*, 2235–2245. [CrossRef] [PubMed]
20. O'Brien, A.C.; Judge, M.M.; Fair, S.; Berry, D.P. High imputation accuracy from informative low-to-medium density single nucleotide polymorphism genotypes is achievable in sheep. *J. Anim. Sci.* **2019**, *97*, 1550–1567. [CrossRef] [PubMed]
21. Wang, Y.; Lin, G.; Li, C.; Stothard, P. Genotype Imputation Methods and Their Effects on Genomic Predictions in Cattle. *Springer Sci. Rev.* **2016**, *4*, 79–98. [CrossRef]
22. Calus, M.P.L.; Bouwman, A.C.; Hickey, J.M.; Veerkamp, R.F.; Mulder, H.A. Evaluation of measures of correctness of genotype imputation in the context of genomic prediction: A review of livestock applications. *Animal* **2014**, *8*, 1743–1753. [CrossRef]
23. Faux, A.-M.; Gorjanc, G.; Gaynor, R.C.; Battagin, M.; Edwards, S.M.; Wilson, D.L.; Hearne, S.J.; Gonen, S.; Hickey, J.M. AlphaSim: Software for Breeding Program Simulation. *Plant. Genome* **2016**, *9*. [CrossRef]
24. Chen, G.K.; Marjoram, P.; Wall, J.D. Fast and flexible Simulation of DNA Sequence Data. *Genome Res.* **2009**, *19*, 136–142. [CrossRef]
25. García, M.L.; Baselga, M. Estimation of Genetic Response to selection in litter size of rabbits using a cryopreserved control population. *Livest. Prod. Sci.* **2002**, *74*, 45–53. [CrossRef]
26. Ragab, M.; Sánchez, J.P.; Mínguez, C.; Vicente, J.S.; Baselga, M. Litter size components in a full diallel cross of four maternal lines of rabbits. *J. Anim. Sci.* **2014**, *92*, 3231–3236. [CrossRef] [PubMed]
27. Argente, M.J. Major Components in Limiting Litter Size. In *Insights from Animal Reproduction*; Payan Carreira, R., Ed.; InTech: London, UK, 2016; pp. 87–114. ISBN 978-953-51-2268-5.
28. R Core Team. *R: A Language and Environmental for Statistical Computing*; R foundation for statistical computing: Vienna, Austria, 2020. Available online: https://www.R-project.org/ (accessed on 15 August 2020).
29. Hickey, J.M.; Kinghorn, B.P.; Tier, B.; van der Werf, J.H.; Cleveland, M.A. A phasing and imputation method for pedigreed populations that results in a single-stage genomic evaluation. *Genet. Sel. Evol.* **2012**, *44*, 9. [CrossRef]
30. Hickey, J.M.; Crossa, J.; Babu, R.; de los Campos, G. Factors Affecting the Accuracy of Genotype Imputation in Populations from Several Maize Breeding Programs. *Crop. Sci.* **2012**, *52*, 654–663. [CrossRef]
31. Legarra, A.; Christensen, O.F.; Aguilar, I.; Misztal, I. Single Step, a general approach for genomic selection. *Livest. Sci.* **2014**, *166*, 54–65. [CrossRef]
32. Legarra, A.; Aguilar, I.; Misztal, I. A relationship matrix including full pedigree and genomic information. *J. Dairy Sci.* **2009**, *92*, 4656–4663. [CrossRef]
33. Christensen, O.F.; Lund, M.S. Genomic prediction when some animals are not genotyped. *Genet. Sel. Evol.* **2010**, *42*, 2. [CrossRef]
34. Calus, M.P.L. Editorial: Genomic selection with numerically small reference populations. *Animal* **2016**, *10*, 1016–1017. [CrossRef] [PubMed]
35. VanRaden, P.M. Efficient methods to compute genomic predictions. *J. Dairy Sci.* **2008**, *91*, 4414–4423. [CrossRef]
36. Garcia-Baccino, C.A.; Legarra, A.; Christensen, O.F.; Misztal, I.; Pocrnic, I.; Vitezica, Z.G.; Cantet, R.J.C. Metafounders are related to Fst fixation indices and reduce bias in single-step genomic evaluations. *Genet. Sel. Evol.* **2017**, *49*, 34. [CrossRef] [PubMed]

37. Aguilar, I.; Fernandez, E.N.; Blasco, A.; Ravagnolo, O.; Legarra, A. Effects of ignoring inbreeding in model-based accuracy for BLUP and SSGBLUP. *J. Anim. Breed. Genet.* **2020**, *137*, 356–364. [CrossRef]
38. Aguilar, I.; Tsuruta, S.; Masuda, Y.; Lourenco, D.A.L.; Legarra, A.; Misztal, I. BLUPF90 suite of programs for animal breeding. In Proceedings of the 11th World Congress of Genetics Applied to Livestock Production, Auckland, New Zealand, 11–16 February 2018; p. 11.751.
39. Pszczola, M.; Strabel, T.; van Arendonk, J.A.M.; Calus, M.P.L. The impact of genotyping different groups of animals on accuracy when moving from traditional to genomic selection. *J. Dairy Sci.* **2012**, *95*, 5412–5421. [CrossRef] [PubMed]
40. Hidalgo, J.; Tsuruta, S.; Lourenco, D.; Masuda, Y.; Huang, Y.; Gray, K.A.; Misztal, I. Changes in genetic parameters for fitness and growth traits in pigs under genomic selection. *J. Anim. Sci.* **2020**, *98*, 1–12. [CrossRef] [PubMed]
41. Li, Y.; Willer, C.; Sanna, S.; Abecasis, G. Genotype imputation. *Annu. Rev. Genom. Hum. Genet.* **2009**, *10*, 387–406. [CrossRef] [PubMed]
42. Gorjanc, G.; Battagin, M.; Dumasy, J.; Antolin, R.; Gaynor, R.C.; Hickey, J.M. Prospects for Cost-Effective Genomic Selection via Accurate Within-Family Imputation. *Crop Sci.* **2017**, *57*, 216–228. [CrossRef]
43. Tsai, H.Y.; Matika, O.; Edwards, S.M.K.; Antolín-Sánchez, R.; Hamilton, A.; Guy, D.R.; Tinch, A.E.; Gharbi, K.; Stear, M.J.; Taggart, J.B.; et al. Genotype imputation to improve the cost-efficiency of genomic selection in farmed Atlantic salmon. *G3 Genes Genomes Genet.* **2017**, *7*, 1377–1383. [CrossRef]
44. Lopez, B.I.; Lee, S.H.; Shin, D.H.; Oh, J.D.; Chai, H.H.; Park, W.; Park, J.E.; Lim, D. Accuracy of genomic evaluation using imputed high-density genotypes for carcass traits in commercial Hanwoo population. *Livest. Sci.* **2020**, *241*, 104256. [CrossRef]
45. Tabor, H.K.; Risch, N.J.; Myers, R.M. Candidate-gene approaches for studying complex genetic traits: Practical considerations. *Nat. Rev. Genet.* **2002**, *3*, 391–397. [CrossRef] [PubMed]
46. Weigel, K.A.; de los Campos, G.; Vazquez, A.I.; Rosa, G.J.M.; Gianola, D.; Van Tassell, C.P. Accuracy of direct genomic values derived from imputed single nucleotide polymorphism genotypes in Jersey cattle. *J. Dairy Sci.* **2010**, *93*, 5423–5435. [CrossRef]
47. Pimentel, E.C.G.; Wensch-Dorendorf, M.; König, S.; Swalve, H.H. Enlarging a training set for genomic selection by imputation of un-genotyped animals in populations of varying genetic architecture. *Genet. Sel. Evol.* **2013**, *45*, 1–12. [CrossRef] [PubMed]
48. Silva, A.A.; Silva, F.F.; Silva, D.A.; Silva, H.T.; Costa, C.N.; Lopes, P.S.; Veroneze, R.; Thompson, G.; Carvalheira, J. Genotype imputation strategies for Portuguese Holstein cattle using different SNP panels. *Czech. J. Anim. Sci.* **2019**, *64*, 377–386. [CrossRef]
49. Zhang, Z.; Druet, T. Marker imputation with low-density marker panels in Dutch Holstein cattle. *J. Dairy Sci.* **2010**, *93*, 5487–5494. [CrossRef] [PubMed]
50. Heidaritabar, M.; Calus, M.P.L.; Vereijken, A.; Groenen, M.A.M.; Bastiaansen, J.W.M. Accuracy of imputation using the most common sires as reference population in layer chickens. *BMC Genet.* **2015**, *16*, 1–14. [CrossRef]
51. Bouwman, A.C.; Hickey, J.M.; Calus, M.P.; Veerkamp, R.F. Imputation of non-genotyped individuals based on genotyped relatives: Assessing the imputation accuracy of a real case scenario in dairy cattle. *Genet. Sel. Evol.* **2014**, *46*, 1–11. [CrossRef]
52. Misztal, I.; Aggrey, S.E.; Muir, W.M. Experiences with a single-step genome evaluation. *Poult. Sci.* **2013**, *92*, 2530–2534. [CrossRef] [PubMed]
53. Zhang, H.; Yin, L.; Wang, M.; Yuan, X.; Liu, X. Factors affecting the accuracy of genomic selection for agricultural economic traits in maize, cattle, and pig populations. *Front. Genet.* **2019**, *10*. [CrossRef]
54. Habier, D.; Fernando, R.L.; Garrick, D.J. Genomic BLUP Decoded: A Look into the Black Box of Genomic Prediction. *Genetics* **2013**, *194*, 597–607. [CrossRef]
55. Fragomeni, B.O.; Lourenco, D.A.L.; Masuda, Y.; Legarra, A.; Misztal, I. Incorporation of causative quantitative trait nucleotides in single-step GBLUP. *Genet. Sel. Evol.* **2017**, *49*, 59. [CrossRef]
56. Zhang, X.; Lourenco, D.; Aguilar, I.; Legarra, A.; Misztal, I. Weighting Strategies for Single-Step Genomic BLUP: An Iterative Approach for Accurate Calculation of GEBV and GWAS. *Front. Genet.* **2016**, *7*, 151. [CrossRef] [PubMed]
57. Fernando, R.L.; Dekkers, J.C.M.; Garrick, D.J. A class of Bayesian methods to combine large numbers of genotyped and non-genotyped animals for whole-genome analyses. *Genet. Sel. Evol.* **2014**, *46*, 50. [CrossRef]
58. Song, H.; Ye, S.; Jiang, Y.; Zhang, Z.; Zhang, Q.; Ding, X. Using imputation-based whole-genome sequencing data to improve the accuracy of genomic prediction for combined populations in pigs. *Genet. Sel. Evol.* **2019**, *51*, 58. [CrossRef] [PubMed]
59. Pimentel, E.C.G.; Edel, C.; Emmerling, R.; Götz, K.-U. How imputation errors bias genomic predictions. *J. Dairy Sci.* **2015**, *98*, 4131–4138. [CrossRef] [PubMed]

Article

A Genome-Wide Association Study Identifying Genetic Variants Associated with Growth, Carcass and Meat Quality Traits in Rabbits

Xue Yang [1,2,†], Feilong Deng [1,3,†], Zhoulin Wu [1], Shi-Yi Chen [1], Yu Shi [1], Xianbo Jia [1], Shenqiang Hu [1], Jie Wang [1], Wei Cao [1] and Song-Jia Lai [1,*]

1 Farm Animal Genetic Resources Exploration and Innovation Key Laboratory of Sichuan Province, Sichuan Agricultural University, Chengdu 611130, China; yangxue790702@163.com (X.Y.); fdeng@uark.edu (F.D.); wzlneil@163.com (Z.W.); sychensau@gmail.com (S.-Y.C.); 18227551690@163.com (Y.S.); jaxb369@sicau.edu.cn (X.J.); sqhu2011@163.com (S.H.); wjie68@163.com (J.W.); caowei_2005@126.com (W.C.)

2 Chengdu Academy of Agriculture and Forestry Sciences, Chengdu 611130, China

3 Special Key Laboratory of Microbial Resources and Drug Development, Research Center for Medicine and Biology, Zunyi Medical University, Zunyi 563000, China

* Correspondence: laisj5794@sicau.edu.cn

† These authors contributed equally to this work.

Received: 21 May 2020; Accepted: 17 June 2020; Published: 20 June 2020

Simple Summary: Rabbit meat has been widely consumed in China and is considered as an ideal food source due to its high protein, low fat, low cholesterol and low sodium contents. The growth rate, carcass characteristics and meat quality are considered economically important traits in the rabbit industry. Genomic selection (GS) could facilitate genetic selection for important economic traits, however, the lack of molecular markers for these traits limits the application of GS in rabbits. Genome-wide association study (GWAS) has the potential to comprehensively identify trait-associated molecular markers and has been applied in animal and plant research. In this study, GWAS was used to examine growth, carcass and meat quality traits of meat rabbits based on specific-locus amplified fragment sequencing (SLAF-seq) technology to identify significantly associated SNPs and functional genes, to be used as a basis for prompting the application of GS in rabbits.

Abstract: Growth, carcass characteristics and meat quality are the most important traits used in the rabbit industry. Identification of the candidate markers and genes significantly associated with these traits will be beneficial in rabbit breeding. In this study, we enrolled 465 rabbits, including 16 male Californian rabbits and 17 female Kangda5 line rabbits as the parental generation, along with their offspring (232 male and 200 female), in a genome-wide association study (GWAS) based on SLAF-seq technology. Bodyweight at 35, 42, 49, 56, 63 and 70 d was recorded for growth traits; and slaughter liveweight (84 d) and dressing out percentage were measured as carcass traits; and cooking loss and drip loss were measured as meat quality traits. A total of 5,223,720 SLAF markers were obtained by digesting the rabbit genome using RsaI + EcoRV-HF® restriction enzymes. After quality control, a subset of 317,503 annotated single-nucleotide polymorphisms (SNPs) was retained for subsequent analysis. A total of 28, 81 and 10 SNPs for growth, carcass and meat quality traits, respectively, were identified based on genome-wide significance ($p < 3.16 \times 10^{-7}$). Additionally, 16, 71 and 9 candidate genes were identified within 100 kb upstream or downstream of these SNPs. Further analysis is required to determine the biological roles of these candidate genes in determining rabbit growth, carcass traits and meat quality.

Keywords: *Oryctolagus cuniculus*; SNPs; SLAF-seq; genome-wide association study; growth trait

1. Introduction

Rabbit meat has a long history of consumption starting from around 1100 BC [1]. It has high nutritional value and is considered a healthy food because of its high protein content and low fat, cholesterol and sodium [2]. Growth rate, carcass characteristics and meat quality are considered important economic traits in rabbit breeding. In the past few years, researchers have worked on improving growth performance and meat quality by advanced molecular breeding methods. Dozens of single-nucleotide polymorphisms (SNPs) have been identified by resequencing gene regions and found to be associated with growth traits in rabbit. Melanocortin receptor 4 (*MC4R*) [3], fat mass and obesity-associated (*FTO*) [4,5], *LEP* [6], *TBC1D1* [4] and *GHR* [7] genes are correlated with growth and carcass traits in rabbit. However, these researchers mainly investigated the correlation between a single SNP present in a specific DNA fragment with a given trait using low-throughput methods [4–6]. For complex traits such as growth performance and meat quality, large-scale analysis is necessary to detect trait-associated SNPs.

Genome-wide association study (GWAS) [8] represents a powerful approach to correlating SNPs and functional genes with quantitative traits. SNPs associated with a specific trait can be considered as molecular markers for application in genomic selection (GS) [9] and as genetic markers [10]. The most important step in GWAS is to acquire high-quality SNPs at the genome-wide level. A high-density SNP array is a high-throughput, cost-effective genotyping assay and is the most widely used genotyping method in GWAS [11,12]. Although there are still disadvantages, for example, that only known SNPs can be detected, there are high costs and great effort involved in establishing an array and that marker distribution is biased [13], it has become possible for researchers to perform GWAS using 10,000 individuals. Whole-genome resequencing is another major genotyping method that has been used over the last 10 years. It is a powerful method for whole-genome SNP discovery [14,15]. However, whole-genome resequencing can be prohibitively expensive in GWAS. Therefore, specific-locus amplified fragment sequencing (SLAF-seq), a high-resolution strategy for large-scale genotyping at the genome level, is a great alternative approach to SNP genotyping in non-popular research species such as rabbits [16]. Compared with high-density SNP arrays and whole-genome resequencing, SLAF-seq is an efficient method for de novo SNP discovery with such advantages as high genotyping accuracy, relatively low cost and a high capacity for large sample sizes. SLAF-seq has been successfully applied in chicken [17,18] and pig [19].

Despite the great success of GWAS in animal science [20], including a recent GWAS study successfully performed by Sosa-Madrid and colleagues to identify genomic regions associated with the intramuscular fat of rabbits based on a high-density SNP array [21], there is still a lack of large-scale research studies linking important economic traits to candidate genes in rabbit. Identification of SNPs associated with economically important traits via GWAS, as a first step, would provide a basis for further improving the breeding efficiency of rabbits. Here, we performed a GWAS study of the growth, carcass and meat quality traits of meat rabbits based on SLAF-seq technology to identify the associated SNPs and to predict functional genes. This study will provide a molecular basis for marker-assisted selection and gene-based selection to improve those traits.

2. Materials and Methods

2.1. Ethics Statement

All animal experiments were approved by the Institutional Animal Care and Use Committee of Sichuan Agricultural University (Permit Number: No. DKY-B20141401). The experiments were performed in accordance with the institutional regulations (no public availability).

2.2. Animals and Phenotypes

For the experiment, 53 Californian female rabbits and 22 Kangda5 line rabbits, a commercial strain of meat rabbit from Qingdao Kangda Foodstuffs Co., Ltd., were selected as the parental generation to generate crossbred F1 offspring. Rabbits with the same day of mating (2 May 2017) and the same day of delivery (2 June 2017), including 16 male Californian rabbits and 17 female Kangda5 line rabbits, were enrolled in the study. All rabbits were fed a pellet diet (10.9 MJ/kg DE, 16.5% CP, 13.3% CF) and housed in cages of 50 cm × 40 cm × 40 cm in size. During day 35–70, two rabbits were placed into one cage block and one rabbit was raised in a separate cage block from 70 days until slaughter. An environmental conditioning control system was used when the conditions were out of the normal range with respect to the ventilation (0.5–4 m^3/h/kg), temperature (18–25 °C) and humidity (20%–70%). A 12 h period of artificial normal environmental lighting was provided. All rabbits of the F1 generation were weaned at the age of 35 days.

Bodyweights were recorded on days 35, 42, 49, 56, 63 and 70. In China, meat rabbits are usually slaughtered at 70 or 84 days old. Therefore, we recorded bodyweight at day 70 and slaughtered at day 84. A total of 432 F1 generation rabbits, including 232 male and 200 female rabbits, was slaughtered by electrical shock after fasting for 24 h. The skin and head were separated from the body by cutting at the level of the third caudal vertebra and of the distal epiphyses of radius–ulna and tibia bones. After bleeding, the hot carcasses were properly handled following procedures described in Reference [22]. The slaughter liveweight (84 d, SLW) and hot carcass weight (HCW) were measured. The dressing out percentage (DoP) was defined as HCW divided by SLW.

Two meat quality traits include cooking loss (CL) and drip loss (DL). The drip loss was quantified by the method of Reference [23]. The cooking loss was measured throughout using the following method—approximately 20 g samples of cube-like raw meat from the biceps femoris muscle of the hind leg were weighed (W1) and steamed for 30 min. Cooked samples were cooled down to room temperature and re-weighed (W2). CL was calculated as follows:

$$\text{CL (\%)} = 100 \times (\text{W2}/\text{W1}) \quad (1)$$

2.3. SLAF-Seq Design

The rabbit genome (GenBank assembly accession: GCA_000003625.1) was used as the reference genome to predict and characterize the presence of putative restriction endonuclease sites. Genome assembly revealed a genome size of 2.74 Gbp and a GC (guanine-cytosine) content of 43.75%. Restriction endonuclease profiles of the reference genome were determined using self-developed software [16]. The most efficient enzyme digestion scheme was selected based on the following criteria—(1) the proportion of restriction fragments located in the repetitive region is as low as possible; (2) zymogenic fragments are distributed evenly in the genome; (3) the number of observed restriction fragments (SLAF tags) meets the expected number of tags according to the primary in silico investigation based on reference genome sequences.

Genomic DNA was extracted from whole blood samples using the phenol–chloroform method and was digested with the selected optimal enzymes. A single "A" nucleotide was added to the 3′ end of the digested fragment (SLAF tag). Dual-index [24] sequencing adapters were ligated to the A-tailed fragments using T4 DNA ligase. Further, PCR was conducted using diluted digested DNA samples, with forward primer (5′-AATGATACGGCGACCACCGA-3′) and reverse primer (5′-CAAGCAGAAGACGGCATACG-3′). PCR products were purified, pooled and separated by electrophoresis using a 2% agarose gel. Fragments of 300–500 bp size were excised and purified using a QIA Gel Extraction Kit (Qiagene, Germany). The library was sequenced on an Illumina HiSeq 2500 (Illumina, Inc., San Diego, CA, USA) platform following the manufacturer's instructions. Raw sequencing reads were identified by dual-indexing and classified to each sample. Clean reads were mapped to the reference genome (GenBank assembly accession: GCA_000003625.1) using SOAP software (http://soap.genomics.org.cn) [25]. The sequences mapped to the reference genome were retained for further analysis.

2.4. Genotyping and Statistical Analysis

Efficiency of RsaI EcoRV-HF® digestion was evaluated using a positive control sample (*Oryza sativa* ssp. *japonica*). SLAF tags were mapped to the reference genome using BWA software (http://bio-bwa.sourceforge.net/) [26] and SNPs were identified using two methods, GATK [27] and Samtools [28]. SNPs identified using both these methods with integrity > 0.8 and MAF > 0.05 were retained for GWAS analysis.

The population structure of the rabbits was evaluated using the ADMIXTURE program [29]. The association analysis between traits and SNPs was performed according to a general linear mixed (GLM) model using PLINK2 software (http://www.cog-genomics.org/plink/2.0/) (–glm) [30,31]. The SNP effects were estimated using the following model:

$$y = Xb + Zu + e \quad (2)$$

where y is the vector of phenotypes; b is the fixed effect of sex; u is the SNP effect with $u \sim N(0, I\sigma_u^2)$; X and Z are each the incidence matrix for b and u, respectively; and e is a vector of residual effects with $e \sim N(0, I\sigma_e^2)$. I is an identity matrix; σ_u^2 is the variance of SNP effects; and σ_e^2 is the residual variance.

Furthermore, Bonferroni correction was applied to determine significance at the genome-wide level. SNPs with an adjusted p-value less than the 10%genome-wide Bonferroni-corrected threshold were annotated using ANNOVAR software [32]. Genome-wide linkage disequilibrium (LD) blocks were estimated using PLINK2 (–ld) and the LD decay is shown in Supplementary Figure S1. Based on the LD decay, the genes within 100 kb of significant associated SNPs were considered as trait-associated candidate genes. We extracted the candidate genes within the ±100 kb region of the associated SNPs according to the genome annotation information in GFF format using a script written in-house in Python3 programming language.

3. Results and Discussion

3.1. Phenotype

Bodyweight at six time points, two carcass traits and two meat quality traits were measured. The descriptive statistics for the measured phenotypic traits are listed in Table 1.

Table 1. Descriptive statistics of phenotypic traits.

Traits	Acronym	Trait Group	N	Mean	SD
35-day BW, g	BW35	Growth	432	787.4213	129.6323
42-day BW, g	BW42	Growth	423	1007.229	124.2199
49-day BW, g	BW49	Growth	419	1234.916	155.4796
56-day BW, g	BW56	Growth	412	1460.34	211.3732
63-day BW, g	BW63	Growth	392	1686.722	279.0648
70-day BW, g	BW70	Growth	382	1927.864	320.2864
Slaughter liveweight (84 d), g	SLW	Carcass	432	2082.037	313.2049
Dressing out percentage, %	DoP	Carcass	430	48.7838	3.9789
Cooking loss, %	CL	Meat quality	432	68.1618	5.383
Drip loss, %	DL	Meat quality	432	4.7754	1.6808

BW: bodyweight; SD: standard deviation.

3.2. SLAF-seq

A total of 3,127,736,005 paired-end reads was generated in this study. Almost 97.07% of the raw reads were successfully mapped to the rabbit genome. The effectiveness of digestion is the key indicator of a successful SLAF-seq. In this study, 96.05% of the sequences were digested normally, which is an ideal digestion rate. A total of 5,223,720 SLAF tags was identified across the whole genome, with sequencing to a 16.57× average depth. A total of 2,498,314 polymorphic SLAF tags was identified and 8,838,009 SNPs were selected for genome-wide association analysis based on the selection criteria (integrity > 0.8; minor allele frequency > 0.05). The density distribution of SNPs was calculated throughout the rabbit genome and is shown in Figure 1. Almost all of the genome's non-overlapped 1 Mbp region contained SNPs, which indicates that the data are reliable.

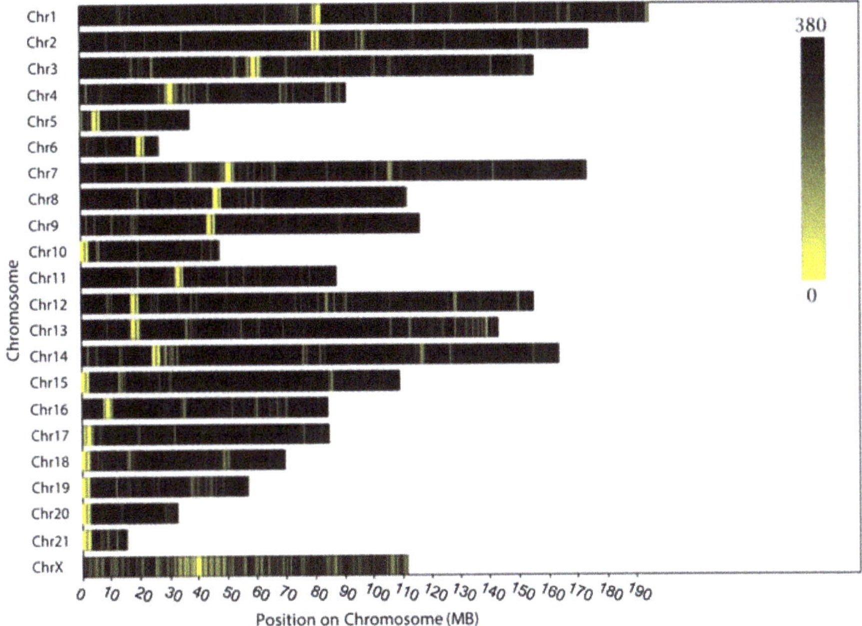

Figure 1. Single-nucleotide polymorphism (SNP) density distribution on chromosomes of the rabbit genome. The horizontal axis (*X*-axis) shows the chromosome length (Mbp). SNP density was calculated per 1 Mbp window. Different colors represent different SNP density levels.

3.3. Population Structure Analysis

The population structure of the rabbits was analyzed using the ADMIXTURE program. The population was first divided into 1–20 subgroups (K). The cross-validation (CV) error of populations was calculated under different k numbers. The number of the Kvalue with the lowest CV is the most suitable. Our analysis revealed that a K value of 18 was most optimal. The samples were divided into 18 subgroups (Figure 2).

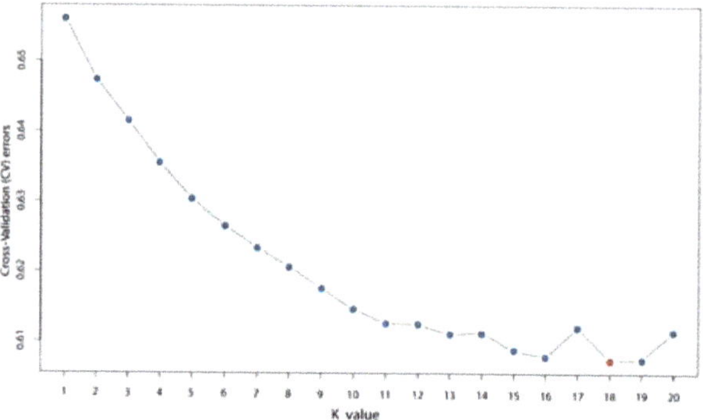

Figure 2. The coefficient of variation for each K value. The point of the lowest cross-validation (CV) error is indicated with red color.

3.4. Genome-Wide Association Analysis

As population stratification might affect GWAS, quantile–quantile (Q-Q) plots of all traits were drawn. The observed p value calculated by the association study fit the expected ones, which suggests that the population stratification was well-corrected and the association analysis using GLM was reliable. The Q-Q plot of each trait is shown following the Manhattan plot of the corresponding trait.

GWAS based on SLAF-seq were successfully used to identify SNPs associated with important economic traits in chickens and pigs [18,19]. In this study, we performed SLAF-seq-based GWAS of six growth, two carcass and two meat quality traits using a General Linear Mixed (GLM) Model in meat rabbits. Table 2 shows the number of SNPs associated with each trait and the number of genes located within 100 kb upstream and downstream. Table S1 shows the location on the genome, the p value and the annotation of all SNPs linked with each trait.

Table 2. Number of SNPs associated with each trait and the number of genes around.

Trait Group	Traits	# SNP	# Gene
growth	35 d BW	1	4
growth	42 d BW	0	0
growth	49 d BW	2	3
growth	56 d BW	4	3
growth	63 d BW	7	N/A
growth	70 d BW	19	9
carcass	Slaughter liveweight (84 d)	1	6
carcass	Dressing out percentage	80	65
meat quality	Cooking loss	5	2
meat quality	Drip loss	10	9

Based on the GLM and Bonferroni correction, 28 SNPs exhibited genome-wide association with the bodyweight trait at 35, 49, 56, 63 and 70 d (Table 2) and 16 genes near or within those SNPs were identified as associated with bodyweight. Manhattan plots for the growth traits are shown in Figures 3 and 4. Interestingly, one SNP on chromosome 8 (Chr8) was found to be associated with bodyweight at both day 49 and 56, which is located within *FGD4* (FYVE, RhoGEF and PH domain containing 4). This gene encodes a protein involved in the regulation of actin cytoskeleton and cell shape [33], which is important for cell growth. The *DNM1L* (dynamin 1 like) gene, located within 100 kb of an SNP significantly associated with BW49 and BW56, encodes a member of the dynamin GTPase superfamily, which is involved in regulating mitochondrial metabolism. The gene expression of *DNM1L* is reported to increase in skeletal muscle following exercise [34]. We thus propose these genes as potential targets for molecular breeding in meat rabbits.

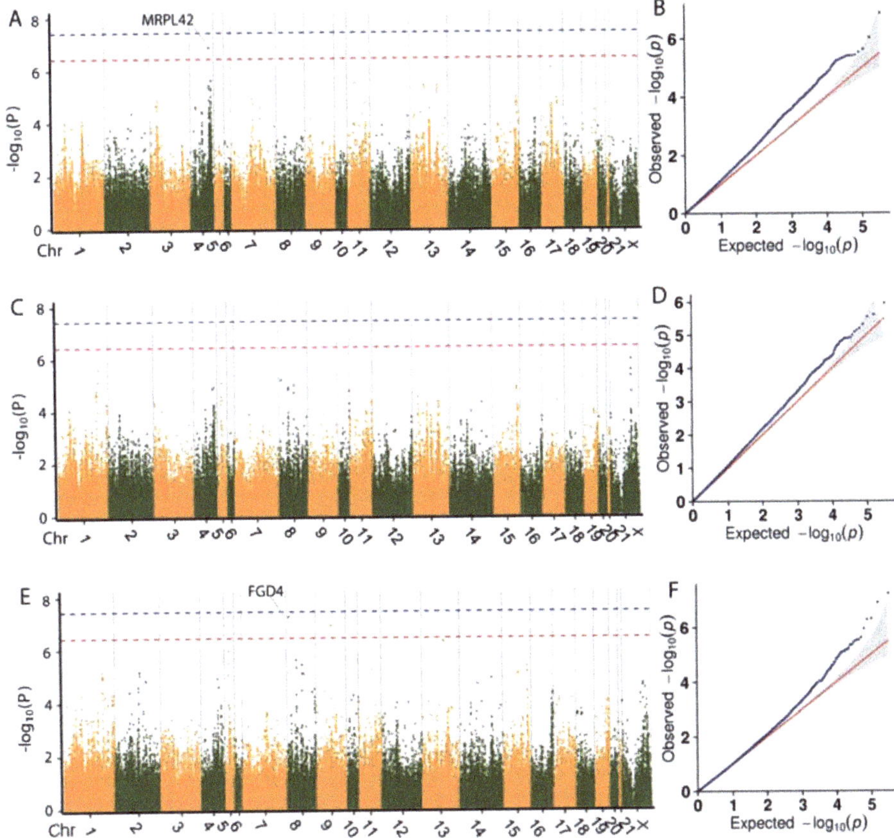

Figure 3. Manhattan plots and quantile–quantile (Q-Q) plots of the general linear mixed (GLM)-based genome-wide association study (GWAS) for bodyweight at day 35 (**A**,**B**), 42 (**C**,**D**) and 49 (**E**,**F**). Negative $\log_{10}$ p values of the filtered high-quality SNPs were plotted against their genomic positions. The dashed lines of orange and blue indicate a 10% and 1% genome-wide Bonferroni-corrected threshold, respectively.

One SNP at position 61,268,427 of Chr9 was identified to be associated with bodyweight at day 49, 56 and 63. The SNP at position 8,681,239 in Chr8 was identified to be associated with bodyweight at day 49 and 56. SNPs at 83,815,488 and 83,815,516 of Chr13 were found to be associated with bodyweight at day 56 and 63. However, no genes were found within 100 kb upstream or downstream of these SNPs. We speculate these SNPs to have an association at two or more time points of bodyweight and be important for the growth of the rabbit and are potential targets for further molecular breeding research.

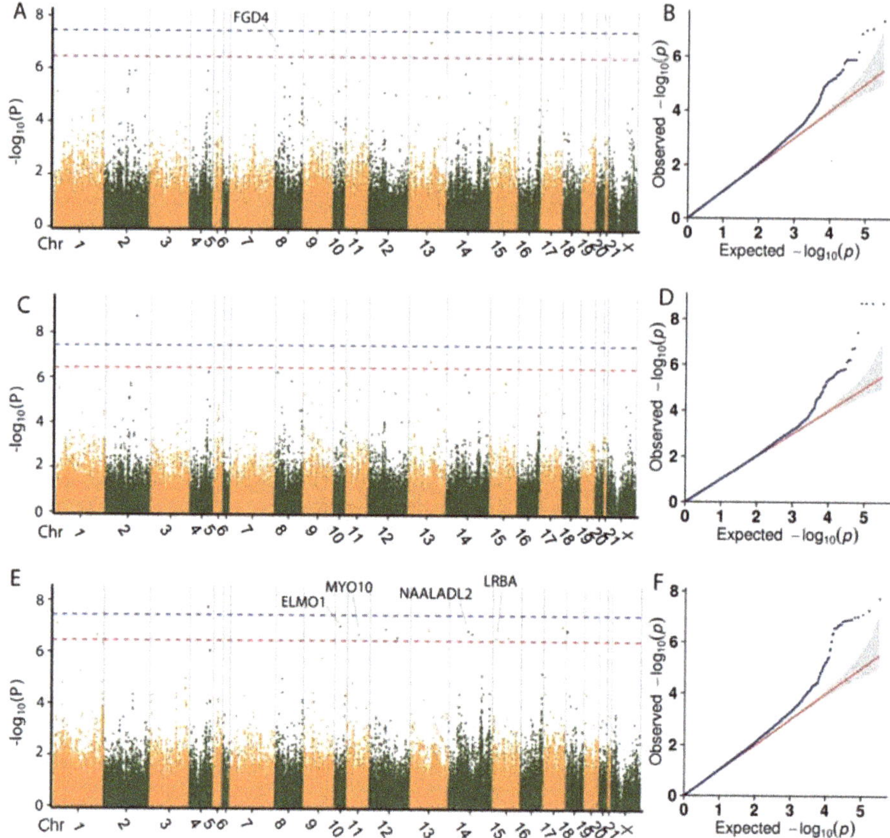

Figure 4. Manhattan plots and Q-Q plots of the GLM-based GWAS for bodyweight at day 56 (**A**,**B**), day 63 (**C**,**D**) and day 70 (**E**,**F**). Negative $\log_{10} p$ values of the filtered high-quality SNPs were plotted against their genomic positions. The dashed lines of orange and blue indicate 10% and 1% genome-wide Bonferroni-corrected threshold, respectively.

A total of 81 SNPs was found to be associated with two carcass traits and 71 genes were found near those SNPs. Manhattan plots for the growth traits are shown in Figure 5. Among them, one SNP located at 37,208,481 of Chr18 was identified to be associated with slaughter liveweight (84 d), with six genes located around the SNP. We identified *LIPA* (lipase A, lysosomal acid type) as the nearest gene, which encodes lipase A, the lysosomal acid lipase (cholesterol ester hydrolase). This enzyme catalyzes the hydrolysis of cholesteryl esters and triglycerides in the lysosome [35]. In addition, 80 SNPs were found associated with carcass rate and 65 genes near these SNPs encoded proteins with various functions (Supplementary Material Table S1). However, there have been no reports of the association between these candidate genes and animal production traits. Further study is required to validate the links between candidate genes and traits.

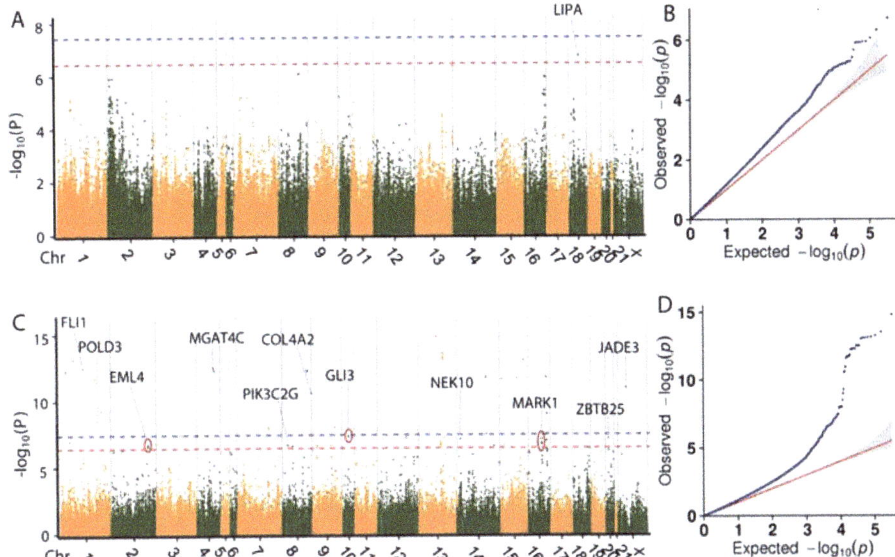

Figure 5. Manhattan plots and Q-Q plots of GLM-based GWAS for the dressing out percentage (**A**,**B**) and slaughter liveweight (**C**,**D**). Negative $\log_{10}$ p values of the filtered high-quality SNPs were plotted against their genomic positions. The dashed lines of orange and blue indicate a 10% and 1% genome-wide Bonferroni-corrected threshold, respectively. Red ellipses indicate that SNPs within the same ellipse share the nearest candidate gene.

For the meat quality phenotype, 15 SNPs were associated with cooking loss and drip loss. Manhattan plots for the growth traits are shown in Figure 6. Five SNPs were associated with cooking loss, with 2 genes located nearby. The nearest gene *OSBPL11* (oxysterol-binding protein like 11) encodes a member of the oxysterol-binding protein (*OSBP*) family, a group of intracellular lipid receptors [36]. Ten SNPs were associated with drip loss, with 9 genes around those SNPs. Two main types of proteins in the human body involved in the maintenance of zinc ion homeostasis—zinc binding protein, which acts as a buffer substance or as a donor of intracellular zinc and zinc transporters, which are involved in the intake and excretion of zinc in cells. Among the five genes found in close proximity to the associated SNPs of the drip loss trait in this study, two encode proteins involved in the maintenance of zinc homeostasis. *ADAM7* (ADAM metallopeptidase domain 7) on chromosome 2 encodes a member of the *ADAMs* family of zinc proteases and *SLC39A6* (solute carrier family 39 member 6) on chromosome 9 encodes a protein with structural characteristics of zinc transporters [37]. These findings indicate that the potential targets for meat quality are the genes relevant to zinc homeostasis.

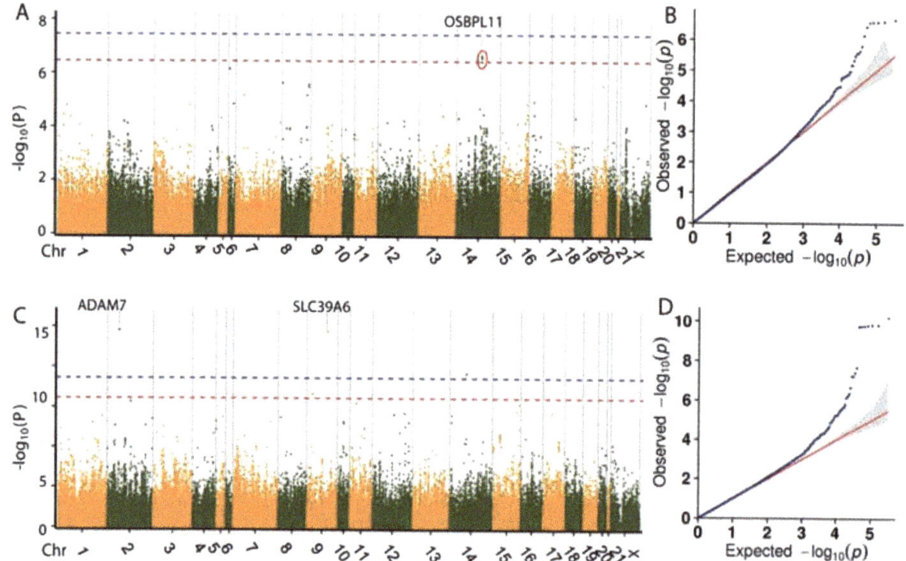

Figure 6. Manhattan plots and Q-Q plots of the GLM-based GWAS for the meat quality traits of cooking loss (**A**,**B**) and drip loss (**C**,**D**) in rabbits. Negative $\log_{10} p$ values of the filtered high-quality SNPs were plotted against their genomic positions. The dashed lines of orange and blue indicate a 10% and 1% genome-wide Bonferroni-corrected threshold, respectively. The red ellipse indicates that SNPs in the same ellipse share the nearest candidate gene.

4. Conclusions

In this study, we performed GWAS of 432 F1 meat rabbits. All samples were genotyped using SLAF-seq technology. The GWAS of 10 economic traits revealed 111 significantly associated SNPs. A total of 98 putative candidate genes were located within 100 kb of significantly associated SNPs. Among them, *FGD4* and *DNM1L* genes, linked with BW49 and BW56 in our study, are reported to be involved in cell growth and mitochondrial metabolism. *ADAM7* on chromosome 2 and *SLC39A6* on chromosome 9 may be associated with drip loss due to its effect on zinc homeostasis that functions in determining meat quality. These findings provide novel insights into the genetic basis of growth, carcass and meat quality traits in rabbits and may contribute to the application of GS in rabbits. However, because only a draft version of the rabbit reference genome is available, there are numerous sequence gaps and missing annotations. These gaps could influence the accurate estimation of LD patterns. Moreover, the missing annotations would inevitably lead to false-negative results when attempting to find causative genes or variants. In addition, relatively high numbers of false positives were observed in the GWAS study. Therefore, a future validation study should be conducted using an independent population.

Supplementary Materials: The following are available online at http://www.mdpi.com/2076-2615/10/6/1068/s1. Figure S1. A schematic representation of genome-wide linkage disequilibrium (LD) decay. Table S1. Significant SNPs for growth, carcass and meat quality traits in rabbits.

Author Contributions: Conceptualization, S.-J.L. and S.-Y.C.; methodology, X.Y. and F.D.; validation, X.Y., and S.H.; formal analysis, X.Y., Z.W., Y.S. and F.D.; investigation, X.Y., X.J., J.W. and W.C.; writing—original draft preparation, X.Y and F.D.; writing—review and editing, S.-J.L.; visualization, X.Y.; project administration, S.-J.L.; funding acquisition, S.-J.L. All authors have read and agreed to the published version of the manuscript.

Funding: This research was funded by the earmarked fund for China Agriculture Research System, grant number CARS-43-A-2.

Conflicts of Interest: All authors declare no conflicts of interest. The funders had no role in the design of the study; in the collection, analyses, or interpretation of data; in the writing of the manuscript, or in the decision to publish the results.

References

1. Bosco, A.D.; Gerencsér, Z.; Szendrő, Z.; Mugnai, C.; Cullere, M.; Kovacs, M.; Ruggeri, S.; Mattioli, S.; Castellini, C.; Zotte, A.D. and et al. Effect of dietary supplementation of Spirulina (Arthrospira platensis) and Thyme (Thymus vulgaris) on rabbit meat appearance, oxidative stability and fatty acid profile during retail display. *Meat Sci.* **2014**, *96*, 114–119. [CrossRef] [PubMed]
2. Zotte, A.D.; Szendrő, Z. The role of rabbit meat as functional food. *Meat Sci.* **2011**, *88*, 319–331. [CrossRef] [PubMed]
3. Fontanesi, L.; Scotti, E.; Cisarova, K.; Di Battista, P.; Dall'Olio, S.; Fornasini, D.; Frabetti, A. A missense mutation in the rabbit Melanocortin 4 Receptor (MC4R) gene is associated with finisching weight in a meat rabbit line. *Anim. Biotechnol.* **2013**, *24*, 268–277. [CrossRef] [PubMed]
4. Yang, Z.-J.; Fu, L.; Zhang, G.-W.; Yang, Y.; Chen, S.-Y.; Wang, J.; Lai, S.-J. Identification and association of SNPs in TBC1D1 gene with growth traits in two rabbit breeds. *Asian-Australas. J. Anim. Sci.* **2013**, *26*, 1529–1535. [CrossRef]
5. Gan, W.; Song, Q.; Zhang, N.; Xiong, X.; Wang, D.; Li, L. Association between FTO polymorphism in exon 3 with carcass and meat quality traits in crossbred ducks. *Genet. Mol. Res.* **2015**, *14*, 6699–6714. [CrossRef]
6. Migdał, Ł.; Koziol, K.; Palka, S.; Migdal, W.; Zabek, T.; Otwinowska-Mindur, A.; Migdal, A.; Kmiecik, M.; Maj, D.; Bieniek, J. and et al. Mutations in Leptin (LEP) gene are associated with carcass and meat quality traits in crossbreed rabbits. *Anim. Biotechnol.* **2017**, *29*, 153–159. [CrossRef]
7. Fontanesi, L.; Sparacino, G.; Utzeri, V.J.; Scotti, E.; Fornasini, D.; Dall'Olio, S.; Frabetti, A. Identification of polymorphisms in the rabbit Growth Hormone Receptor (GHR) gene and association with finishing weight in a commercial meat rabbit line. *Anim. Biotechnol.* **2016**, *27*, 77–83. [CrossRef]
8. McCarthy, M.I.; Abecasis, G.R.; Cardon, L.R.; Goldstein, D.B.; Little, J.; Ioannidis, J.P.A.; Hirschhorn, J.N. Genome-wide association studies for complex traits: Consensus, uncertainty and challenges. *Nat. Rev. Genet.* **2008**, *9*, 356–369. [CrossRef]
9. Hayes, B.; Goddard, M. Genome-wide association and genomic selection in animal breeding. *Genome* **2010**, *53*, 876–883. [CrossRef]
10. Shu, X.-O.; Long, J.; Lu, W.; Li, C.; Chen, W.Y.; Delahanty, R.; Cheng, J.; Cai, H.; Zheng, Y.; Shi, J.; et al. Novel genetic markers of breast cancer survival identified by a genome-wide association study. *Cancer Res.* **2012**, *72*, 1182–1189. [CrossRef]
11. Tsai, H.-Y.; Hamilton, A.; Tinch, A.E.; Guy, D.R.; Gharbi, K.; Stear, M.J.; Matika, O.; Bishop, S.C.; Houston, R.D. Genome wide association and genomic prediction for growth traits in juvenile farmed Atlantic salmon using a high density SNP array. *BMC Genom.* **2015**, *16*, 969. [CrossRef] [PubMed]
12. Gonzalez-Pena, D.; Gao, G.; Baranski, M.; Moen, T.; Cleveland, B.M.; Kenney, P.B.; Vallejo, R.; Palti, Y.; Leeds, T.D. Genome-wide association study for identifying loci that affect fillet yield, carcass, and body weight traits in Rainbow Trout (Oncorhynchus mykiss). *Front. Genet.* **2016**, *7*, 743. [CrossRef]
13. Mason, A.S.; Higgins, E.E.; Snowdon, R.J.; Batley, J.; Stein, A.; Werner, C.; Parkin, I.A.P. A user guide to the Brassica 60K Illumina Infinium™ SNP genotyping array. *Theor. Appl. Genet.* **2017**, *130*, 621–633. [CrossRef] [PubMed]
14. Cao, J.; Schneeberger, K.; Ossowski, S.; Günther, T.; Bender, S.; Fitz, J.; Koenig, D.; Lanz, C.; Stegle, O.; Lippert, C.; et al. Whole-genome sequencing of multiple Arabidopsis thaliana populations. *Nat. Genet.* **2011**, *43*, 956–963. [CrossRef] [PubMed]
15. Yano, K.; Yamamoto, E.; Aya, K.; Takeuchi, H.; Lo, P.-C.; Hu, L.; Yamasaki, M.; Yoshida, S.; Kitano, H.; Hirano, K.; et al. Genome-wide association study using whole-genome sequencing rapidly identifies new genes influencing agronomic traits in rice. *Nat. Genet.* **2016**, *48*, 927–934. [CrossRef]
16. Sun, X.; Liu, N.; Zhang, X.; Li, W.; Liu, H.; Hong, W.; Jiang, C.; Guan, N.; Ma, C.; Zeng, H.; et al. SLAF-seq: An efficient method of large-scale de novo SNP discovery and genotyping using high-throughput sequencing. *PLoS ONE* **2013**, *8*, e58700. [CrossRef]

17. Wang, W.; Zhang, T.; Zhang, G.; Wang, J.; Han, K.; Wang, Y.; Zhang, Y. Genome-wide association study of antibody level response to NDV and IBV in Jinghai yellow chicken based on SLAF-seq technology. *J. Appl. Genet.* **2015**, *56*, 555. [CrossRef]
18. Li, F.; Han, H.; Lei, Q.; Gao, J.; Liu, J.; Liu, W.; Zhou, Y.; Li, H.; Cao, D. Genome-wide association study of body weight in Wenshang Barred chicken based on the SLAF-seq technology. *J. Appl. Genet.* **2018**, *59*, 305–312. [CrossRef]
19. Li, Z.; Wei, S.; Li, H.; Wu, K.; Cai, Z.; Li, N.; Wei, W.; Li, Q.; Chen, J.; Liu, H.; et al. Genome-wide genetic structure and differentially selected regions among Landrace, Erhualian, and Meishan pigs using specific-locus amplified fragment sequencing. *Sci. Rep.* **2017**, *7*, 10063. [CrossRef]
20. Zhang, H.; Wang, Z.; Wang, S.; Li, H. Progress of genome wide association study in domestic animals. *J. Anim. Sci. Biotechnol.* **2013**, *3*, 26. [CrossRef]
21. Sosa-Madrid, B.S.; Hernández, P.; Blasco, A.; Haley, C.S.; Fontanesi, L.; Santacreu, M.A.; Pena, R.N.; Navarro, P.; Ibáñez-Escriche, N. Genomic regions influencing intramuscular fat in divergently selected rabbit lines. *Anim. Genet.* **2019**, *51*, 58–69. [CrossRef] [PubMed]
22. Blasco, A.; Ouhayoun, J. Harmonization of criteria and terminology in rabbit meat research. *World Rabbit Sci.* **2010**, *1*, 4. [CrossRef]
23. Traoré, S.; Aubry, L.; Gatellier, P.; Przybylski, W.; Jaworska, D.; Kajak-Siemaszko, K.; Santé-Lhoutellier, V. Higher drip loss is associated with protein oxidation. *Meat Sci.* **2012**, *90*, 917–924. [CrossRef]
24. Kozich, J.J.; Westcott, S.L.; Baxter, N.; Highlander, S.K.; Schloss, P.D. Development of a dual-index sequencing strategy and curation pipeline for analyzing amplicon sequence data on the miseq illumina sequencing platform. *Appl. Environ. Microbiol.* **2013**, *79*, 5112–5120. [CrossRef] [PubMed]
25. Li, R.; Yu, C.; Li, Y.; Lam, T.-W.; Yiu, S.M.; Kristiansen, K.; Wang, J. SOAP2: An improved ultrafast tool for short read alignment. *Bioinformatics* **2009**, *25*, 1966–1967. [CrossRef]
26. Li, H.; Durbin, R. Fast and accurate short read alignment with Burrows-Wheeler transform. *Bioinformatics* **2009**, *25*, 1754–1760. [CrossRef]
27. McKenna, A.; Hanna, M.; Banks, E.; Sivachenko, A.; Cibulskis, K.; Kernytsky, A.; Garimella, K.; Altshuler, D.; Gabriel, S.; Daly, M.; et al. The genome analysis toolkit: A MapReduce framework for analyzing next-generation DNA sequencing data. *Genome Res.* **2010**, *20*, 1297–1303. [CrossRef]
28. Li, H.; Handsaker, B.; Wysoker, A.; Fennell, T.; Ruan, J.; Homer, N.; Marth, G.; Abecasis, G.R.; Durbin, R.; Genome Project Data Processing Subgroup; et al. The sequence alignment/map format and SAMtools. *Bioinformatics* **2009**, *25*, 2078–2079. [CrossRef]
29. Alexander, D.H.; Novembre, J.; Lange, K. Fast model-based estimation of ancestry in unrelated individuals. *Genome Res.* **2009**, *19*, 1655–1664. [CrossRef]
30. Price, A.L.; Patterson, N.J.; Plenge, R.M.; Weinblatt, M.E.; Shadick, N.A.; Reich, D. Principal components analysis corrects for stratification in genome-wide association studies. *Nat. Genet.* **2006**, *38*, 904–909. [CrossRef]
31. Chang, C.C.; Chow, C.C.; Tellier, L.C.A.M.; Vattikuti, S.; Purcell, S.M.; Lee, J.J. Second-generation PLINK: Rising to the challenge of larger and richer datasets. *GigaScience* **2015**, *4*, 7. [CrossRef] [PubMed]
32. Wang, K.; Li, M.; Hakonarson, H. Annovar: Functional annotation of genetic variants from high-throughput sequencing data. *Nucleic Acids Res.* **2010**, *38*. [CrossRef] [PubMed]
33. Ikeda, W.; Nakanishi, H.; Takekuni, K.; Itoh, S.; Takai, Y. Identification of splicing variants of Frabin with partly different functions and tissue distribution. *Biochem. Biophys. Res. Commun.* **2001**, *286*, 1066–1072. [CrossRef]
34. Moore, T.M.; Zhou, Z.; Cohn, W.; Norheim, F.; Lin, A.J.; Kalajian, N.; Strumwasser, A.R.; Cory, K.; Whitney, K.; Ho, T.; et al. The impact of exercise on mitochondrial dynamics and the role of Drp1 in exercise performance and training adaptations in skeletal muscle. *Mol. Metab.* **2018**, *21*, 51–67. [CrossRef]
35. Anderson, R.A.; Rao, N.; Byrum, R.S.; Rothschild, C.B.; Bowden, N.W.; Hayworth, R.; Pettenati, M. In situ localization of the genetic locus encoding the Lysosomal Acid Lipase/Cholesteryl Esterase (LIPA) deficient in Wolman disease to chromosome 10q23.2-q23.3. *Genomics* **1993**, *15*, 245–247. [CrossRef] [PubMed]

36. Taylor, F.R.; Kandutsch, A.A. Oxysterol binding protein. *Chem. Phys. Lipids* **1985**, *38*, 187–194. [CrossRef]
37. Taylor, K.M.; Morgan, H.E.; Johnson, A.; Hadley, L.J.; Nicholson, R.I. Structure–function analysis of LIV-1, the breast cancer-associated protein that belongs to a new subfamily of zinc transporters. *Biochem. J.* **2003**, *375*, 51–59. [CrossRef]

Article

Myeloperoxidase and Lysozymes as a Pivotal Hallmark of Immunity Status in Rabbits

Rafał Hrynkiewicz, Dominika Bębnowska and Paulina Niedźwiedzka-Rystwej *

Institute of Biology, University of Szczecin, Felczaka 3c, 71-412 Szczecin, Poland; rafal.hrynkiewicz@gmail.com (R.H.); bebnowska.d@wp.pl (D.B.)

* Correspondence: paulina.niedzwiedzka-rystwej@usz.edu.pl

Received: 15 August 2020; Accepted: 3 September 2020; Published: 4 September 2020

Simple Summary: Rabbit breeding is a very important element in the context of broadly understood industrial breeding, as rabbits are one of the main and most frequently chosen economic directions. Effective rabbit breeding, however, requires full control over the health of these animals, which is particularly related to the orientation regarding their immune status. There are many indicators that can be used to assess the immune system, but the greatest attention should be paid to those that change rapidly over time and reflect the body's first line of defense. Peripheral blood granulocytes contain enzymes with strong antimicrobial properties, the level of which changes as a result of various external factors, e.g., viral infection, which was assessed in this study. The aim of the study was to evaluate the dynamics of myeloperoxidase (MPO) and lysozyme (LZM) in the experimental infection of rabbits with the *Lagovirus europaeus*/GI.1a virus, which is a pathogen causing high mortality, decimating rabbit farms all over the world in a short time. The results obtained in the dynamic system show that the levels of assessed enzymes significantly change in the blood during infection. Assessing the immune system using these indicators could therefore be a potential biomarker for the immune status of rabbits.

Abstract: Infectious diseases, due to their massive scale, are the greatest pain for all rabbit breeders. Viral infections cause enormous economic losses in farms. Treating sick rabbits is very difficult and expensive, so it is very important to prevent disease by vaccinating. In order to successfully fight viral infections, it is important to know about the immune response of an infected animal. The aim of this study was to analyze the immune response mediated by antimicrobial peptides (myeloperoxidase (MPO) and lysozyme (LZM)) in peripheral blood neutrophils and rabbit serum by non-invasive immunological methods. The study was carried out on mixed breed rabbits that were experimentally infected with two strains (Erfurt and Rossi) of the *Lagovirus europaeus*/GI.1a virus. It has been observed that virus infection causes changes in the form of statistically significant increases in the activity of MPO and LZM concentration, while in the case of LZM activity only statistically significant decreases were noted. Additionally, clinical symptoms typical for the course of the disease were noted, and the probability of survival of the animals at 60 h p.i. (post infection) was 30% for the Erfurt strain, and −60% for the Rossi strain. The obtained results of MPO and LZMs suggest that these enzymes, especially MPO, may serve as a prognostic marker of the state of the immune system of rabbits.

Keywords: myeloperoxidase; lysozyme; rabbits; viral infection; rabbit hemorrhagic disease

---

1. Introduction

Rabbit breeding is an important and future-oriented direction of animal production, especially from an economic point of view. Apart from being a key element of the economy of many countries, rabbits also exist as an important link in the trophic chains of the Mediterranean ecosystems [1–5]. As laboratory animals, they are widely used in various types of diagnostic tests and analyses [2,6–8].

Currently, rabbit breeding is becoming increasingly popular. The data concerning the world production of these animals indicate that annually about 2 million of them are slaughtered exclusively for meat [9].

The loss of rabbits, especially due to various diseases, is very problematic for breeders, because even the best farms, by not following the rules of disease prevention and improper treatment, face huge problems. The treatment of sick rabbits is very difficult, as it requires cumbersome, long-lasting and systematic procedures [10,11], and the causes of rabbits' diseases can be very different. The most common diseases of rabbits are infectious diseases, which can occur en masse, thus leading to serious losses [10,11].

One of the most dangerous diseases that can decimate rabbit breeding is rabbit haemorrhagic disease (RHD) [10–12]. RHD is a highly infectious and deadly viral disease manifested by acute viral hepatitis in rabbits, caused by single-stranded RNA (ssRNA) virus ssRNA virus *Lagovirus (L.) europaeus* [13–15]. The genus *L. europaeus* is divided into two main gene groups related to rabbit hemorrhagic disease virus (RHDV): GI or European hare syndrome virus (EBHSV): GII [14]. Within the GI gene group, four genotypes—GI.1, GI.2, GI.3, and GI.4—were distinguished. The GI.1 genotype includes classical RHDV strains (*L. europaeus*/GI.1), while the GI.2 genotype was classified as RHDV2, discovered in France in 2010 (*L. europaeus*/GI.2) [16]. Genotype GI.3 is represented by RCV-E1 (*L. europaeus*/GI.3), and GI.4 by RCV-A1 (*L. europaeus*/GI.4) and RCV-E2 (*L. europaeus*/GI.4d). The strains present in these genotypes are called nonpathogenic rabbit calicivirus (RCV). The strains of the GI.1 and GI.2 genotypes are known infectious agents responsible for the development of RHD. According to the recently proposed nomenclature [14], GI.1 is further subdivided into different antigenic variants (GI.1a-d) based on phylogeny and genetic distance [14].

Due to its features, RHD has been included in the list of notifiable diseases to the World Organisation for Animal Health (OIE). Every animal in which RHD is suspected or confirmed is immediately euthanized and cremated [10,11]. The first data of RHD were reported in Wuxi, Jiangsu Province in China in 1984 [17]. The disease was observed in the population of European rabbits (*Oryctolagus cuniculus*) of the Angora breed, which were imported from the former German Democratic Republic to China for breeding purposes [17,18]. In China, within one year, the disease contributed to the deaths of over 140 million domestic rabbits and spread over about 50,000 km^2 [18]. The Chinese epidemic of RHD caused enormous losses in the economic sphere of the country [18,19]. The strong expansion of the virus soon led to its rapid spread throughout Europe [2]. Since 2010, the RHDV2 virus (*L. europaeus*/GI.2) has also spread to countries in Europe, Australia, America, and Africa [20]. The appearance of RHD on the Iberian Peninsula led to huge losses in the wild rabbit population [1–3,16,21]. There was a great interest in this problem due to the rabbit's key contribution to the Mediterranean ecosystems, as it provides food for several endangered endemic species, such as the Iberian lynx (*Lynx pardinus*) and the Iberian imperial eagle (*Aquila heliaca*) [3]. The reduction of the wild rabbit population therefore has a negative impact on the persistence of these predators [3,22,23].

Currently, the disease is present in Europe, Asia, Africa, and Australia. Single epizooties are recorded from time to time in North America (United States, Canada, Cuba). According to the current information in the WAHID (World Animal Health Information Database) system [11], in the years 2005–2018, the occurrence of RHD was reported or suspected in 50 countries, of which more than half of the reports were reported in European countries [11]. Unfortunately, so far there is no effective cure that can save infected rabbits. The only correct method of fighting RHD is its prevention through prophylactic vaccinations [1,3,20,24].

Moreover, the key to administering proper vaccination, as well as efficient fighting against the virus, is the knowledge of the immune system response of the infected animal. In light of the above, the aim of the study was to analyze with a noninvasive immunological method, the immunological response mediated by antimicrobial peptides (myeloperoxidase and lysozyme) in neutrophils of peripheral blood and serum of rabbits infected experimentally with *L. europaeus*/GI.1a.

2. Materials and Methods

2.1. Animals

The study was conducted on a group of 25 Polish mixed-breed rabbits of various sexes, of the *Oryctolagus cuniculus* species, weighing in the range of 2.50–3.80 kg, marked as conventional animals, from licensed breeding under constant veterinary and zoo-hygienic supervision [25]. The rabbits were not vaccinated against *Lagovirus europaeus*, were in good health, and did not show any disease symptoms. Throughout the experiment, the animals stayed in a vivarium belonging to the Institute of Biology, the Faculty of Mathematical, Physical and Natural Sciences, the University of Szczecin. During the tests, appropriate zoo-technical conditions were ensured, in accordance with the recommended Polish standards developed in line with the European Union Directive with regards temperature and humidity, as well as lighting and size of cages for animals [26]. After transporting to the vivarium, the animals were subjected to a two-week adaptation period and tested for the presence of anti-RHDV antibodies using a ready-made ELISA (Institutio Zooprofilattico Sperimentale, Italy). Each of the rabbits had a separate metal cage, was fed with complete feed for rabbits (16% Rabbit, Motycz, Poland) at an amount of 0.15–0.20 kg per day, and had unlimited access to fresh drinking water. All studies were conducted with the approval of the Local Ethics Committee for Experiments on Animals in Poznań (license no. 1/2009).

2.2. The Scheme of the Experiment

Two antigenic variants of *Lagovirus europaeus*/GI.1a with positive hemagglutination (HA+) were selected for the study:

- *L. europaeus*/GI.1a/Erfurt (Germany, 2000),
- *L. europaeus*/GI.1a/Rossi (Germany, 2002).

The rabbits that were used in the experiment were divided into three groups. The animals were classified into groups based on randomness. The first group ($n = 10$) were rabbits (60% female; 40% male) infected with *L. europaeus*/GI.1a/Erfurt. The second group ($n = 10$) were rabbits (50% female; 50% male) infected with *L. europaeus*/GI.1a/Rossi ($n = 10$). The third group ($n = 5$) were rabbits (40% female; 60% male) that were classified as the control group.

The virus infection occurred after the first blood sampling (0 hour of the experiment). The virus was infected by intramuscular injection of *L. europaeus*/GI.1a antigens into the lower limb muscle. Control group ($n = 5$) rabbits were given a placebo in the form of PBS (phosphate-buffered saline) in the same way. Subsequent blood samples were taken in hours: 8, 12, 24, 36, 48, 52, 56, and 60 h post infection (p.i.). Myeloperoxidase (MPO) activity was determined in peripheral blood samples, whereas in blood serum the concentration and activity of lysozyme (LZM) was determined.

2.3. Virus Preparation and Administration

Viruses were obtained from animals that died under natural conditions in different parts of Germany (strains were obtained in lyophilized form from Dr. Horst Schirrmeier, Friedrich Loeffler Institute Greifswald, Germany). Virus identification was performed using the real-time PCR method under the conditions previously described by Niedźwiedzka-Rystwej et al. [27]. Liver homogenate was prepared from the recovered livers, and the animals were experimentally infected in order to increase the amount of virus. After their death, the liver was prepared as a 20% homogenate, which was purified by centrifugation (3000 rpm), treatment with 10% chloroform for 60 min, and centrifugation again. Then, a suspension in glycerol was prepared in a 1:1 ratio [28,29]. All the antigens prepared in this way had the same number of virus particles, with the density from 1.310 to 1.340 g/cm^3. Additionally, the virus titer was determined using the HA hemagglutination test, which was 1:1280 for both strains.

2.4. Myeloperoxidase (MPO) Activity

Determination of myeloperoxidase (MPO) activity in neutrophils was performed according to Graham's method, as described by Zawistowski [30]. This method is based on making a smear of edetate blood, collected in a sample with an anticoagulant in the form of heparin, and the use of a color reaction with myeloperoxidase (MPO) in PMN (polymorphonuclear) cells, based on the reaction of benzidine, which in the presence of hydrogen peroxide and peroxidase passes into brown oxybenzidine. The reagent used in the color reaction with MPO was made from 250 mL of 40% ethyl alcohol, in which 350 mg of benzidine was dissolved; then the reagent was filtered and 0.35 mL of hydrogen peroxide was added. All blood smears were fixed with a mixture of formalin and ethyl alcohol, and then, for at least 3 min, were stained with a benzidine reagent. At a later stage, the preparations were stained with a 7% Giemsa solution for 7 min to stain the PMN cell nucleus. Then, the preparations prepared in this way were viewed under an optical microscope at 1000-fold magnification using immersion. In the preparations, the number of granulocytes was calculated, and the degree of color of the granules was determined according to the adopted scale (Figure 1). MPO activity was expressed as a factor, according to the intensity of the granule color in 100 PMN cells, and calculated on the basis of this formula according to Afanasyev and Kolot [31]:

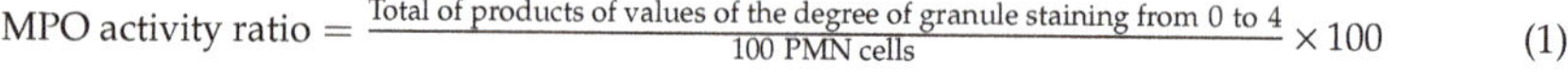

$$\text{MPO activity ratio} = \frac{\text{Total of products of values of the degree of granule staining from 0 to 4}}{\text{100 PMN cells}} \times 100 \quad (1)$$

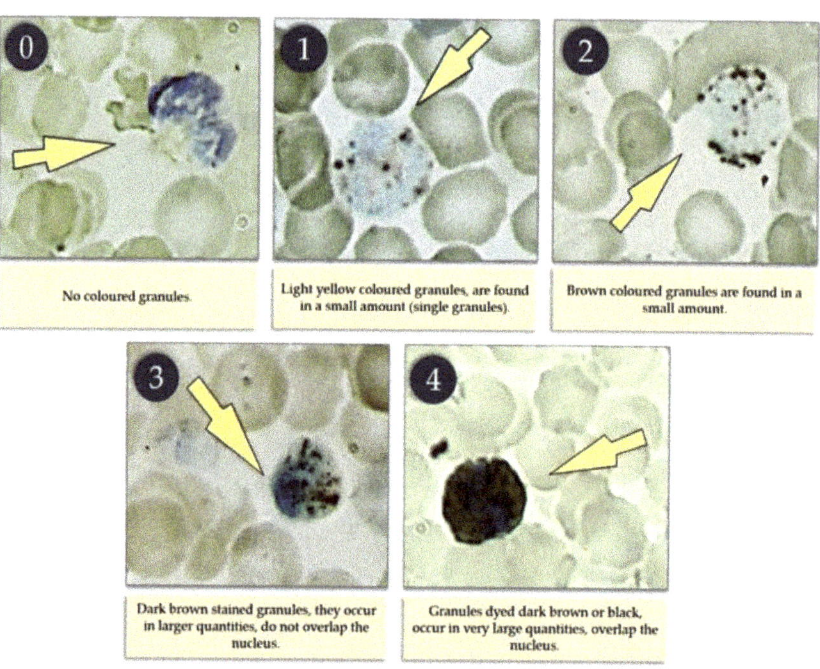

Figure 1. Scale showing the intensity of granule staining in PMN cells.

2.5. Concentration and Activity of Lysozyme (LZM)

The concentration of LZM was determined in the blood serum towards *Micrococcus (M.) lysodeikticus* using the platelet diffusion method, according to the method of Hankiewicz [32].

The substrate was a 1% agarose gel prepared from 1 g of agarose dissolved in 100 mL of 0.067 M phosphate buffer at pH 6.2, with the addition of 1 g of NaCl, in which *M. lysodeikticus* (Sigma, Saint Louis, MO, USA) was suspended (150 mg of bacteria in 15 mL phosphate buffer, at 0.067 M pH

6.2). The agarose gel prepared in this way was poured between glass plates. The agarose gel plates were allowed to set, and then 17 μL wells were carved into the gel.

The LZM standard with a concentration of 100 mg/L was obtained by dissolving 10 mg of hen egg white (Sigma, Saint Louis, MO, USA) with activity of 50,000 IU/mg in 100 mL of 0.067 M phosphate buffer at pH 6.2. A subsequent dilution at a concentration of 64 mg/L was obtained by transferring 6.4 mL of LZM standard solution (100 mg/L) into 3.6 mL of 0.067 M phosphate buffer at pH 6.2. Subsequent concentrations of 32, 16, 8, 4, 2, 1, 0.5, 0.25, and 0.125 mg/L were obtained by serial dilution by transferring 1 mL of standard from each higher concentration to 1 mL of buffer. The thus-prepared LZM standard (17 μL) and the test serum (17 μL) were spotted into the grooved wells in an agar medium, and then the plates with the medium were incubated in a humid chamber and placed in an incubator at 37 °C for 18 h. After incubation, the diameter of radiolucency zones around individual wells was measured; during this process, the size of the diameter depends on the concentration of LZM in the blood serum. The results of the LZM concentration are read from the calibration curve made on the basis of standard solutions.

In turn, the activity of LZM was calculated on the basis of the formula presented by Szmigielski [33] and presented as an indicator:

$$\text{LZM activity ratio} = \frac{\text{LZM concentration in serum (mg/L)}}{\text{absolute number of neutrophils}^{*}} \tag{2}$$

The absolute neutrophil count (*) was calculated by multiplying the total leukocyte count (in 1000) by the percentage of neutrophils from the blood quality picture (leukogram).

2.6. Clinical Studies of Experimental Animals

Throughout the experiment, clinical symptoms and the survival of rabbits infected with *L. europaeus*/GI.1a were recorded. Rabbit survival is presented as the percentage of survival analyzed using the Kaplan–Meier method.

2.7. Statistical Analysis

All results were statistically analyzed using the Student's *t*-test with Cochran–Cox correction, assuming $p = 0.05$. The analysis was performed using in the *Statistica* software, ver. 13.1 (Statsoft, Poland) and Microsoft Excel (Microsoft 365, Redmond, WA, USA).

3. Results

3.1. Myeloperoxidase (MPO) Activity

The values of MPO activity in the group of rabbits infected with *L. europaeus*/GI.1a/Erfurt ranged from 1.15 to 2.13, with the standard deviation (SD ±) within the range of 0.08–0.29 (Figure 2). On the other hand, in the case of animals infected with *L. europaeus*/GI.1a/Rossi, MPO activity ranged from 1.07 to 2.34, with the standard deviation (SD ±) from 0.16 to 0.65 (Figure 2). In the group of control rabbits, MPO activity values ranged from 1.01 to 1.27, with the standard deviation (SD ±) from 0.02 to 0.15 (Figure 2).

Both results of MPO activity in rabbits infected with *L. europaeus*/GI.1a/Erfurt and *L. europaeus*/GI.1a/Rossi throughout the entire digestion of the experiment showed a clear upward trend of the parameter under study, compared to the results of MPO activity obtained in the studies of rabbits from the control group. The obtained results were subjected to statistical analysis, which showed that both *L. europaeus*/GI.1a/Erfurt and *L. europaeus*/GI.1a/Rossi strains caused eight statistically significant changes in the form of increases in the parameter under study. These changes were observed after 8, 12, 24, 36, 48, 52, 56, and 60 h p.i. (Figure 2).

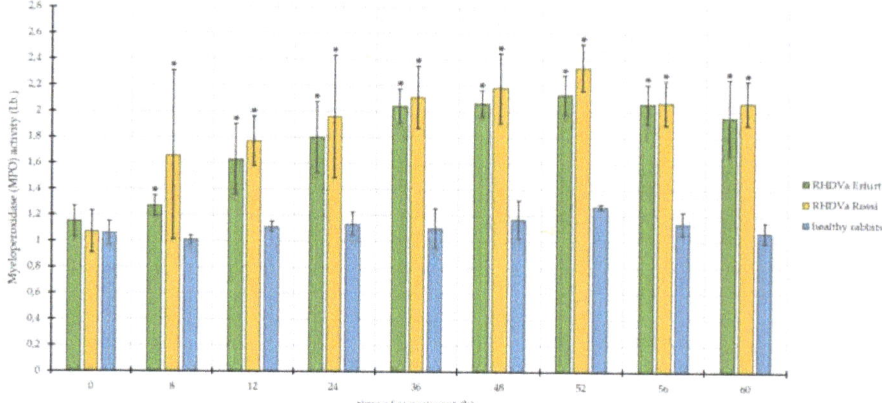

Figure 2. Values of myeloperoxidase (MPO) activity in rabbits infected with *L. europaeus*/GI.1a/Erfurt and *L. europaeus*/GI.1a/Rossi, as well as control rabbits. * statistically significant change with respect to the control group ($p < 0.05$).

3.2. Lysozyme (LZM) Concentration

The parameters of LZM concentration for *L. europaeus*/GI.1a/Erfurt infected rabbits ranged from 4.22 mg/L to 6.40 mg/L, with standard deviation (SD ±) from 0.78 to 1.91 (Figure 3). On the other hand, the results recorded for *L. europaeus*/GI.1a/Rossi infection ranged from 4.03 mg/L to 5.49 mg/L, with the standard deviation (SD ±) from 0.54 to 2.12. (Figure 3). The concentration of LZM in control rabbits ranged from 3.60 mg/L to 4.50 mg/L, with the standard deviation (SD ±) ranging from 0.20 to 0.90 (Figure 3).

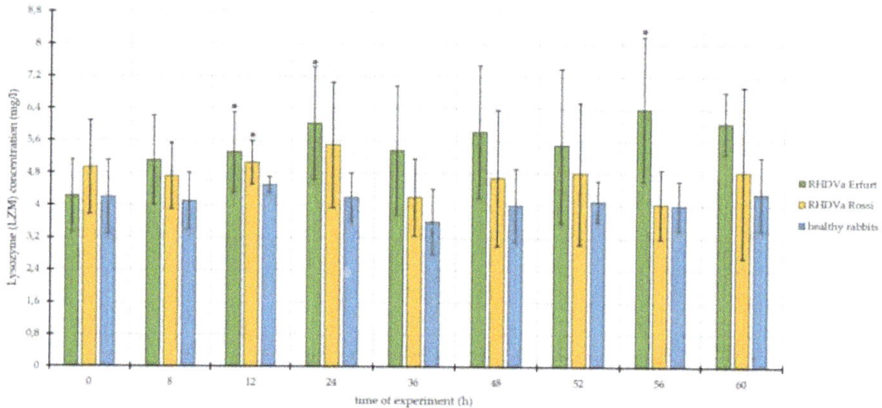

Figure 3. Values of lysozyme (LZM) concentration in rabbits infected with *L. europaeus*/GI.1a/Erfurt and *L. europaeus*/GI.1a/Rossi, as well as control rabbits. * statistically significant change respect to the control group ($p < 0.05$).

Both of the parameters for LZM concentration in rabbits infected with *L. europaeus*/GI.1a/Erfurt and *L. europaeus*/GI.1a/Rossi, in relation to the results obtained in the study of rabbits from the control group, showed an upward trend throughout the duration of the study. Nevertheless, they were accompanied by significant standard deviation. The obtained results were analyzed statistically, which showed that in the case of infection with *L. europaeus*/GI.1a/Erfurt, three statistically significant changes were noted, which fell at 12, 24, and 56 h p.i., while the *L. europaeus*/GI.1a/Rossi strain in within the examined parameter showed only one change of statistical significance; this change was observed at 12 h p.i. (Figure 3). High standard deviations and relatively low numbers of statistically relevant changes may be caused by the rapid losses of rabbits and the methodical character of the analysis, which depends on many outside factors.

3.3. Lysozyme (LZM) Activity

The results of studies on LZM activity in rabbits infected with *L. europaeus*/GI.1a/Erfurt ranged from 0.0015 to 0.0057, with the standard deviation (SD ±) from 0.0002 to 0.0047 (Figure 4). In *L. europaeus*/GI.1a/Rossi-infected animals, LZM activities ranged from 0.0004 to 0.0014, with a standard deviation (SD ±) of 0.0001 to 0.0008 (Figure 4). In the group of control rabbits, LZM activity values ranged from 0.0013 to 0.0023, with standard deviation (SD ±) from 0.0002 to 0.0008 (Figure 4).

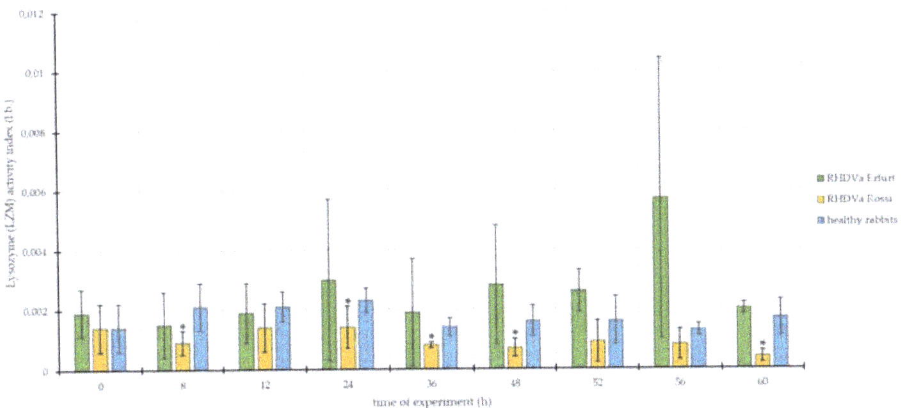

Figure 4. Values of lysozyme (LZM) activity in rabbits infected with *L. europaeus*/GI.1a/Erfurt and *L. europaeus*/GI.1a/Rossi, as well as control rabbits. * statistically significant change respect to the control group ($p < 0.05$).

In the case of rabbits infected with *L. europaeus*/GI.1a/Erfurt, the parameters of LZM activity, in relation to the results obtained in the study of rabbits from the control group, showed a slight upward trend at 0 hours. Then at 12 and 24 h p.i., there was a decrease in the tested parameter; from 24 to 60 h p.i., the tested parameter showed only an upward trend. Analyzing the values of LZM activity in rabbits infected with *L. europaeus*/GI.1a/Rossi, it was observed that the parameter values, in relation to the control group, showed only a downward trend throughout the duration of the experiment. The obtained result was subjected to statistical analysis, which showed that in the case of infection with *L. europaeus*/GI.1a/Erfurt, all changes within the value of the examined parameter are not statistically significant, while in the group of rabbits infected with *L. europaeus*/GI.1a/Rossi, there were five statistically significant changes in the form of decreases in the activity of LZM. These changes were observed at 8, 24, 36, 48, and 60 p.i. (Figure 4). It needs to be added that similar to LZM concentration, LZM activity is characterized by noticeably high standard deviations, which was probably caused by individual animals with relatively higher or lower results. This may lead to an overall conclusion on LZM that it is not enough stable to serve as a prognostic factor.

3.4. Clinical Signs Infection of Lagovirus Europaeus/GI.1a.

During the experiment, many clinical symptoms were noted, including sneezing, thickening of the blood, no response to external stimuli, rapid breathing, general body stiffness, lethargy, increased thirst, increased blood clotting, and lack of appetite. All observed clinical symptoms were characteristic of the course of RHD. The first symptoms in animals were observed as early as 8 h p.i. In the following hours of the experiment, symptoms began to worsen, and continued until the animal died or until the end of the experiment. Using the Kaplan–Meier method, we calculated the probability of survival of animals from both groups of infected rabbits and plotted them in the graph (Figure 5). In the group of animals infected with *L. europaeus*/GI.1a/Erfurt, there were seven deaths from 10 tested animals during the experiment.

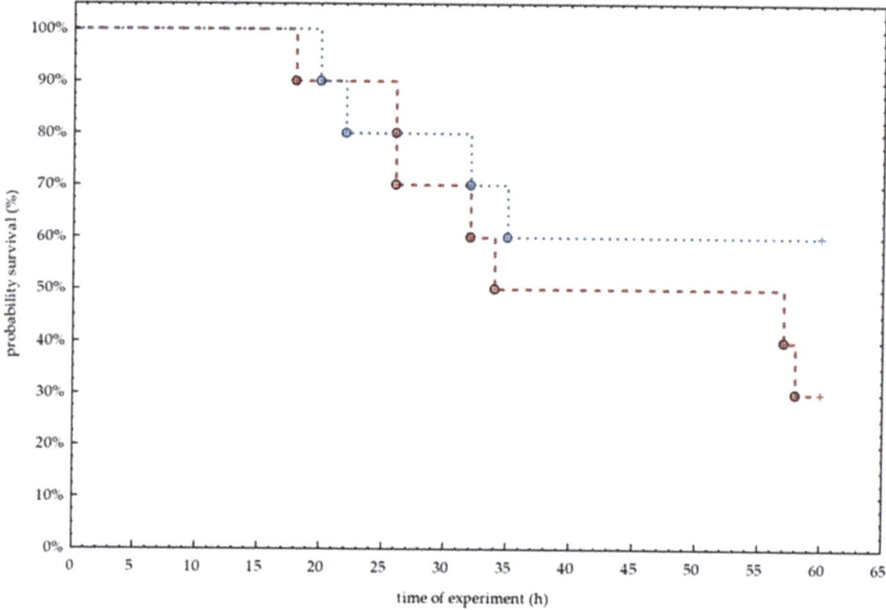

Figure 5. Percentage of rabbit survival recorded for the two tested strains (*L. europaeus*/GI.1a/Erfurt and *L. europaeus*/GI.1a/Rossi), analyzed by Kaplan–Meier method.

The first death was recorded between 12 and 24 h p.i., then between 24 and 36 h p.i. another four deaths were recorded. The last fall of the animals was observed between 56 and 60 h after the start of the experiment. The probability of survival 60 h after infection with *L. europaeus*/GI.1a/Erfurt was 30% (Figure 5).

In the case of the group of animals infected with *L. europaeus*/GI.1a/Rossi, there were four deaths from 10 infected animals. The first two deaths, as in the case of *L. europaeus*/GI.1a/Erfurt, were recorded between 12 and 24 from the beginning of the experiment. Another two deaths were observed between 24 and 36 h p.i. The probability of survival for 60 h after infection in the group of animals infected with *L. europaeus*/GI.1a/Rossi was 60% (Figure 5).

4. Discussion

Results show that MPO seems to be a sensitive marker of the condition of immune system activity of infected rabbits, as it by definition increases over the course of infection. It was previously described that MPO in viral infections in increasing, and the elevated level of MPO results from the possible lysis of leukocytes, being the first line of antimicrobial defense [34,35]. After entering of the virus into the host cells, phagocytosis is performed, and the leukocytes lead to increased reactive oxygen species through MPO and NADPH oxidase, meaning that the MPO level and activity is directly correlated with the activation of phagocytes [36]. Moreover, the level of MPO enzyme is also discussed as being a prompt indicator of endothelial dysfunction, inflammation, atherosclerosis, and oxidative stress [37]. Our results show that MPO activity levels seems to be stable and behave similarly in both studied strains of *Lagovirus europaeus*/GI.1a levels of MPO—in the Erfurt strain levels were between 1.15 and 2.13, and in case of the Rossi strain the range was 1.07–2.34. The trend is similarly increasing from 8 to 60 h p.i. Also, no impact was noted and considered as significant as far as sex of the animals is concerned. Taking all of the above into consideration, it may be stated that MPO activity may be a more useful prognostic tool, compared to LZM, for defining the status of the immune system in the breed animals, in order to avoid the unexpected losses of the farmed rabbits.

Among animal models, the role of myeloperoxidase has been emphasized mainly in mice, where experiments with MPO-deficient mice showed their susceptibility to *Candida albicans* [38] and *Klebsiella pneumoniae* [39] infection. Also, in rats, infection causes decreases in MPO levels [40]. Keeping in mind that animal models are not ideally reflecting the human condition, it is worth noticing that in humans, several types of tissue injuries and pathogenesis of many diseases, like rheumatoid arthritis, cardiovascular diseases, liver disease, diabetes, and cancer, have been linked with MPO-derived oxidants [41]. Due to the above, it can be stated that elevated levels of MPO activity may be one of the best diagnostic tools of inflammatory and oxidative stress biomarkers. Finally, latest research shows that MPO may not only be pivotal for innate immunity, but seems to have a potential role in adaptive immunity [42] and autoimmune diseases [43].

Lysozyme is an enzyme hydrolyzing glycosidic linkages in bacterial peptidoglycan (PGN); due to the low toxicity of lysozyme, it is used as a natural preservative to control bacteria in meat products [44]. What is more, this enzyme may act as chitinase and activate bacterial autolysins, and it also exhibits antiviral activity against several human and animal viruses [45]. Also, since the natural substrate for lysozyme is PGN, it has similar functions as PGN recognition molecules, such as CD14, TLR2, or NOD-like proteins [46]. The antibacterial, antiviral, and antifungal roles of LZM has been claimed mainly in human-orientated studies [47]; its role in animal models is modestly represented. Yet, an extremely interesting study that corresponds to ours was found related to influenza, in which it was registered that the level of LZM is inhibited by this virus, whereas simultaneously, the virus does not influence the level of MPO [48]. So far, apart from ours, this is the only noted correlation between those potential biomarking substances of the immune system's condition.

The mortality of the infected animals showed that the probability of survival differs in the studied viruses of *L. europaeus*/GI.1a; however, it is important to add that this condition was probably affected by several other factors impacting immune system status.

5. Conclusions

The results on the roles of MPO and LZM as important hallmark and prognostic factors of survival of rabbits suffering from infection with *L. europaeus*/GI.1a suggests that those antimicrobial enzymes located in neutrophils not only play a significant role in defending the host from the infection, by activating the phagocytes to actively fight with the virus, but also may serve as a prognostic marker of immune system status. Basing on the results, MPO may be a more reliable indicator of inflammatory response than LZM. Further, more extensive studies on the subject are required, especially to check the correlation between the enzymes' activity and viral loads, in order to better understand the mechanism and use it in the protection of rabbits. Also, considering the fact that currently, *Lagovirus europaeus*/GII is widespread all around the world, it would be interesting to check if the role of MPO and LZM is repeatable.

Author Contributions: Conceptualization, P.N.-R. and R.H.; methodology, P.N.-R and R.H.; software, R.H.; validation, R.H. and D.B.; formal analysis, R.H., D.B., and P.N.-R.; investigation, R.H. and D.B.; resources, P.N.-R.; data curation, R.H.; writing—original draft preparation, R.H., D.B., and P.N.-R.; writing—review and editing, R.H., D.B., and P.N.-R.; visualization, R.H.; supervision, P.N.-R.; project administration, P.N.-R.; funding acquisition, P.N.-R. All authors have read and agreed to the published version of the manuscript.

Funding: This research received no external funding.

Conflicts of Interest: The authors declare no conflict of interest.

References

1. Alda, F.; Gaitero, T.; Suárez, M.; Merchán, T.; Rocha, G.; Doadrio, I. Evolutionary history and molecular epidemiology of rabbit haemorrhagic disease virus in the Iberian Peninsula and Western Europe. *BMC Evol. Biol.* **2010**, *10*, 347. [CrossRef]
2. Abrantes, J.; van der Loo, W.; Le Pendu, J.; Esteves, P.J. Rabbit haemorrhagic disease (RHD) and rabbit haemorrhagic disease virus (RHDV): A review. *Vet. Res.* **2012**, *43*, 12. [CrossRef]
3. Abrantes, J.; Lopes, A.M.; Dalton, K.P.; Parra, F.F.; Esteves, P.J. Detection of RHDVa on the Iberian Peninsula: Isolation on an RHDVa strain from a Spanisch rabbitry. *Arch. Virol.* **2014**, *159*, 321–326. [CrossRef]
4. Rouco, C.; Abrantes, J.; Serronha, A.; Lopes, A.M.; Maio, E.; Magalhaes, M.J.; Blanco, E.; Barcena, J.; Esteves, P.J.; Santos, N.; et al. Epidemiology of RHDV2 (*Lagovirus europaeus*/GI.2) in free-living wild European rabbits in Portugal. *Transbound. Emerg. Dis.* **2018**, *65*, e373–e382. [CrossRef]
5. Delibes-Mateos, M.; Smith, A.T.; Slobodchikoff, C.N.; Swenson, J.E. The paradox of keystone species persecuted as pests: A call for the conservation of abundant small mammals in their native range. *Biol. Conserv.* **2011**, *144*, 1335–1346. [CrossRef]
6. Niedźwiedzka-Rystwej, P.; Tokarz-Deptuła, B.; Abrantes, J.; Esteves, P.J.; Deptuła, W. B and T lymphocytes in rabbits change according to the sex and throughout the year. *Pol. J. Vet. Sci.* **2020**, *23*, 37–42. [CrossRef]
7. Esteves, P.J.; Abrantes, J.; Baldauf, H.M.; BenMohamed, L.; Chen, Y.; Christensen, N.; González-Gallego, J.; Giacani, L.; Hu, J.; Kaplan, G. The wide utility of rabbits as models of human diseases. *Exp. Mol. Med.* **2018**, *50*, 1–10. [CrossRef] [PubMed]
8. Mage, R.G.; Esteves, P.J.; Rader, C. Rabbit models of human diseases for diagnostics and therapeutics development. *Dev. Comp. Immunol.* **2019**, *92*, 99–104. [CrossRef] [PubMed]
9. Food and Agriculture Organization of the United Nations (FAOSTAT). Available online: http://www.fao.org/faostat/en/#data/QL (accessed on 9 August 2020).
10. World Organisation for Animal Health (OIE). Rabbit Haemorrhagic Disease. Available online: https://www.oie.int/fileadmin/Home/eng/Animal_Health_in_the_World/docs/pdf/Disease_cards/RHD.pdf (accessed on 9 July 2020).
11. World Animal Health Information Database (WAHIS OIE). Available online: https://www.oie.int/wahis_2/public/wahid.php/Diseaseinformation/Diseasetimelines (accessed on 9 August 2020).
12. Niedźwiedzka-Rystwej, P.; Deptuła, W. Phagocytosis of neutrophils in rabbit infected with antigenic variant of RHD virus. *Pol. J. Vet. Sci.* **2012**, *15*, 143–150. [CrossRef] [PubMed]
13. International Committee on Taxonomy of Viruses (ICTV). Available online: https://talk.ictvonline.org/ictv-reports/ictv_online_report/positive-sense-rna-viruses/w/caliciviridae/1163/genus-lagovirus (accessed on 31 July 2020).
14. Le Pendu, J.; Abrantes, J.; Bertagnoli, S.; Guitton, J.S.; Le Gall-Reculé, G.; Lopes, A.M.; Marchandeau, S.; Alda, F.; Almeida, T.; Célio, A.P.; et al. Proposal for a unified classification system and nomenclature of lagoviruses. *J. Gen. Virol.* **2017**, *98*, 1658–1666. [CrossRef] [PubMed]
15. Vinjé, J.; Estes, M.K.; Esteves, P.; Green, K.Y.; Katayama, K.; Knowles, N.J.; L'Homme, Y.; Martella, V.; Vennema, H.; White, P.A.; et al. ICTV Virus Taxonomy Profile: Caliciviridae. *J. Gen. Virol.* **2019**, *100*, 1469–1470. [CrossRef] [PubMed]
16. Le Gall-Recule, G.; Zwingelstein, F. Detection of a new variant of rabbit haemorrhagic disease virus in France. *Vet. Rec.* **2011**, *168*, 137–138. [CrossRef] [PubMed]
17. Liu, S.J.; Xue, H.P.; Pu, B.Q.; Quian, N.H. A new viral disease in rabbits. *Anim. Hus. Vet. Med.* **1984**, *16*, 253–255.
18. Huang, H.B. Vaccination against and immune response to viral haemorrhagic disease of rabbits: A review of research in the People's Republic of China. *Rev. Sci. Tech. Off. Int. Epiz.* **1991**, *10*, 481–498.
19. Capucci, L.; Frigoli, G.; Ronshold, L.; Lavazza, A.; Brocchi, E.; Rossi, C. Antigenicity of the rabbit hemorrhagic disease virus studied by its reactivity with monoclonal antibodies. *Virus Res.* **1995**, *37*, 221–238. [CrossRef]
20. Bębnowska, D.; Niedźwiedzka-Rystwej, P. Characteristics of a new variant of rabbit haemorrhagic disease virus—RHDV2. *Acta Biol.* **2019**, *15*, 83–97. [CrossRef]
21. Abrantes, J.; Lopes, A.M.; Dalton, K.P.; Melo, P.; Correia, J.J.; Ramada, M.; Alves, P.C.; Parra, F.; Esteves, P.J. New variant of rabbit hemorrhagic disease virus, Portugal, 2012–2013. *Emerg. Infect. Dis.* **2013**, *19*, 1900–1902. [CrossRef]

22. Carvalho, C.L.; Silva, S.; Gouveia, P.; Costa, M.; Duarte, E.L.; Henriques, A.M.; Barros, S.S.; Luis, T.; Ramos, F.; Fagulha, T.; et al. Emergence of rabbit haemorrhagic disease virus 2 in the archipelago of Madeira, Portugal (2016–2017). *Virus Genes* **2017**, *53*, 922–926. [CrossRef]
23. Silverio, D.; Lopes, A.M.; Melo-Ferreira, J.; Magalhães, M.J.; Monterroso, P.; Serronha, A.; Maio, E.; Alves, P.C.; Esteves, P.J.; Abrantes, J. Insights into the evolution of the new variant rabbit haemorrhagic disease virus (GI.2) and the identification of novel recombinant strains. *Transbound. Emerg. Dis.* **2018**, *65*, 983–992. [CrossRef]
24. Müller, C.; Ulrich, R.; Franzke, K.; Müller, M.; Köllner, B. Crude extracts of recombinant baculovirus expressing rabbit hemorrhagic disease virus 2 VLPs from both insect and rabbit cells protect rabbits from rabbit hemorrhagic disease caused by RHDV2. *Archiv. Virol.* **2019**, *164*, 137–148. [CrossRef]
25. Information and training materials of the Laboratory Animals Section, General Assembly of the Association of Agriculture Engineers and Technicians. In *Materiały Informacyjno-Szkoleniowe Sekcji ds. Zwierząt Laboratoryjnych*; ZG Stowarzyszenia Inżynierów i Techników Rolnictwa: Warsaw, Poland, 1987; pp. 26–77. (In Polish)
26. Regulation of the Minister of Agriculture and Rural Development of 10 March 2006 on Detailed Conditions for Maintenance of Laboratory Animals in Experimental Units, Breeding Units and Suppliers (Polish Journal of Laws of 2006, No 50, Item 368). Available online: http://prawo.sejm.gov.pl/isap.nsf/DocDetails.xsp?id=WDU20060500368 (accessed on 9 July 2020).
27. Niedźwiedzka-Rystwej, P.; Deptuła, W. Lymphocyte subpopulations and apoptosis of immune cells in rabbits experimentally infected with a strain of the RHD virus having a variable haemagglutination capacity. *Pol. J. Vet. Sci.* **2012**, *15*, 43–49. [CrossRef] [PubMed]
28. Niedźwiedzka-Rystwej, P.; Deptuła, W. Non-specyfic immunity in rabbits infected with 10 strain of the rabbit haemorrhagic disease virus with different biological properties. *Centr. Europ. J. Biol.* **2012**, *5*, 613. [CrossRef]
29. Fitzner, A.; Kęsy, A.; Niedbalski, W.; Paprocka, G.; Walkowiak, B. Identification of the dominating VP60 polypeptide in domestic isolates of the RHD virus. *Med. Wet.* **1996**, *52*, 303–305.
30. Zawistowski, S. *Histology—Techniques and Basics*; PZWL: Warsaw, Poland, 1976. (In Polish)
31. Afanasjew, W.I.; Kolot, L.I. Izmienienije peroksydazy i cytochromaksydazy w krowi Karpow *Cyprinus carpio* L. pri krasnuchie. *Woprosy Ichtiologii* **1971**, *11*, 940–943.
32. Hankiewicz, J.; Świerczek, E. Comparison analysis of lysozyme evaluation with agar diffusion and nefelometric method. *Przegl. Lek.* **1975**, *32*, 376–378. (In Polish)
33. Szmigielski, S. The Analysis of up and down Regulated Granulocytes. Ph.D. Thesis, Wojskowy Instytut Medycyny Lotniczej, Warsaw, Poland, 1972. (In Polish).
34. Guven, F.M.K.; Aydin, H.; Yildiz, G.; Engin, A.; Celik, V.K.; Bakir, D.; Deveci, K. The importance of myeloperoxidase enyme activity in the pathogenesis of Crimean-Congo haemorrhagic fever. *J. Med. Microbiol.* **2013**, *62*, 441–445. [CrossRef]
35. Kulander, L.; Pauksens, K.; Venge, P. Soluble adhesion mlecules, cytokines and cellular markers in serum in patients with acute infections. *Scand. J. Infect. Dis.* **2001**, *33*, 290–300. [CrossRef]
36. Altinyazar, H.C.; Gurel, A.; Koca, R.; Armutcu, F.; Unalacak, M. The status of oxidants and antioxidants in the neutrophils of patients with recurrent aphthous stomatitis. *Turk. J. Med. Sci.* **2006**, *36*, 87–91.
37. Michowitz, Y.; Kisil, S.; Guzner-Gur, H.; Rubinstein, A.; Wexler, D.; Sheps, D.; Keren, G.; George, J. Usefullness of serum myeloperoxidase in prediction of mortality in patients with severe heart failure. *Isr. Med. Assoc. J.* **2008**, *10*, 884–888.
38. Aratani, Y.; Kouama, H.; Nyui, S.; Suzuki, K.; Kura, F.; Maeda, N. Severe impairment in early host defense agains Cancida albicans in mice deficient in myeloperoxidase. *Infect. Immun.* **1999**, *67*, 1828–1836. [CrossRef]
39. Hirche, T.O.; Gaut, J.P.; Heinecke, J.W.; Belaaouaj, A. Myeloperoxidase plays critical roles in killing *Klebsiella pneumoniae* and inactivating neutrophil elastase: Effects on host defense. *J. Immunol.* **2005**, *174*, 1557–1565. [CrossRef] [PubMed]
40. Christensen, R.D.; Rothstein, G. Neutrophil myeloperoxidase concentration: Changes with development and during bacterial infection. *Pediatr. Res.* **1985**, *19*, 1278–1282. [CrossRef] [PubMed]
41. Khan, A.A.; Alsahli, M.A.; Rahmani, A.H. Myeloperoxidase as an active disease biomarker: Recent biochemical and pathological perspectives. *Med. Sci.* **2018**, *6*, 33. [CrossRef] [PubMed]
42. Odobasic, D.; Kitching, A.R.; Holdsworth, S.R. Neutrophil-mediated regulation of innate and adaptive immunity: The role of myeloperoxidase. *J. Immunol. Res.* **2016**, *2016*, 2349817. [CrossRef] [PubMed]
43. Strzepa, A.; Pritchard, K.A.; Dittel, B.N. Myeloperoxidase: A new player in autoimmunity. *Cell. Immunol.* **2017**, *317*, 1–8. [CrossRef]

44. Ly-Chatain, M.H.; Moussaoui, S.; Vera, A.; Rigobello, V.; Demarigny, Y. Antiviriral effect of cationic compounds on bacteriophages. *Front. Microbiol.* **2013**, *4*, 46. [CrossRef]
45. Villa, T.G.; Feijoo-Siota, L.; Rama, J.L.R.; Ageitos, J.M. Antivirals against animal viruses. *Biochem. Pharmacol.* **2017**, *133*, 97–116. [CrossRef]
46. Dziarski, R.; Gupta, D. Peptidoglycan recognition in innate immunity. *J. Endotoxin Res.* **2005**, *11*, 304–310. [CrossRef]
47. Sahoo, N.R.; Kumar, P.; Bhusan, B.; Bhattacharya, T.K.; Dayal, S.; Sahoo, M. Lysozyme in livestock: A guide to selection for disease resistance: A review. *J. Anim. Sci. Adv.* **2012**, *2*, 347–360.
48. Pang, G.; Clancy, R.; Cong, M.; Ortega, M.; Zhigang, R.; Reeves, G. Influenza virus inhibits lysozyme secretion by sputum neutrophils in subjects with chronic bronchial sepsis. *Am. J. Respir. Crit. Care Med.* **2000**, *161*, 718–722. [CrossRef]

MDPI
St. Alban-Anlage 66
4052 Basel
Switzerland
Tel. +41 61 683 77 34
Fax +41 61 302 89 18
www.mdpi.com

Animals Editorial Office
E-mail: animals@mdpi.com
www.mdpi.com/journal/animals

www.ingramcontent.com/pod-product-compliance
Lightning Source LLC
LaVergne TN
LVHW070859170726
843515LV00005B/680

* 9 7 8 3 0 3 6 5 4 4 9 7 7 *